Lecture Notes in Civil Engineering

Volume 582

Lecture Notes in Civil Engineering (LNCE) publishes the latest developments in Civil Engineering—quickly, informally and in top quality. Though original research reported in proceedings and post-proceedings represents the core of LNCE, edited volumes of exceptionally high quality and interest may also be considered for publication. Volumes published in LNCE embrace all aspects and subfields of, as well as new challenges in, Civil Engineering. Topics in the series include:

- Construction and Structural Mechanics
- Building Materials
- Concrete, Steel and Timber Structures
- Geotechnical Engineering
- Earthquake Engineering
- Coastal Engineering
- Ocean and Offshore Engineering; Ships and Floating Structures
- Hydraulics, Hydrology and Water Resources Engineering
- Environmental Engineering and Sustainability
- Structural Health and Monitoring
- Surveying and Geographical Information Systems
- Indoor Environments
- Transportation and Traffic
- Risk Analysis
- Safety and Security

To submit a proposal or request further information, please contact the appropriate Springer Editor:

- Pierpaolo Riva at pierpaolo.riva@springer.com (Europe and Americas);
- Swati Meherishi at swati.meherishi@springer.com (Asia—except China, Australia, and New Zealand);
- Wayne Hu at wayne.hu@springer.com (China).

All books in the series now indexed by Scopus and EI Compendex database!

S. M. Shiva Nagendra · Praveen C. Ramamurthy ·
Sotiris Vardoulakis · Kraichat Tantrakarnapa ·
Richard J. Ball
Editors

Select Proceedings of the 8th Indian International Conference on Air Quality Management (IICAQM 2023)

Springer

Editors
S. M. Shiva Nagendra
Department of Civil Engineering
Indian Institute of Technology Madras
Chennai, Tamil Nadu, India

Praveen C. Ramamurthy
Department of Materials Engineering
Indian Institute of Science Bangalore
Bengaluru, Karnataka, India

Sotiris Vardoulakis
Department of Environmental Public Health, Faculty of Public Health, HEAL Centre
University of Canberra
Canberra, ACT, Australia

Kraichat Tantrakarnapa
Department of Social and Environmental Medicine, Faculty of Tropical Medicine
Mahidol University
Payathai, Thailand

Richard J. Ball
Department of Architecture and Civil Engineering
University of Bath
Bath, UK

ISSN 2366-2557 ISSN 2366-2565 (electronic)
Lecture Notes in Civil Engineering
ISBN 978-981-96-2358-7 ISBN 978-981-96-2359-4 (eBook)
https://doi.org/10.1007/978-981-96-2359-4

This Springer imprint is published by the registered company Springer Nature Singapore Pte Ltd.
The registered company address is: 152 Beach Road, #21-01/04 Gateway East, Singapore 189721, Singapore

Preface

The Indian International Conference on Air Quality Management (IICAQM) serves as an important forum that offers opportunities to scientists, engineers, educators, and practitioners to share information on recent progress in theory, technology, modelling, and policy in the area of air quality management and health.

Air pollution is a cross-sectoral issue with missions originating from diverse sources. Despite technological advances and continuous efforts from the research and scientific community and the government, air pollution remains one of India's most significant environmental threats. Particulate matter poses the biggest challenge, with emission levels continuously exceeding standards, particularly in urban areas. Also, ambient and household air pollution contributed significantly to millions of premature deaths in the country. Air pollutants, released into the atmosphere, undergo multiple atmospheric transformation processes as they are transported over long distances from their sources. They are responsible for adverse effects on human health, visibility, building materials, and ecosystems over spatial scales ranging from local to global. The impact of air pollutants on human health is governed by the toxicity of pollutants, exposure time, and exposure pathway. Therefore, there is an increasing need to manage air quality to minimize our exposure to toxic air pollutants.

The 1st (IICAQM 2016), 3rd (IICAQM 2018), and 7th (IICAQM 2022) conferences were held at IIT Madras, the 2nd (IICAQM 2017) was held at IIT Delhi, the 4th (IICAQM 2019) at IIT Bombay, the 5th (IICAQM 2020) was hosted online by IIT Madras, and the 6th (IICAQM 2021) was hosted online by IIT Guwahati. We are delighted that this time, the 8th IICAQM (2023) was hosted by IISc Bengaluru.

More than 35 technical papers were submitted, and after a rigorous peer-review process, 25 papers have been accepted for presentation. The conference, combining both air and water, includes 11 keynote addresses and 24 invited talks, in addition to paper, poster, and industry presentations. The conference proceedings will be published as a dedicated book chapter in Springer Nature India Private Limited.

The IICAQM 2023 would not have been possible without the support from the collaborating institutions, including the Indian Institute of Technology Madras, Indian Institute of Science Bengaluru, Indian Institute of Technology Guwahati, Indian Institute of Technology Bombay, Indian Institute of Technology Kanpur,

Australian National University, University of Bath, Asian Institute of Technology, Mahidol University and University of California, Riverside, in association with CE4HEAL and CEPHA Networks, AQMA, IIT Delhi, IIT Palakkad, Indian Institute of Technology (Indian School of Mines) Dhanbad, Ministry of Earth Sciences, Ministry of Environment, Forest and Climate Change, Greater Chennai Corporation, Tiruchirappalli City Corporation, Madurai City Corporation, Tamil Nadu Pollution Control Board (TNPCB), Central Pollution Control Board (CPCB) and National Environmental Engineering Research Institute (NEERI). We are also delighted to have received support and generous sponsorship from industries including TSI Instruments India Private Limited, Jawaharlal Nehru Port Authority, Kerala State Pollution Control Board, AlfaTech Services, Envirotech Instruments Private Limited, SENSurAir Private Limited, and the enthusiastic help from our colleagues, friends, staff, and students.

Finally, we would like to thank all the speakers, experts, delegates, and authors for their valuable contributions. We acknowledge the help of the eminent members of our scientific advisory board, the IICAQM 2023 organizing committee, and all the reviewers who supported the peer-review process. We hope that all the fellow participants will have fruitful and productive interactions during and after the conference.

Chennai, India	Prof. Dr. S. M. Shiva Nagendra
Bengaluru, India	Prof. Dr. Praveen C. Ramamurthy
Canberra, Australia	Prof. Dr. Sotiris Vardoulakis
Salaya, Thailand	Dr. Kraichat Tantrakarnapa
Bath, UK	Dr. Richard J. Ball

Contents

Air Quality Measurement and Monitoring

Air Quality Health Impacts and Policy Interventions

Editors and Contributors

About the Editors

Dr. S. M. Shiva Nagendra is Professor in the Department of Civil Engineering and Chairman of the Graduate Aptitude Test in Engineering (GATE) and Joint Admission test for Masters (JAM) at the Indian Institute of Technology Madras (IITM), Chennai, India. He has more than 25 years of experience in research, teaching, consultancy, and community engagement. He has published more than 110 research publications in international and national refereed journals, two reference books, and more than 100 papers in conferences. He has been awarded 6 Indian patents. He is Co-Author of books titled *Urban Air Quality Monitoring, Modelling and Human Exposure Assessment* (ISBN: 978-981-15-5511-4) and *Artificial Neural Networks in Vehicular Pollution Modelling* (SCI-41, ISBN-10:3-540-37417-5) published by Springer. He is Director of the Clean Environment for Planetary Health in Asia (CEPHA) Network supported by UKRI. He is Founder-Chairman of the Indian International Conference on Air Quality Management (IICAQM) series.

Dr. Praveen C. Ramamurthy is Professor at the Department of Materials Engineering and Chair of the Interdisciplinary Centre for Water Research at the Indian Institute of Science, Bengaluru, Karnataka, India. He completed M.Sc. in polymer science from the University of Mysore and Ph.D. from Clemson University in 2004. He worked as R&D Engineer in Therm-O-Disc, Ohio, the USA, and as Research Scientist in Hoku Scientific, Honolulu, Hawaii, the USA. He joined IISc in the year 2007 and works as Professor in the Department of Materials Engineering. His current research focuses on organic electronics and sensors for analytes like VOCs, heavy metal ions, nitrate, and biosensors. He has more than 270 peer-reviewed papers published and 21 patents to his credit.

Dr. Sotiris Vardoulakis is Professor of Environmental Public Health and Director of the Healthy Environment and Lives (HEAL) Global Research Centre at the University of Canberra, Australia, and Adjunct Professor at the Indian Institute of Technology Madras. He is Co-director of the Clean Air and Planetary Health in Asia (CEPHA) Network, the Clean Energy for Healthy Environments and Lives (CE4HEAL) partnership, and the International Consortium for Urban Environmental Health and Sustainability (Healthy-Polis). Previously, he was Director of Research and Head of the WHO Collaborating Centre on Occupational Health at the Institute of Occupational Medicine in Edinburgh, and before that, Head of the Environmental Change Department at Public Health England. He also held academic positions at the London School of Hygiene and Tropical Medicine and the University of Birmingham. Professor Vardoulakis' main research interests include climate change, health impact assessment, environmental epidemiology, and public health communication and policy.

Dr. Kraichat Tantrakarnapa is currently working at the Faculty of Tropical Medicine, MAHIDOL University, Thailand. He has more than 25 years of experience in many natural resources and environmental areas in Thailand and neighbouring countries. He has participated in Consulting Units for many organizations and local and international agencies. He has worked in environmental health, environmental health impact assessment, climate change and health impact, occupational health and safety management system, and environmental planning. He holds a bachelor of science in statistics from Chiang Mai University in Thailand and a master of science in technology of environmental management from Mahidol University, also in Thailand. Further, he completed Ph.D. in environmental engineering from Suranaree University of Technology in Thailand. Additionally, he completed post-graduate training in Occupational Health and Safety in the Workplaces at the International Training Centre of the International Labour Organization (ITC-ILO) and the University of Torino in Italy.

Dr. Richard J. Ball is Reader in the Department of Architecture and Civil Engineering at the University of Bath. He graduated from the University of Bath with B.Eng. in materials science and engineering and later completed Ph.D. in materials science, also at Bath. He has published more than 100 papers in refereed international journals covering batteries, fuel cells, solar cells, ceramics, composites, NDT, sensors, sustainable construction materials, and air quality. Recent research has focused on the development of novel coatings for buildings to help improve air quality and the study of air quality in refugee camps around the world. He is Chartered Engineer, Chartered Scientist, and Fellow of the Institute of Materials, Minerals, and Mining.

Contributors—IICAQM 2023

A. Beig Department of Civil Engineering, Indian Institute of Technology Delhi, New Delhi, India

S. Adarsh Department of Civil Engineering, TKM College of Engineering, Kollam, India

Aishi Nath Department of Civil Engineering, National Institute of Technology Meghalaya, India

Amit Chakrabarti Centre for Ageing and Mental Health (I-CAM), Indian Council of Medical Research (ICMR), Kolkata, India

Aniruddha Mukhopadhayay Department of Environmental Science, University of Calcutta, Kolkata, India

Anupa Yadav Centre for Ageing and Mental Health (I-CAM), Indian Council of Medical Research (ICMR), Kolkata, India; Department of Environmental Science, University of Calcutta, Kolkata, India

Argha Kamal Guha Department of Civil Engineering, Indian Institute of Technology Guwahati, Guwahati, India

Asim Saha Centre for Ageing and Mental Health (I-CAM), Indian Council of Medical Research (ICMR), Kolkata, India

B. Chandra Mohan Department of ECE, Bapatla Engineering College, Bapatla, India

B. S. Shilpa Department of Civil Engineering, Vidyavardhaka College of Engineering, Mysuru, India

B. T. Shivendra Department of Civil Engineering, Dayananda Sagar College of Engineering, Bengaluru, India

Bhanu Chaudhary Department of Civil Engineering, Sardar Vallabhbhai National Institute of Technology, Surat, India

Bhaven N. Tandel Department of Civil Engineering, Sardar Vallabhbhai National Institute of Technology, Surat, India

Brajesh Singh Department of Civil Engineering, Indian Institute of Technology Ropar, Rupnagar, India

D. Maskell Department of Architecture and Civil Engineering, University of Bath, Claverton Down, UK

Digambar S. Londhe Department of Civil Engineering, Visvesvaraya National Institute of Technology, Nagpur, India

Dishant M. Pandya Department of Mathematics, Pandit Deendayal Energy University Gandhinagar, Gujarat, India

G. Habib Department of Civil Engineering, Indian Institute of Technology Delhi, New Delhi, India

G. P. Shivashankara Department of Civil Engineering, Dayananda Sagar College of Engineering, Bengaluru, India

Ganesh Chandra Dhal Department of Civil Engineering, National Institute of Technology Meghalaya, India

H. K. Ramaraju Department of Civil Engineering, Dayananda Sagar College of Engineering, Bengaluru, India

H. N. Sowmya Department of Civil Engineering, Dayananda Sagar College of Engineering, Bengaluru, India

Hsi-Hsien Yang Department of Environmental Engineering and Management, Chaoyang University of Technology, Taichung, Taiwan

Indramani Dhada Department of Civil Engineering, Indian Institute of Technology Ropar, Rupnagar, India

J. T. Wilson Department of Architecture and Civil Engineering, University of Bath, Claverton Down, UK

K. Saran Kumar Department of Electrical Engineering, Indian Institute of Technology Madras, India

Kanagaraj Rajagopal Department of Environmental Engineering, Delhi Technological University, New Delhi, India

Karigowda Department of Civil Engineering, Indian Institute of Technology Delhi, New Delhi, India

M. Imran Department of Civil Engineering, Indian Institute of Technology Delhi, New Delhi, India

M. Nabi Department of Civil Engineering, Indian Institute of Technology Delhi, New Delhi, India

Madhumita Chakraborty Department of Civil Engineering, National Institute of Technology Surat, India

Manan Shah Department of Chemical Engineering, Pandit Deendayal Energy University, Gandhinagar, Gujarat, India

Manish Naja Atmospheric Division, Aryabhatta Research Institute of Observational Sciences (ARIES), Nainital, Uttarakhand, India

Manoj Kumar Environmental Science and Biomedical Metrology Division, CSIR-National Physical Laboratory, New Delhi, India; Academy of Scientific and Innovative Research (AcSIR), Ghaziabad, India

Monika Sharma Department of Environmental Engineering, Delhi Technological University, New Delhi, India

Mukesh Khare Department of Civil Engineering, Indian Institute of Technology Delhi, New Delhi, India

B. Muraleekrishnan Department of Civil Engineering, TKM College of Engineering, Kollam, India

Nikki Choudhary Environmental Science and Biomedical Metrology Division, CSIR-National Physical Laboratory, New Delhi, India; Academy of Scientific and Innovative Research (AcSIR), Ghaziabad, India

Nilesh J. Vasa Department of Electrical Engineering, Indian Institute of Technology Madras, India

R. Prajwal Department of Civil Engineering, Dayananda Sagar College of Engineering, Bengaluru, India

Pralhad Walvekar Department of Civil Engineering, Zeal College of Engineering and Research, Pune, Maharashtra, India

Preeti Tiwari Environmental Science and Biomedical Metrology Division, CSIR-National Physical Laboratory, New Delhi, India; Academy of Scientific and Innovative Research (AcSIR), Ghaziabad, India

Prince Vats Department of Civil Engineering, Indian Institute of Technology Ropar, Rupnagar, India

Pritha Bhattacharjee Department of Environmental Science, University of Calcutta, Kolkata, India

Priyanka Srivastava Atmospheric Division, Aryabhatta Research Institute of Observational Sciences (ARIES), Nainital, Uttarakhand, India

Richard J. Ball Department of Architecture and Civil Engineering, University of Bath, Claverton Down, UK

R. N. Patel Department of Electrical Engineering, National Institute of Technology Raipur, India

Rajeev Kumar Mishra Department of Environmental Engineering, Delhi Technological University, New Delhi, India

Ramya Selvaraj Department of Electrical Engineering, National Institute of Technology Raipur, India

Robin A. Christian Department of Civil Engineering, National Institute of Technology Surat, India

S. M. Shiva Nagendra Department of Civil Engineering, Indian Institute of Technology Madras, Chennai, India

S. P. S. Prabhakaran Department of Energy and Environment, National Institute of Technology, Tiruchirappalli, India

S. Ramachandran Space and Atmospheric Sciences Division, Physical Research Laboratory, Ahmedabad, Gujarat, India

S. Satyanarayanan Department of Applied Mechanics, Indian Institute of Technology Madras, Chennai, India

Sailee Redekar Department of Civil Engineering, Sinhgad College of Engineering, Pune, Maharashtra, India

Sajeed I. Ghanchi Department of Mathematics, Pandit Deendayal Energy University, Gandhinagar, Gujarat, India

Saket Ranjan Department of Civil Engineering, Indian Institute of Technology Madras, Chennai, India

Sakshi Gupta Environmental Science and Biomedical Metrology Division, CSIR-National Physical Laboratory, New Delhi, India; Academy of Scientific and Innovative Research (AcSIR), Ghaziabad, India

Sharad Gokhale Department of Civil Engineering, Indian Institute of Technology Guwahati, Guwahati, India

R. Shruthi Department of Civil Engineering, BGS Institute of Technology, Adichunchanagiri University, Mandya, India

Smaranika Panda Department of Civil Engineering, National Institute of Technology Surat, India

Sudhir Kumar Sharma Environmental Science and Biomedical Metrology Division, CSIR-National Physical Laboratory, New Delhi, India; Academy of Scientific and Innovative Research (AcSIR), Ghaziabad, India

Sunil Kumar Gupta Department of Environmental Engineering and Management, Chaoyang University of Technology, Taichung, Taiwan

Susan Mariam Rajesh Department of Civil Engineering, TKM College of Engineering, Kollam, India

Vignesh Mohan Department of Environmental Engineering, Delhi Technological University, New Delhi, India

Vijay Kumar Soni Indian Meteorological Department, New Delhi, India

Vivek Bargate Department of Electrical Engineering, National Institute of Technology Raipur, India

X. Ke Department of Architecture and Civil Engineering, University of Bath, Claverton Down, UK

Yashwant B. Katpatal Department of Civil Engineering, Visvesvaraya National Institute of Technology, Nagpur, India

Contributors—IICAQM 2022

A. Shukla Transport Planning and Environment Division, CSIR-Central Road Research Institute, New Delhi, India

Abhishek Penchala Department of Mining Engineering, Indian Institute of Technology Kharagpur, India

Aditya Kumar Patra Department of Mining Engineering, Indian Institute of Technology Kharagpur, India; School of Environmental Science and Engineering, Indian Institute of Technology Kharagpur, India

Amit Passi Department of Civil Engineering, Indian Institute of Technology Madras, Chennai, India

Chirashree Ghosh Department of Environmental Studies, University of Delhi, New Delhi, India

P. Hema Department of Civil and Environmental Engineering, Indian Institute of Technology Tirupati, Andhra Pradesh, India

Kanagaraj Rajagopal, Department of Environmental Engineering, Delhi Technological University, New Delhi, India

M. P. Maiya Department of Mechanical Engineering, Indian Institute of Technology Madras, Chennai, India

Manuj Sharma Department of Civil and Environmental Engineering, Indian Institute of Technology Tirupati, Andhra Pradesh, India

N. Dogra Environment and Health, The Energy and Resources Institute (TERI), New Delhi, India

N. Manesha Manoon Mani Department of Civil and Environmental Engineering, Indian Institute of Technology Tirupati, Andhra Pradesh, India

Namrata Mishra School of Environmental Science and Engineering, Indian Institute of Technology Kharagpur, India

Pooja Saroj Department of Environmental Studies, University of Delhi, New Delhi, India

R. Singh Transport Planning and Environment Division, CSIR-Central Road Research Institute, New Delhi, India

Rajeev Kumar Mishra Department of Environmental Engineering, Delhi Technological University, New Delhi, India

Ravish Dubey School of Environmental Science and Engineering, Indian Institute of Technology Kharagpur, India

S. Padma Transport Planning and Environment Division, CSIR-Central Road Research Institute, New Delhi, India

S. Ram Department of Civil Engineering, Gautam Buddha University, Greater Noida, Uttar Pradesh, India

S. Ramachandran Space and Atmospheric Sciences Division, Physical Research Laboratory, Ahmedabad, Gujarat, India

Samrat Santra School of Environmental Science and Engineering, Indian Institute of Technology Kharagpur, India

Sarup Das Department of Civil Engineering, Indian Institute of Technology Madras, Chennai, India

S. M. Shiva Nagendra Department of Civil Engineering, Indian Institute of Technology Madras, Chennai, India

Siddhartha Singh India Meteorological Department, New Delhi, India

Sri Harsha Kota Department of Civil Engineering, Indian Institute of Technology Delhi, New Delhi, India

Suresh Jain Department of Civil and Environmental Engineering, Indian Institute of Technology Tirupati, Andhra Pradesh, India

Thirunagari Baby Keerthi Department of Civil Engineering, Indian Institute of Technology Delhi, New Delhi, India

U. Gupta Department of Civil Engineering, Gautam Buddha University, Greater Noida, Uttar Pradesh, India

Vijay Kumar Soni India Meteorological Department, New Delhi, India

List of Figures

Traffic-Induced Ultrafine Particle Concentration in Megacity Delhi

Impact of Solid Waste Dump Yards Fire on Air Quality of National Capital Delhi

Effect of Sulfur Content on Filterable and Condensable $PM_{2.5}$ Emissions from Diesel Generator

Variation of Atmospheric Carbon Dioxide Over Maharashtra Using Remote Sensing Data

Initial Results of the Degradation of α-Pinene Utilising a Metakoalin-Based Geopolymer

Measurement and Evaluation of Driving Patterns and Real-World Emissions from Heavy Tractor Vehicles Under Varied Load Conditions

Road Traffic-Borne $PM_{2.5}$ and Noise Indices in Urban Roadside Microenvironments

Seasonal and Temporal Distribution of Urban Fine Particulates and Associated Health Risk Analysis: A Long-Term Measurement (2018–2020) Analysis in Megacity Delhi

Elemental Composition, Sources and Health Impact of PM_{10} Over Nainital, India

Monitoring of Personal Exposure Due to Particulate Matter

Probable Health Risk of Pollutants and Noise Found Inside Bottling Beverage Plant of Himachal Pradesh, India

Assessment of Benzene Exposure Through Urinary S-Phenylmercapturic Acid (SPMA) in the Adult Population of West Bengal Emphasizing on Sociodemographic and Lifestyle Factor

Assessment of Source Effects and Action Plan for Controlling Air Pollution in Non-attainment Areas of Ribhoi District, Meghalaya: A Traffic Survey Report

Investigation of Occupational Thermal Comfort in Urban Subway Stations

Measurement of Particulate Matter Levels in and Around the Mining Area Using Low-cost Sensors

Measurements of Size-Resolved Nanoparticle Concentration and Its Distribution in Polluted Urban Environment During Induced Firework Event

Emissions Assessment and Health Impacts of PM2.5 Due to Burning Sugarcane Residue in Mandya District, Karnataka

Chemical Characteristics, Source Apportionment and Health Risk Analysis of Atmospheric Particulates at the Urban Traffic Site in Vijayawada City, Andhra Pradesh India

A Review of Problems and Directions in Creating a Region-Specific Emission Inventory for the Agricultural Sector

Assessment of Crop Yield Loss Due to Surface Ozone Pollution Over Delhi NCR and Indo-Gangetic Plain

Appendix A: Impact of Solid Waste Dump Yards Fire on Air Quality of National Capital Delhi

Appendix B: Seasonal and Temporal Distribution of Urban Fine Particulates and Associated Health Risk Analysis: A Long-Term Measurement (2018–2020) Analysis in Megacity Delhi

List of Tables

Elemental Composition, Sources and Health Impact of PM_{10} Over Nainital, India

Probable Health Risk of Pollutants and Noise Found Inside Bottling Beverage Plant of Himachal Pradesh, India

Assessment of Benzene Exposure Through Urinary S-Phenylmercapturic Acid (SPMA) in the Adult Population of West Bengal Emphasizing on Sociodemographic and Lifestyle Factor

Assessment of Source Effects and Action Plan for Controlling Air Pollution in Non-attainment Areas of Ribhoi District, Meghalaya: A Traffic Survey Report

Investigation of Occupational Thermal Comfort in Urban Subway Stations

Emissions Assessment and Health Impacts of PM2.5 Due to Burning Sugarcane Residue in Mandya District, Karnataka

Chemical Characteristics, Source Apportionment and Health Risk Analysis of Atmospheric Particulates at the Urban Traffic Site in Vijayawada City, Andhra Pradesh India

A Review of Problems and Directions in Creating a Region-Specific Emission Inventory for the Agricultural Sector

Assessment of Crop Yield Loss Due to Surface Ozone Pollution Over Delhi NCR and Indo-Gangetic Plain

Appendix B: Seasonal and Temporal Distribution of Urban Fine Particulates and Associated Health Risk Analysis: A Long-Term Measurement (2018–2020) Analysis in Megacity Delhi

Appendix C: Measurements of Size Resolved Nano Particle Concentration and Its Distribution in Polluted Urban Environment During Induced Firework Event

Air Quality Modelling, Analysis and Sensor Applications

Low-Cost Dust Sensors: Contemporary Approach of Construction Dust Monitoring

Sailee Redekar and Pralhad Walvekar

Abstract Construction sector is major contributor to economic growth and productive livelihood. According to the pollution control board, PM_{10} and $PM_{2.5}$ are the main pollutants from construction sites and are responsible for the air pollution in developing nations. In order to update the real-time data from PM studies and pertinent papers, this study uses a scientific literature review methodology. The work signifies contemporary monitoring approach for air quality monitoring from construction sites, its utilization, working principle of equipment, equipment cost, etc. During the construction process, air-monitoring sensors with improved saturation point and sensitivity are utilized to measure PM. The low-cost sensor possesses major benefits than conventional monitoring equipment, and it found to be effective for utilization in developing country like India. Additionally, the research will support strategies to reduce particulate matter emission from the construction industry by changing the way that monitoring approaches are used to measure particulate matter. The updated building regulations can solve the problems caused by PM emissions and result in a green and sustainable construction method.

Keywords Emission · Air pollutants · Particulate matter · Construction industry · Sustainable construction practices

1 Introduction

Infrastructure is the foundation of economic development. The construction sector contributes an average of 6.5% GDP, ranking second only to the agriculture sector. However, Jones Lang Lasalle Meghraj [20] found that India's sustainable

S. Redekar (✉)
Department of Civil Engineering, Sinhgad College of Engineering, Pune, Maharashtra, India
e-mail: saileeredekar99@gmail.com

P. Walvekar
Department of Civil Engineering, Zeal College of Engineering and Research, Pune, Maharashtra, India

S. M. Shiva Nagendra et al. (eds.), *Select Proceedings of the 8th Indian International Conference on Air Quality Management (IICAQM 2023)*, Lecture Notes in Civil Engineering 582, https://doi.org/10.1007/978-981-96-2359-4_1

building programme is now getting little attention. Construction-related environmental contamination comes in 36 different forms. Depending on the type, pollution has different effects on the environment. Environmental effect is now a key consideration when assessing building projects, alongside time and budget constraints since pollution linked with construction projects created upstream of the development process until building deconstruction and removal [29]. There are numerous varieties of construction projects, including heavy construction, infrastructure projects, specialized industrial construction, institutional and commercial buildings, and residential constructions [40]. Monitoring air quality involves measuring ambient air contaminants in a methodical manner with the goal of determining the cause of the air pollution (causal analysis) or determining the exposure of sensitive receptors (such as people, animals, plants and works of art). Continuously detecting particular air pollutants, also known as "criteria air pollutants," is what is referred to as "air quality monitoring" [46]. PM, or the total amount of suspended particles and dust, is one of the primary air pollutants created by construction sites [45].

Particles having a diameter of 10 microns or smaller, $PM_{2.5}$ and PM_1 are particularly hazardous for human inhalation since they can lead to major health problems such as silicosis, cancer, early death and asthma [8].

30% of air pollution, according to representatives of the Delhi Pollution Control Committee (DPCC), is caused by dust from construction projects [41]. Construction workers and adjacent residents both are put in danger because of the high amount of particles that are released into the environment during the procedure. Construction workers' occupational health is in danger due to dust exposure and may have detrimental direct or indirect effects on the labour-intensive construction industry. Construction workers have an 11% probability of developing COPD, whereas never-smokers could experience a possibility as high as 52.6% due to exposure to PM [7]. Rapid infrastructure development in emerging nations like India is the cause of big construction projects. Even though it indicates a higher exposure rate, the PM concentration that is released around a building site is generally ignored. Based on these dangerous quantities, the exposure is not quantified. Still, traditional techniques and tools are used for field monitoring. There have been no developments in field monitoring methods in the nation. Monitoring procedures and mitigating recommendations for lowering PM concentration in the construction sector are lacking. The construction industry does not currently have a systematic framework for emitted particulate matter mitigation measures or its monitoring near building sites. The monitoring of PM and exposure brought on by construction projects is examined in this study.

2 Literature Review

In this work, the parameters of the literature review before coming up with a list of search terms (such as "particulate matter," "construction dust," "dust sensors," and "PM monitoring"). The study's principal goal was taken into consideration when

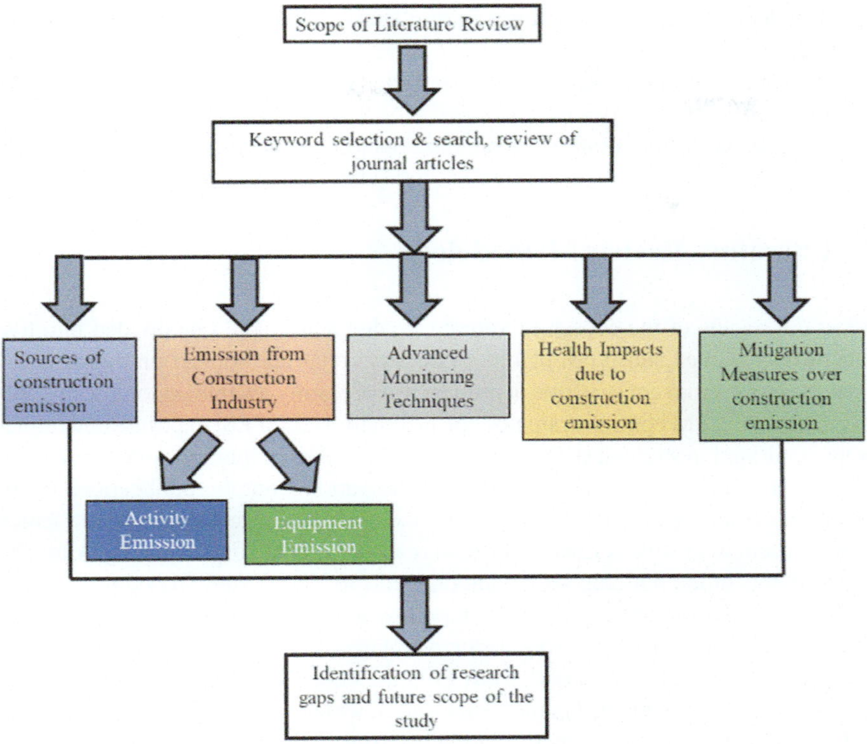

Fig. 1 Techniques for reviewing and evaluating literature

choosing the keywords. A careful keyword search found relevant study articles. The primary academic databases Scopus, Web of Science, Science Direct and Google Scholar were utilized in the search. The articles' breadth and suitability for inclusion or exclusion based on the inclusion and exclusion criteria stated below were evaluated after a brief content review. There were not many more publications, though, because of the limitations of this study. Consequently, it is employed to verify a keyword search's accuracy. Techniques employed in the review procedure, as depicted in Fig. 1.

3 Criteria for Insertion and Removal

1. Publications addressing PM, sophisticated monitoring methods and exposure from construction activities were taken into consideration for this research.
2. The literature on the health effects of PM exposure on construction workers was excluded.

3. Articles on construction dust that don't take place at building sites are not included.
4. However, measurements of PM from other industries are not included in the research.
5. Articles on PM monitoring for building operations were inserted in the review.

4 Literature Review Methodology

An initial review of the chosen literature revealed that it could be divided into five categories: sources and types of other emissions that are noticed on construction sites; emission from construction activities; use of readily available monitoring tools and techniques; PM control strategies, the health impacts of PM exposure and implemented mitigation strategies. The five groups of the chosen literary works were identified after rigorous examination, inductive derivation from the publications themselves and grouping into five categories. Utilization of these techniques because they provide them with superior instruments to reach a rational conclusion regarding the requirements of the building sector and the paths for future study based on a wealth of data from the literature. A concept-based review approach is one of these tools, along with content analysis using deductive reasoning. The findings of a literature study done over a 24-year period, from 1999 to 2023, as depicted in Fig. 1.

Total 150 papers screened in the review out of which 49 selected for literature and 101 excluded from review as per stated in insertion and removal criteria.

There are 26 papers covering sources of PM_{10} to $PM_{2.5}$ particles among the 49 literatures evaluated. According to the graph, there is not much written on building emissions. From 1999 until 2015, the issue of construction emissions was ignored, but in 2016, researchers started to pay attention to it. In contrast to recent years 2017, 2019 and 2020, and beyond, when the issue was taken into account by researchers, no research on the specified topic was conducted in 2008 or 2009. As per Fig. 2, the PM research focused rigorously after 2017. From the year 1999 to 2010, 1 or 0 publication is available in the first 10 years of span. After 2011 to 2015 publications increased to 2 or 3 in the year 2013. In the years 2016 and 2018, available publications are 4. In the year 2017, 7 publications are available. In the years 2019 and 2020, the number comes to 6. In the years 2021, 2022 and 2013, the available publications are 5, 3 and 4, respectively. Table 1 lists the different categories of the studied literature as well as the publication and citation data.

In order to address contemporary issues, the study uses a review technique including both scientific and grey literature. Because the field of air sensors is continuously increasing and new advancements and information frequently emerge outside of the typical peer-reviewed publications, a wider search was required to comprehend properly the current state of the subject. The academic databases Scopus, Web of Science, IEEE Xplore Digital Library and a general Google (Scholar) search were used in the search, which had publication dates from 1999 through 2023.

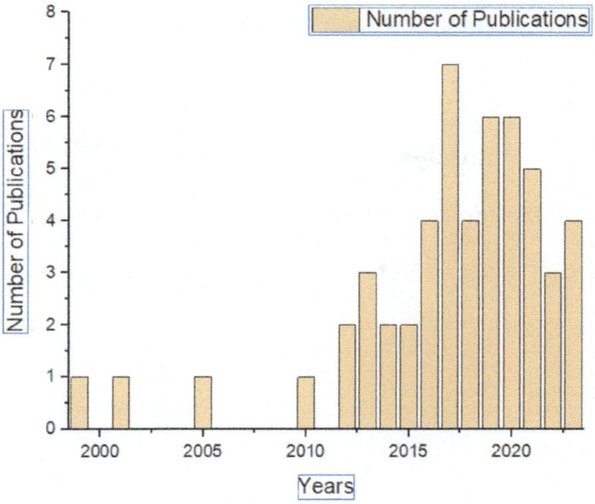

Fig. 2 Year-wise publication details of available literature

5 Construction Industry Emissions

One of the main types of air pollutants is particulate matter released from building sites, according to the USEPA [45]. As per [17], particulate pollution from construction activities has the potential to be quite significant. Among the most common construction activities are site preparation, earthmoving (including material trucking), paving of roadways, building and other structure construction, demolition and the application of architectural coatings.

6 Emission from Construction Activities

Marshall et al. [28] focused on non-road construction activities carried out on-site. Construction activities considered in the study are from both substructure and superstructure. Building construction activity and emissions are directly correlated,according to the study's methodology, [33] discussed here is straightforward, using freely available, cost-free data, and necessitates little image preprocessing or processing. For the purpose of tracking and reducing fugitive dust, it identifies places with high levels of dust-generating activities.

de Moraes et al. [13] calculated the concentrations of particle matter 10 μm (PM_{10}), and total suspended particles (TSP) produced at construction sites, with an emphasis on concrete and masonry works and their chemical constituents, were calculated. This study builds the framework for enhancing the PM collection procedure on building sites and offers knowledge that is applicable to numerous real-world scenarios. Sandanayake et al. [38] carried out a study with the following goals was conducted: (1) to estimate and compare the variation in emissions while constructing

Table 1 Survey of particular research articles

Category	Incites	References
Emissions	PM	[34]
	PM	[17]
	PM, RCS, toxic pollutants	[13]
Emission from construction activities	Construction activities from commercial building are considered	[28]
	Emission from CLT work taken for study	[1]
	Major substructure and superstructure activities are considered	[8]
	Direct and indirect emissions from construction sites are considered	[38]
	Equipment and machinery emissions are taken into account	[37]
	PM emission taken for study	[13]
	Truck emission taken for study	[44]
Monitoring Techniques	Focus of study is applicability of Sensors	[31]
	Sensor selection techniques determined for real-time monitoring	[10]
	Types of sensors are mentioned	[26]
	Low-cost dust sensors are used to find out emission concentration	[8]
	Sensor selection and data validation techniques determined	[36]
	Preliminary monitoring techniques with sensors are suggested	[39]
	Sensors in combination with GRIMM for emission estimation used	[22]

a building's foundation and structure,and (2) to critically examine the variation in direct and indirect emissions depending on various resource usage and consumption scenarios. The findings show the necessity of identifying significant emission sources and affect groups at various developmental phases.

Marzouk et al. [29] calculated six various types of emissions, including smog, eutrophication particles, particulate matter, greenhouse gases and sulphur dioxide. A model built on BIM has been created. The total direct and indirect emissions from construction operations are calculated across all project life cycle phases, including manufacture, transportation, construction, operation, maintenance and deconstruction/demolition. Ahmed & Arocho [1] conducted a study with the intention of calculating the PM concentration, monitoring the situation and categorizing buildings based on the rate of construction emission.

Sandanayake et al. [37] designated the construction stage (foundation, structure or entire building) that is used for emission analysis. This makes it possible for the user to identify which devices emit the most polluting compounds and at what rates. The authors want to improve the present version of CEET so that they can choose from a variety of emission factor standards when calculating emissions. Cheriyan & Choi [7] emphasized the requirements of the construction sector in terms of workplace air quality, health consequences, and progress and lag areas of research in this sector,a comprehensive study was conducted to incorporate existing research papers. Activities like excavation, foundation, mixing, masonry tasks, concreting, destruction, on-site electricity and fuel consumption are considered during substructure and superstructure construction [43]. The study's findings presented. In Jianghan Plain, China, it examined dust dispersion during the construction of an urban commercial-residential complex. The study's research subjects were each section's $PM_{2.5}$ and TSP concentration values. Estimates of the crucial variables relating to dust dispersion during earthwork building supplied by the study. These variables could be used to show how dust is distributed and changes during the construction of earthworks. Romina Paolucci, Marianna Rotilio, Stefano Ricci [35] represent a system for detecting tiny dust that is electronic. The apparatus particularly detects levels of $PM_{2.5}$ and PM_{10} emitted during the demolition of a L'Aquila structure that was significantly damaged by the earthquake in 2009.

7 Utilization of Cost-Effective Monitoring Tools and Techniques

Monitoring is a huge concern in developing countries like India; since equipment costs are very high, trained supervision is required for monitoring, and transportation issues may emerge owing to larger equipment, making the monitoring process time-consuming. Monitoring of PM is necessary to reduce the negative effects of particulate pollution produced by construction sites on the local community, human health, ambient air quality and environmental pollution.

8 Monitoring Scenario (National and Global)

Chung et al. [9] utilized continuous real time airborne particle detectors to measure PM in his experiment and adviced to use PM detectors for PM monitoring. The system consists of a beta attenuation monitor and integrated nephelometer, a Dusttrak sampler and a persistent aerosol mass monitor (CAMM, Andersen Instruments). As per Kinsey et al. [25], at sampling heights of 1.3, 2.7, 4.1 and 6 m for each air sampling station, four high-volume air samplers that were vertically oriented and outfitted with Sierra Model 235CP cyclone precipitators were set up. According to the study, the exposure profiling was completed. A sample of PM_{10} was taken using the Model 235CP. Engel-Cox et al. [14] presented a US-based satellite data monitoring system, and it was a visualization method to combine data from satellite sensors, data from the ground and information from back trajectory analysis. A new regulation governing the movement of particulate matter across state lines needed this information. Kinsey & Cowherd [24] employed a hybrid sampler to monitor construction in the USA. It displays the outcomes of component operation testing performed at real building sites between 1998 and 2001.

Pianalto & Yool [33] utilized the temporal fluctuations in the mid-infrared spectral response to dust sources from neighbouring construction were linked with the use of an image processing technique. Image differencing of the TM band 5 (mid-infrared) precisely pinpoints the place and area impacted by construction-related soil disturbance with a change threshold of 5 SD of the mean. Azarmi et al. [6] examined the release of particles, especially ultrafine particles, during the mixing of fresh concrete (with Portland cement and GGBS, PFA or pulverized fuel ash), as well as the subsequent drilling and cutting of hardened concrete, with the help of a fast response differential mobility spectrometer (DMS50) and a GRIMM particle spectrometer.

Hassan et al. [18] employed, a repairable dust sampler used, and an experimental campaign was carried out in Doha, Qatar, between April and May of 2014 [16]. The dust concentration was measured using low-cost sensors and HVS to follow the situation while flagging urgent issues and alternative remedies. The number of real-time air quality monitoring devices being created and enhanced has substantially increased the spatial and temporal resolution of datasets that were previously accessible, enhancing air pollution maps and forecast models [50]. The author investigated that beta attenuation monitors continuous real-time airborne particle monitors are useful.Continuous aerosol mass monitor (CAMM), two integrated nephelometers, Dusttrak sampler and tapered element oscillating microbalance. Continuous aerosol mass monitor (CAMM), two integrated nephelometers, Dusttrak sampler and tapered element oscillating microbalance. According to (M. [11], in order to describe the behaviour and size of PM_{10} and $PM_{2.5}$, a portable wind erosion device known as PI-SWERL was employed in this study of Nanjing, a typical metropolitan city in the Yangtze River. Ahmed & Arocho (2019b) used TSI DustTrak II 8530EP monitors which are used to measure and compare the emission of PM in construction sites. Using various materials in order to determine the difference in PM emission levels between construction sites and normal areas analyses the compliance of PM concentration between various construction sites with EPA standard and determines correlation between four different PM sizes.

Welsh [48] developed techniques for monitoring samples: (1) passive sampling methods (diffusion tubes), (2) active (semi-automatic) sampling approaches, (3) automatic point ozone, carbon monoxide, sulphur dioxide and particulate matter. Araújo & Costa [4] performed gravimetric sampler with filters, Construction-related PM research is still in its early phases. Because there is still no consensus on established experimental methodologies for assessing PM in building sites, this hinders the development of knowledge on this subject.

Wojcikowski et al. [49] recommended to use a modular, portable Internet of Things sensor to measure PM. A two-step technique is employed to enhance the measurement results due to the PM detector's subpar accuracy. Prior to employing a multiplicative surrogate model to adjust the responses nonlinearly, non-physical signal spikes are first deleted. Anlimah et al. [3] when assessing the effectiveness of a low-cost IoT particulate matter (PM_{10} and $PM_{2.5}$) sensor in the urban setting of Santiago, Chile, it discovered that the interventions used water, surfactant, foam and air currents in a variety of designs and combinations to decrease dust exposure.

These methods provide variable degrees of RCS and respirable dust control efficacy. Although there is evidence that interventions considerably reduce dust concentration levels, it is possible that occupational overexposure to RCS will still occur despite the controls in place.

Cui [12], in order to ensure a smooth implementation of dust monitoring, the stats necessity and development benefits of urban building site dust monitoring are discussed. Following the major points of dust monitoring technology, some suggestions were made for potential issues with the current technology and presented a plan that included three components—dust treatment and control from the construction unit, construction unit and supervision unit—to assist reduce construction-related dust pollution. The air quality of the surroundings is revealed by monitoring. Measurements are essential for identifying and prioritizing the sources and regions that require the greatest attention, for benchmarking against national or international norms and guidelines, for gauging the efficacy of emission reduction measures and for tracking overall progress. Different monitoring methods are used to find particle matter and here are some modern approaches along with their benefits and drawbacks.

9 Contemporary Monitoring Approach

Affordable sensor systems low-cost sensor systems often come in the form of compact, low-power gadgets that cost between hundred pounds for a single pollutant to thousand pounds for a full multi-pollutant system. They used handheld for mobile pollution investigations or at fixed places (such as affixed to a light). They are normally battery-powered. According to study by Atzori et al. [5], sensing RFID systems will make it possible to create RFID networks, which are comprised of miniature RFID-based computing and sensing components and RFID readers, which act as data sinks for the information produced by the sensing RFID tags and supply the network with power.

As per Naticchia et al. [32], the RFDs contained a Sharp Inc. nephelometer sensor with the sensor code GP2Y1010AU0F. Gewiss TM kept all of the electronic components, including a tilt sensor and the communication devices, in a watertight plastic box. The maximum current requirement for this optically based sensor, which monitors dust concentration, is 20 mA,therefore, it has no impact on the overall device's power consumption.

Wang et al. [47] utilized Guidelines from the US EPA's 2013 Air Sensor Workshop to evaluate three inexpensive particle sensors based on light scattering.

Gozzi et al. [16] used mobile monitoring of the various PM qualities, professional devices have been utilized on a variety of platforms (including automobile, van, pedestrian, bicycle, tram and airship). Mobile air quality monitoring is mainly carried out using motor vehicles, some of which may be equipped with large and weighty instruments, DMPS, SMS, etc.

Kelly et al. [22] compared the PMS sensor's performance to that of two high-end, light-scattering devices. A beta attenuation monitor, a tapered element oscillating microbalance and the gravimetric federal reference method (GRIMM) were used to compare the sensor's performance in ambient tests. Clements et.al [10] proposed sensor selection criteria for monitoring. A critical first step in any data collection effort is selecting the appropriate sensor. Sensors are commonly tested in actual outside field settings, where they are placed alongside more typical FRM/FEM equipment and the resulting data is analysed. When seeking for a suitable sensor, these findings should be taken into account. According to Clements et al. [10], the environmental impact becomes a very important signal when evaluating construction projects with schedule and cost constraints because it is formed from the upstream of the development process till building deconstruction and removal [29]. Kelly et al [22] signified that in both a controlled wind tunnel and in the real world, researchers evaluated a low-cost particulate matter sensor (Plantower PMS 1003/3003) used by a community air quality network during a variety of winter cold-pool occurrences associated to high ambient PM levels [22].

Yang et al. [50] stated "Paper" is a real-time tracker based on a robust to occlusions and deformations region-based fully convolutional Siamese network (R-FCSN) in this study. Smaoui et al. [42] created a method for real-time dust monitoring utilizing building information modelling (BIM) and a network of inexpensive mobile dust sensors. Dust sensors and single-board processors combined into field-deployable systems. Based on an accurate ground truth sensor, inaccurate sensors are automatically calibrated. A real-time presentation of the data gathered by dust sensors using a BIM-based visualization system. In a controlled setting, a prototype system is created and tested.

According to Morawska et al. [31], the goal of the study was to the need for and viability of sensor technology for exposure assessment in the construction sector for construction pollution monitoring. A thorough literature search was performed to address this problem, including both scientific and grey literature. Smaoui et al. [42] presented the data gathered from dust sensors and created a BIM-based visualization. In a controlled setting, a prototype system is created and tested. Schwarz et.al [39] contributed to a new field of use for inexpensive PM sensors. This study offers a preliminary evaluation of the uses and functions of inexpensive PM sensors for monitoring emissions continuously [39].

As per Amit Kumar et.al (2019), the dust sensor and its differences from conventional sensors are examined first in the study. These sensors provide developing nations with a wide range of advantages, such as the potential to expand the scope of current monitoring networks and the cost viability of setting up networks in rural areas or smaller towns (A. [26].

An air quality and dust level monitoring system are given in Akshatha [2]. The amount of formaldehyde, dust, temperature and humidity pollution is detected in an area. Cheriyan & Choi [8] stated that using both static and dynamic PM monitoring, Alphasense sensors were the subject of a comparative investigation,their additional specifications were presented. Through the USB interface, the Alphasense directly sends the recorded data to the computer. Because it is less expensive than most of the other dust sensors, a Sharp dust sensor is a cost-effective choice among the dust sensors that were chosen. The Sharp dust sensor requires communication with an external microcontroller and displays in order to track PM concentration.

Li et al. [27] assessed the performance of nine inexpensive PM monitors. Both a SidePak and a GRIMM were used as reference tools. According to their operational philosophies and data reporting formats, the monitors split into two groups, and for each group, a linear correlation factor is for the mass concentration measurement of $PM_{2.5}$. In Jo et al. [19], the creation of an IoT-based indoor dust monitoring platform is discussed. Based on a technique recommended by the Korean Ministry of Environment, experiments were conducted to confirm the air quality measurement device utilized in the platform. Cheriyan & Choi [7] identified two different sensor types for investigation, and findings have already been compared. This post's main objective is to employ inexpensive dust sensors like Sharp and Alphasense to track PM emissions and exposure from a building site.

Sahu et al. [36] used a PM Sensor New Plantower PM sensors (model PMS7003) that were the low-cost new sensors investigated in this experiment. A Plantower PMS7003 sensor is used by the Atmos device to measure the amounts of PM_1, $PM_{2.5}$ and PM_{10}. Using a DHT22 sensor, temperature and relative humidity were detected. Kim et al. [23] adopted using an IoT-based PM monitoring system that entails offering services to clients and transmitting data obtained from a measuring equipment through a network. As per Giordano et al. [15] stated dust sensors must be collocated with one another, as well as with a standard procedure or other instrument of preference, in order to achieve network precision and measurement accuracy. Every potential combination of weather and PM concentrations that the sensors are likely to function in should be included in the ideal calibration periods. Kane et al. [21] utilized, sensors (Sharp GP2Y1010AU0F) in various other air quality studies. A microcontroller from Adafruit (New York City, NY, USA), which includes clock on a daughter board and a microSD card interface, is used to capture the sensor data. Zafra-Pérez et al. [51] stated that an innovative strategy for using portable, low-cost sensors (LCSs) to monitor dust concentrations in open-pit mines is suggested. The three-year study (2019–2021) took place in the renowned Riotinto mine. It included a complete mobile LCS calibration that met the requirements of the USA and EU. Molina Rueda et al. [30], with little to no scaling capability or measurement reliability outside of this size range, accumulation mode PM (PM_{10}) sensors are the best classification for low-cost PM sensors frequently employed in crowdsourced measurement networks.

10 Discussion

As per the current construction scenario, contractors should focus more on the immediate environmental effects if they wish to maintain a sustainable construction site. The sensor board installed on buses includes modern technologies like GPRS and GPS antenna, data loggers, chemical sensors, temperature and humidity sensors, and these may all be used for real-time monitoring. Due to their low price, extensive coverage and capacity to record both temporal and spatial heterogeneities in air pollution concentrations, these sensors have been hailed as a potential remedy for the lack of air quality monitoring in developing countries. In developing nations like India, monitoring is a major concern due to expensive equipment prices, the need for experienced supervision, potential transportation problems brought on by larger equipment and the lengthy nature of the monitoring process. Real-time monitoring with inexpensive dust sensors done on the spot to fill the gap, surveillance in underdeveloped nations.

By (i) expanding the scope and area covered by air quality monitoring in developing countries and (ii) extending the current air quality monitoring programmes to underrepresented areas for rich countries, modern sensors may be able to overcome the limitations of traditional monitoring. The low-cost sensors are more likely to employ in more locations than their more expensive equivalent. The ability to undertake high-resolution exposure assessment using the obtained data is one benefit of this approach (Table 2).

These sensors also gather local meteorological information, which allows for proper input conditions for air quality modelling and reliable results. These sensors are also used for other things including emission inventories, spotting pollution hotspots and providing the public with real-time exposure assessments. Sensors take very little power.

11 Summary

In emerging countries like India, rapid infrastructure development is a substantial source of particle pollution. This report did a detailed analysis of research publications published since 2008 to identify current research advances and suggest future research areas. According to the review, there are not many research articles in this topic according to the time frame (1999 to 2023). However, as more people are becoming aware of PM exposure and its harmful effects on health, there is now more research available. Even though it indicates a greater exposure rate, the amount of PM that emitted near a building site is often ignored. However, it is challenging to address this issue and grow towards becoming a more sustainable and environmentally friendly sector because the regulations now in place that oversee building practices do not take into account all aspects of construction operations. Although the study's findings led to the publication of more research articles, experts continue to

Table 2 Details of contemporary monitoring approach

Sr. no.	Equipment	Working principle	Specifications	References
1	RFID Sensor Networks	Uses radio waves to provide signals that cause the tag to activate	Operating Frequency13.56 MHz EEPROM Size256 byte (2048 bit)512 bit user memory Unique Serial Number (UID)8 byte (64 bit)8 byte (64 bit)	[5]
2	Nephelometer Sensor	Instead of absorbing light, particles in a solution will scatter it as it passes through	Display: 2 line, 16-character LCD Dimensions: 3.75W × 6.75H × 2.0D inches 9.5W × 17.2H × 5.1D cm Weight: 24 oz. (0.67 kg)	[47]
3	GRIMM Spectrophotometer	Light scattering inside the measurement cell allows for the detection of the particles in the sampled air. Every particle's scattering light pulse is counted, and the strength of its signal is categorized according to its particle size	Grimm spectrometer 1.109 with 31 size channels and integrated PFTE-filter GRIMM Specifications 1.109 Particle counting up to 2 000 000 PA Dust range [ug/m^3] 0.1.100.000 Size Range [um] 0 25. > 32 battery 12VDC (up to 8 h) or 110.230VAC Weight/Dimensions 24 kg (5 31b) with battery/24 × 13 × 7 cm	[22]
4	Sharp Sensor GP2Y1010AU0F	A "optical spot" will develop on the PSD when the light beam is reflected by an object and reaches the light detector	Maximum range:80 cm Minimum range:10 cm Sampling rate:26 Hz1 Minimum operating voltage:4.5 V Maximum operating voltage:5.5 V	[8]

(continued)

Table 2 (continued)

Sr. no.	Equipment	Working principle	Specifications	References
5	Alphasence Sensor	Utilize laser beams to identify particles with a diameter between 0.38 micron and 17 microns and 16 bins (0.35 to 40 m and 24 bins for OPC-N3). Employing embedded algorithms, count readings were transformed into mass concentrations of PM1.0, $PM_{2.5}$ and PM_{10}	Dimensions: 75 (L) × 60 (W) × 63.5 (H) (mm)Weight: < 105 g Battery: No Power supply: 175 mA (~ 5 V DC) (180 mA for OPC-N3) Sensor lifetime: N/A Clock function: No Sampling mechanism: Fan Environmental operating conditions: -10 to 50 C Temp, 0 to 95% RH	[8]
6	Plantower Sensor PMS7003	The sensor makes use of the laser scattering concept, which entails utilizing a laser to disperse light from airborne suspended particles before harvesting it to a certain extent to form the scattering light variation curve	Total Response Time < 10 Second(s) DC Power Supply Type:5.0 Min:4.5 Max:5.5 Volt(V) Active Current ≤ 100 Milliampere(mA) Standby Current ≤ 200 Milliampere(mA)	[36]

Abbreviation PFTE Filter- polytetrafluoroethylene Filter

emphasize the need for more knowledge on PM creation during various building operations and dust monitoring methodologies. It is important to consider the main factors that influence PM generation from various construction-related emission sources, as well as the different types of construction activities, features of different PM size distributions, materials and tools utilized, and the time of the activity. The standard monitoring approach needs the necessary staff, field and lab infrastructure, equipment and financial resources, among other things, in order to use for measuring air quality.

Experts emphasize the need for more information on PM production during various building operations and dust monitoring approaches notwithstanding the study that was completed and the surge in research articles. It is important to consider the primary variables affecting the production of PM from diverse construction emission sources as well as the many different types of building operations. These elements include the distributional characteristics of various PM sizes, the supplies and equipment used and the duration of the activity.

The conventional monitoring approach used to obtain air quality measurements has a fundamental requirement for resources, including personnel, field and lab infrastructure, equipment and cash. The study provides a detailed picture of PM emissions from building operations using cutting-edge monitoring techniques. The study's findings might encourage researchers to carry out more research in this area, which might lead to a change in the rules governing the construction industry. The problem of PM emissions is solved by the updated building standards, which will also result in a durable, ecologically friendly construction technique, etc.

References

1. Ahmed S, Arocho I (2019) Emission of particulate matters during construction: a comparative study on a Cross Laminated Timber (CLT) and a steel building construction project. J Build Eng 22:281–294
2. Akshatha J (2019) Air quality and dust level monitoring using IoT. Int Res J Eng Technol, pp 1410–1414. www.irjet.net
3. Anlimah F, Gopaldasani V, MacPhail C, Davies B (2023) A systematic review of the effectiveness of dust control measures adopted to reduce workplace exposure. Environ Sci Pollut Res 30(19):54407–54428
4. Araújo IPS, Costa DB (2022) Measurement and monitoring of particulate matter in construction sites: guidelines for gravimetric approach. Sustainability (Switzerland) 14(1)
5. Atzori L, Iera A, Morabito G (2010) The internet of things: a survey. Comput Netw 54(15):2787–2805
6. Azarmi F, Kumar P, Mulheron M (2014) The exposure to coarse, fine and ultrafine particle emissions from concrete mixing, drilling and cutting activities. J Hazard Mater 279:268–279
7. Cheriyan D, Choi JH (2020) A review of research on particulate matter pollution in the construction industry. J Clean Product 254
8. Cheriyan D, Choi JH (2020) Estimation of particulate matter exposure to construction workers using low-cost dust sensors. Sustain Cities Society 59:102197
9. Chung A, Chang DPY, Kleeman MJ, Perry KD, Cahill TA, Dutcher D, McDougall EM, Stroud K (2001) Comparison of real-time instruments used to monitor airborne particulate matter. J Air Waste Manag Assoc 51(1):109–120
10. Clements AL, Griswold WG, Abhijit RS, Johnston JE, Herting MM, Thorson J, Collier-Oxandale A, Hannigan M (2017) Low-cost air quality monitoring tools: from research to practice (A workshop summary). Sensors (Switzerland) 17(11):1–20
11. Cui M, Lu H, Etyemezian V, Su Q (2019) Quantifying the emission potentials of fugitive dust sources in Nanjing, East China. Atm Environ 207:129–135
12. Cui T (2023) Development of dust monitoring in Urban construction sites and suggestions on dust control. J Innovat Develop 2(2):18–21
13. de Moraes RJB, Costa DB, Araújo IPS (2016) Particulate matter concentration from construction sites: concrete and masonry works. J Environ Eng 142(11):05016004
14. Engel-Cox JA, Young GS, Hoff RM (2005) Application of satellite remote-sensing data for source analysis of fine particulate matter transport events. J Air Waste Manag Assoc 55(9):1389–1397
15. Giordano MR, Malings C, Pandis SN, Presto AA, McNeill VF, Westervelt DM, Beekmann M, Subramanian R (2021) From low-cost sensors to high-quality data: a summary of challenges and best practices for effectively calibrating low-cost particulate matter mass sensors. J Aerosol Sci 158(July):105833
16. Gozzi F, Della Ventura G, Marcelli A (2016) Mobile monitoring of particulate matter: State of art and perspectives. Atmos Pollut Res 7(2):228–234
17. Guide NAAQ (2013) Construction-generated criteria air pollutant and precursor emissions
18. Hassan H, Tsiouri V, Kakosimos KE (2015) Developing emission factors of fugitive particulate matter emissions for construction sites in the middle east area. XIII Int Con Air Pollut Control 9(2):50–54

19. Jo J, Jo B, Kim J, Kim S, Han W (2020) Development of an IoT-based indoor air quality monitoring platform. J Sen 2020:13–15
20. Meghraj JLL (2016) Sustainable buildings and construction for India: policies practice and performance. Energy Build 92(14):324–337
21. Kane F, Abbate J, Landahl EC, Potosnak MJ (2022) Monitoring particulate matter with wearable sensors and the influence on student environmental attitudes. Sensors 22(3)
22. Kelly KE, Whitaker J, Petty A, Widmer C, Dybwad A, Sleeth D, Martin R, Butterfield A (2017) Ambient and laboratory evaluation of a low-cost particulate matter sensor. Environ Pollut 221:491–500
23. Kim H, Tae S, Zheng P, Kang G, Lee H (2021) Development of IoT-based particulate matter monitoring system for construction sites. Int J Environ Res Public Health 18(21)
24. Kinsey JS, Cowherd C (2012) Particulate emissions from construction activities. J Air Waste Manag Assoc 55(6):772–783
25. Kinsey JS, Linna KJ, Squier WC, Muleski GE, Cowherd C (2004) Characterization of the fugitive particulate emissions from construction mud/dirt carryout. J Air Waste Manag Assoc 54(11):1394–1404
26. Kumar A, Gurjar BR (2019) Low-cost sensors for air quality monitoring in developing countries–A critical view. Asian J Water Environ Pollut 16(2):65–70
27. Li J, Mattewal SK, Patel S, Biswas P (2020) Evaluation of nine low-cost-sensor-based particulate matter monitors. Aerosol Air Qual Res 20(2):254–270
28. Marshall SK, Rasdorf W, Lewis P, Frey HC (2012) Methodology for estimating emissions inventories for commercial building projects. J Archit Eng 18(3):251–260
29. Marzouk M, Abdelkader EM, Al-Gahtani K (2017) Building information modeling-based model for calculating direct and indirect emissions in construction projects. J Clean Prod 152:351–363
30. Molina Rueda E, Carter E, L'Orange C, Quinn C, Volckens J (2023) Size-resolved field performance of low-cost sensors for particulate matter air pollution. Environ Sci Technol Lett 10(3):247–253
31. Morawska L, Thai PK, Liu X, Asumadu-Sakyi A, Ayoko G, Bartonova A, Bedini A, Chai F, Christensen B, Dunbabin M, Gao J, Hagler GSW, Jayaratne R, Kumar P, Lau AKH, Louie PKK, Mazaheri M, Ning Z, Motta N, Williams R (2018) Applications of low-cost sensing technologies for air quality monitoring and exposure assessment: how far have they gone? Environ Int 116(April):286–299
32. Naticchia B, Fava G, Carbonari A, Quaquero E (2014) Preliminary tests on a wireless sensor network for pervasive dust monitoring in construction sites. Open Environ Eng J 7(1):10–18
33. Pianalto FS, Yool SR (2013) Monitoring fugitive dust emission sources arising from construction: a remote-sensing approach. GIScience Remote Sens 50(3):251–270
34. Borana R (2021) Pollution and construction sites. Blog, pp 1–4
35. Paolucci R, Rotilio M, Ricci S, Pelliccione A, Ferri G (2022) A sensor-based system for dust containment in the construction site. The Routledge Companion to Architectural Pedagogies of the Global South, pp 162–173
36. Sahu R, Dixit KK, Mishra S, Kumar P, Shukla AK, Sutaria R, Tiwari S, Tripathi SN (2020) Validation of low-cost sensors in measuring real-time PM10 concentrations at two sites in Delhi national capital region. Sensors (Switzerland) 20(5):1–19
37. Sandanayake M, Zhang G, Setunge S (2019) Estimation of environmental emissions and impacts of building construction–A decision making tool for contractors. J Build Eng 21:173–185
38. Sandanayake M, Zhang G, Setunge S, Luo W, Li CQ (2017) Estimation and comparison of environmental emissions and impacts at foundation and structure construction stages of a building–A case study. J Clean Prod 151:319–329
39. Schwarz AD, Meyer J, Dittler A (2018) Opportunities for low-cost particulate matter sensors in filter emission measurements. Chem Eng Technol 41(9):1826–1832
40. Shaikh ZG, Bhavay Sharma Manager DASDVADKW (2020) Handbook of Clean Construction

41. Singh S (2021) Construction dust pollution. 1–2. Schwarz AD, Meyer J, Dittler A (2018) Opportunities for low-cost particulate matter sensors in filter emission measurements. Chem Eng Technol 41(9):1826–1832
42. Smaoui N, Kim K, Gnawali O, Lee YJ, Suh W (2018) Respirable dust monitoring in construction sites and visualization in building information modeling using real-time sensor data. Sens Mater 30(8):1775–1786
43. Sung WT, Hsiao SJ (2021) Building an indoor air quality monitoring system based on the architecture of the Internet of Things. EURASIP J Wirel Commun Netw 2021(1):1–41
44. Thomas CM (2014) Proceedings of the 18th international symposium on advancement of construction management and real estate. In: Proceedings of the 18th international symposium on advancement of construction management and real estate
45. USEPA (1999) National air quality and emissions trends report, 1999 EPA 1999 annual mean PM 2.5 concentrations (μg/m^3). March, 258
46. USEPA (2020) Basic information about air emissions monitoring. United States environmental protection agency, pp 4. Wang Y, Li J, Jing H, Zhang Q, Jiang J, Biswas P (2015) Laboratory evaluation and calibration of three low-cost particle sensors for particulate matter measurement. Aerosol Sci Technol 49(11):1063–1077
47. Wang Y, Li J, Jing H, Zhang Q, Jiang J, Biswas P (2015) Laboratory evaluation and calibration of three low-cost particle sensors for particulate matter measurement. Aerosol Sci Technol 49(11):1063–1077
48. Welsh LC (2021) Monitoring methodologies. Air Quality in Wales
49. Wojcikowski M, Pankiewicz B, Bekasiewicz A, Cao TV, Lepioufle JM, Vallejo I, Odegard RPHH (2022) A surrogate-assisted measurement correction method for accurate and low-cost monitoring of particulate matter pollutants. Meas: J Int Meas Confed 200:111601
50. Yang L, Jiang P, Wang F, Wang X (2017) Region-based fully convolutional Siamese networks for robust real-time visual tracking institute of artificial intelligence and robotics 710049, 28 Xianning Road, Xi ' an, China. IEEE Int Conf Image Process (ICIP) 2017:1–5
51. Zafra-Pérez A, Boente C, de la Campa AS, Gómez-Galán JA, de la Rosa JD (2023) A novel application of mobile low-cost sensors for atmospheric particulate matter monitoring in open-pit mines. Environ Technol Innov 29:102974

Multifractal Cross-Correlation Analysis of Particulate Matter Concentrations

Adarsh Sankaran, Susan Mariam Rajesh, and Muraleekrishnan Bahuleyan

Abstract Assessing and managing air quality can be made easier for policymakers by examining changes in PM concentrations over time. Fractal characterization of $PM_{2.5}$ and PM_{10} of Thiruvananthapuram City in Kerala is performed using concentration at various locations using multifractal detrended fluctuation analysis (MFDFA). The cross-correlation of PM concentration in a multifractal standpoint is proposed in this study for the first time. It was observed that $PM_{2.5}$ is more multifractal and complex than PM_{10}. Knowledge about time series behavior can be used to make prediction models for PMs. The correlation patterns are found to be different for PM with different variables. Declining trend and negative correlation across the scales are noted for PM-relative humidity and PM-temperature relationships. PM-solar radiation relationship showed an increasing pattern and positive correlations across time scales. All the time series displayed multifractality on considering pair-wise analysis using MFCCA, but there is a reduction of multifractality over individual cases, during PM-meteorological variable paired analysis.

Keywords PM · Correlation · Multifractal · Meteorological factors

1 Introduction

A major problem observed in cities around the world is air pollution which is governed by anthropogenic emissions and meteorological factors. The air pollutant series is characterized by its complex behavior and fluctuations which makes its predictions challenging [2]. Therefore, proper understanding of air pollution time series is essential for improving the predictability and early warning. Air pollutants in the form of particulate matter (PM) of different sizes bring even more concerns on

A. Sankaran · S. M. Rajesh (✉) · M. Bahuleyan
Department of Civil Engineering, TKM College of Engineering Kollam, Kollam, India
e-mail: susanmrajesh@gmail.com

A. Sankaran
e-mail: adarsh1982@tkmce.ac.in

S. M. Shiva Nagendra et al. (eds.), *Select Proceedings of the 8th Indian International Conference on Air Quality Management (IICAQM 2023)*, Lecture Notes in Civil Engineering 582, https://doi.org/10.1007/978-981-96-2359-4_2

controlling air quality standards of cities. PM with diameter lower than 2.5 and 10 has a drastic impact on human health, because of which proper understanding on its fluctuation characteristics is highly warranted. Since we can extract shorter time scale data from longer time scale data, scale invariance is the essential quality that aids in prediction [13]. Additionally, understanding the scaling behavior of PM data is useful for managing air pollution [23]. Several techniques like fractal analysis prove useful for examining the concept of scale invariance [17]. Nevertheless, it was shown that a single fractal exponent was insufficient to explain the intricate geophysical processes,as a result, multifractal detrended fluctuation analysis (MFDFA), introduced by Kantelhardt et al. [11], became widely utilized to evaluate the scaling behavior of such systems. Due to the influence of multiple associated factors like meteorological variables, diffusion, dilution, and anthropogenic forcing, PM time series are often complex with many scaling exponents. Several studies have been done to understand the scaling properties of air pollutant concentrations around the world [3, 6–9, 13–16, 19–21, 27, 29, 30]. The assessment of air pollution concentrations is significantly impacted by several elements of air pollution systems, such as solar radiation, atmospheric self-purification, and meteorological conditions [26]. Thus, investigating the scale-dependent correlations in a multifractal standpoint may help to get useful insights on the predictability of PM concentration. Unfortunately, such studies are rare in Global context, along with the scarcity of studies on multifractality of air pollutants in the Indian context. On considering the Indian conditions, only few studies attempted multifractal analysis of PMs [4, 16]. Maninaran and Narayana [18] analyzed the multifractal properties and cross-correlation characteristics of Air Pollution Index (API) data using MF detrended cross-correlation analysis (MFDCCA) method. They have considered API data of 9 air pollutants present in the campus of University of Hyderabad for 2013–2016 period. They reported anti-correlation and multifractality in behavior for all the 36 cases. PM2.5 and PM10 were found to behave in an anti-correlated manner with all other air contaminants. Chelani and Gautam [4] evaluated the multifractality of wet and dry seasonal PM data of six cities in India using MFDFA. Detrended cross-correlation analysis (DCCA) is a widely used technique to investigate the relationship between numerous related geophysical variables [24]. For conducting such scale-dependent correlation studies, its multifractal extension, known as MFDCCA [31], and the improved alternative multifractal cross-correlation analysis (MFCCA), are dependable possibilities Oświęcimka et al. [22], Wątorek et al. [28], Sankaran et al. [1]. The key objectives of this inquiry include:

(i) Analyze the multifractal properties of $PM_{2.5}$, PM_{10} and four of the meteorological variables of Thiruvananthapuram City in Kerala, India;
(ii) Analyze the scale-dependent correlations of PM and meteorological variables in a multifractal standpoint, using the sign preserved MFCCA approach.

2 Materials and Methods

2.1 *Data and Study Area*

Thiruvananthapuram (8.5241°N, 76.9366°E) is the capital of Kerala, in southern India. Plamoodu is the monitoring site operated by Central Pollution Control Board (CPCB) that is located about 10 km from the coast for ambient air quality. The measurements of the daily PM_{10} and $PM_{2.5}$ levels and meteorological parameters during the period 2018–2021 are used for this study. Less than 10% of the data was found to be missing, according to the analysis of the data. Using the Predictive Mean Matching (PMM) method in R programming, the missing data was determined [25].

2.2 *MFDFA*

The algorithm of MFDFA is described well in the literature [10, 11]. It involves (i) the computation of the accumulated deviation of the 'profile' of the time series, which is the series of deviation from its mean (ii) division of the profile into a number of non-overlapping segments of length *s* (referred as scale), *Ns*, (iii) least square fitting of each non-overlapping segment using a polynomial *m*, to remove the local trends. Here, the deviation of the polynomial fit from the 'profile' provides the detrended time series; (iv) finding the variance of the series, called as 'fluctuation function ($F_q(s)$)', is raised for different moment orders 'q' and the averaging over all segments. The scaling and multifractal behavior are confirmed if $F_q(s)$ follows a power law with s.

2.3 *Multifractal Parameters*

Then the slope of $F_q(s)$ versus plot at logarithmic scale gives the generalized Hurst exponent (H). In terms of the series' long/short memory persistence, the value of $h(2)$ is comparable to the traditional H. The derivative of q-order mass exponent ($\tau(q) = qh(q) - 1$) gives singularity exponent which helps comment on the multifractality of time series. The singularity spectrum is defined by the plot between $f(\alpha)$ versus α, where $f(\alpha) = q\alpha - \tau(q)$ is a useful measure to comment on the strength of multifractality of the series. Larger the base width (W), larger will be the multifractality. In a multifractal time series, the singularity spectrum takes the form of an inverted parabola, where the left wing represents negative q-orders, and the right wing corresponds to positive q-orders. The shape of the singularity spectrum $f(\alpha)$ curve holds important information about the distribution characteristics of the analyzed dataset, providing insights into the singular nature of the temporal data series. Asymmetry index (R) is defined as the ratio of the difference between ($\Delta\alpha_L - \Delta\alpha_R$) and ($\Delta\alpha_L + \Delta\alpha_R$) where $\Delta\alpha_L$ and $\Delta\alpha_R$ refer the width of left and right wings of the singularity spectrum. A positive R value suggests a leftward shift in the multifractal spectrum, which is likely due to the presence of local high fluctuations. In contrast, a negative R value indicates a rightward shift with local low fluctuations. The complexity of the time series is shown by the value of α for zero moment order, also referred to as the Holder exponent (α_0).

2.4 MFCCA

When estimating generalized moments, MFCCA can take into account the sign of the fluctuation function. This feature makes it possible to compute the detrended covariance function and makes it easier to evaluate the joint and individual persistence between the two series. In the MFCCA algorithm, the fluctuation function in each segment reads as:

$$\mathrm{ff}^2_{\mathrm{XY}}(\nu, S) = \left\{ \frac{1}{S} \sum_{k=i}^{S} \left[\begin{pmatrix} X\left((\nu - 1) + k\right) - p^m_{X,\upsilon}(k)\right) \\ \times\left(Y((\nu - 1) + k) - p^m_{Y,\upsilon}(k)\right) \end{pmatrix} \right] \right\} \tag{1}$$

$$F^q_{\mathrm{XY}}(s) = \frac{1}{2N_s} \sum_{\upsilon=0}^{2N_s-1} \mathrm{sign}\left[\mathrm{ff}^2_{\mathrm{XY}}(\upsilon, s)\right] |\mathrm{ff}^2_{\mathrm{XY}}(\upsilon, s)|^{\frac{q}{2}} \tag{2}$$

The cross-correlation coefficient is defined as the ratio between the detrended covariance function F_{XY} and the detrended variance functions F_X and F_Y [12]

$$\rho_{\mathrm{XY}} = \frac{F^q\mathrm{XY}}{\sqrt{F^q\mathrm{XF}^qY}} \tag{3}$$

Theoretically, the value of ρ_{XY} lies in the range [−1, 1]. With the use of cross-correlation, one may estimate the scale-dependent correlation between two candidate time series, providing a more thorough comprehension of the actual physical link between the variables.

3 Results and Discussion

In this analysis, firstly the data is examined in detail. In order to prevent any bias, the datasets are subjected to seasonal trend decomposition (STL) (Cleveland [5]. The deseasonalized datasets of PM and related parameters are estimated by considering minimum scale of a week and maximum scale as L/4 [7]. The fluctuation functions, multifractal spectrum, GHE plot, and mass exponent plots of all the series are results of MFDFA analysis as presented in Figs. 1, 2, 3 and 4.

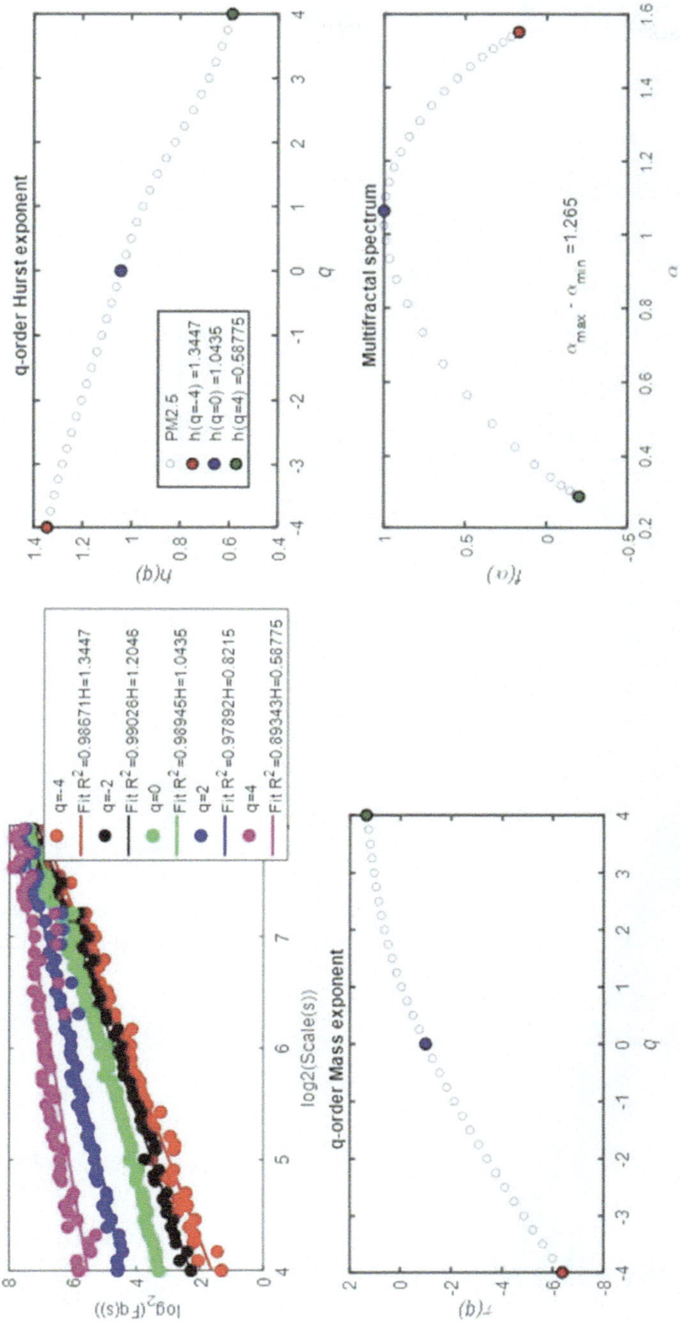

Fig. 1 Fluctuation function and multifractal plots of $PM_{2.5}$

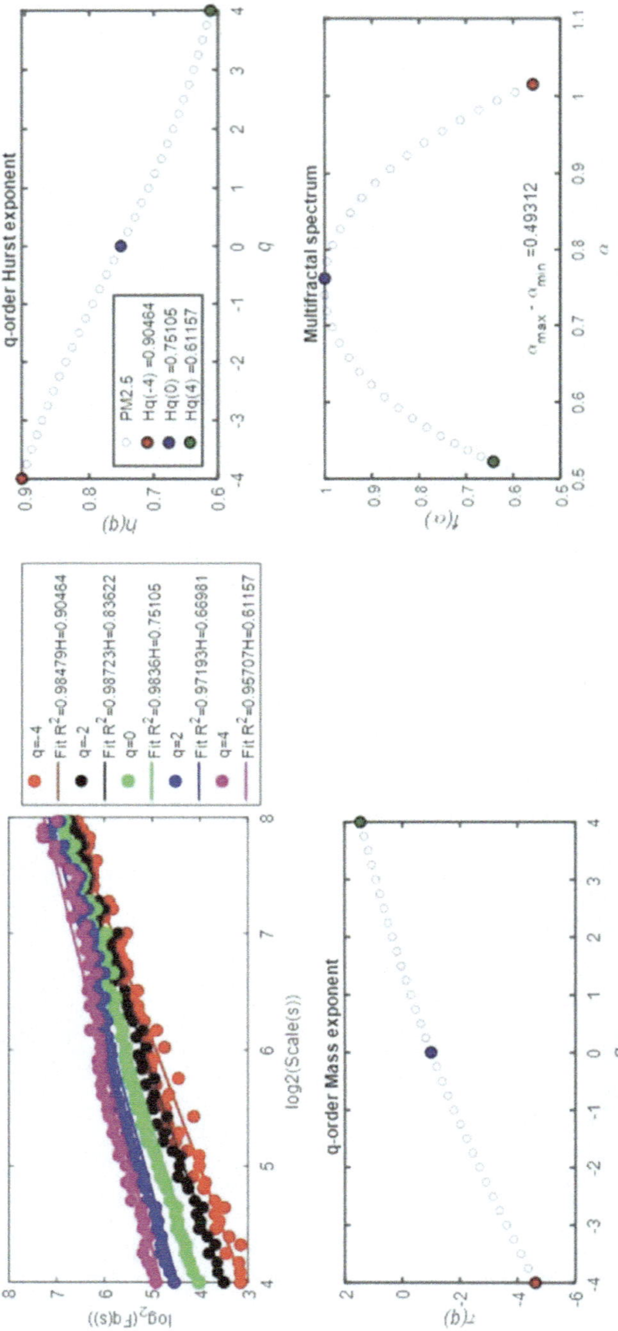

Fig. 2 Fluctuation function and multifractal plots of PM_{10}

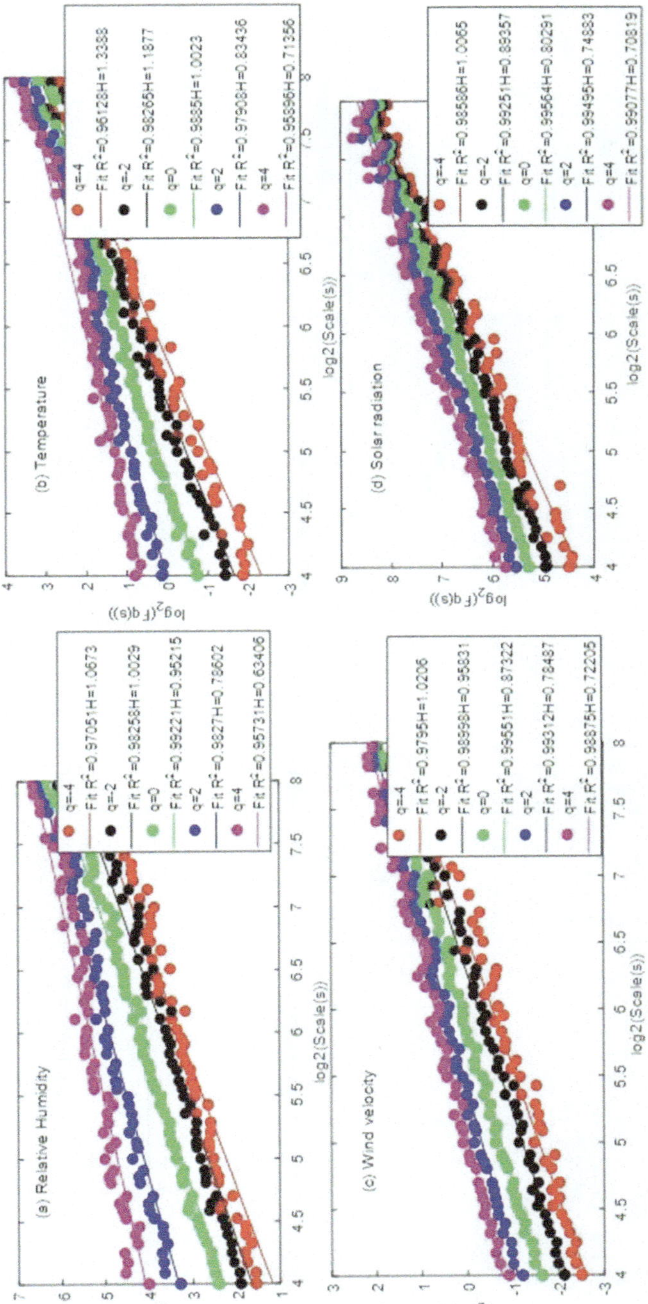

Fig. 3 Fluctuation functions of meteorological variables **a** RH; **b** T; **c** U; **d** R

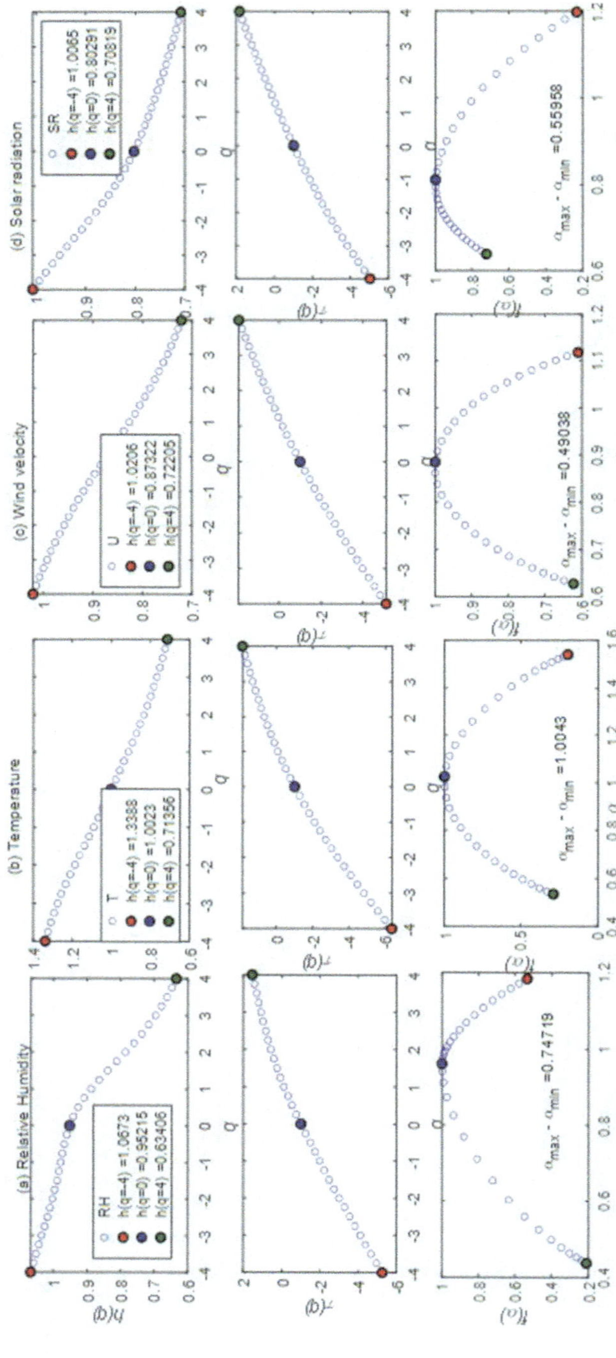

Fig. 4 Multifractal analysis plots of meteorological variables **a** RH; **b** T; **c** U; **d** R. The GHE plot, the mass exponent plot, and the multifractal spectrum are displayed in the upper, middle, and lower panels, respectively

Table 1 Multifractal parameters of PM and meteorological variables

Variable	H	WI	R	Δh(q)	ΔH	α_0
$PM_{2.5}$	0.822	1.265	−0.165	−0.375	0.757	1.023
PM_{10}	0.670	0.493	0.115	0.084	0.293	0.740
Relative humidity (RH)	0.786	0.747	−0.349	−0.322	0.433	0.941
Temperature (T)	0.834	1.004	0.114	0.092	0.625	0.980
Wind velocity (U)	0.785	0.490	0.046	0.012	0.299	0.862
Solar radiation (R)	0.749	0.560	0.441	0.487	0.298	0.795

By assessing the cross-correlations among various PM components, policymakers can gain insights into the sources and dynamics of air pollution in Indian cities. This information can aid in the development of targeted and effective policies to address the specific challenges of air pollution, such as identifying pollution sources, establishing emission control measures, and improving air quality management strategies, ultimately contributing to better public health and environmental outcomes.

Table 1 shows that $PM_{2.5}$ shows more multifractality and complex behavior when compared with PM_{10}. For this observatory station, wind velocity time series possess less multifractality, which is in contrast to the notion of complex behavior of wind velocity. This might be because of the stable wind pattern in the location of Plamoodu observatory of Thiruvananthapuram City. Thus, it can be implied that the complexity of the series might be depending on location and other climatic parameters. All the series are found to be possessing strong long memory persistence and the highest persistence was noted to be for Temperature series. One can see a good similarity between the spectral width (W) and ΔH, both are useful measures for detecting the multifractal strength. A right-skewed spectrum is indicated by a positive R value, and a left-skewed spectrum is indicated by a negative R value. These values indicate the presence of complexity, which is characterized by large fluctuations and multifractality at a larger scale, while showing insensitivity to smaller amplitude fluctuations. $PM_{2.5}$ was found to have negative asymmetry while PM_{10} was found to have a positive asymmetry. Identical nature (in terms of sign) was also noted in $\Delta h(q)$ values. There exists a good correlation of 0.92 between H and α_0.

MFCCA application was performed by considering the scaling range of weekly to annual scale. The values of $q = 2$ are known to be a good estimate to analyze the multifractality, for which the properties are extracted. Figures 5 and 6 represent MFCCA plots for $PM_{2.5}$ and PM_{10}, respectively.

The combined impact of PMs' multifractality and MVs is summed up in Table 2, and it is clear from the table that the series are persistent even when the joint influence is taken into account. It is found that the combined persistence is roughly 50% that of the individual persistence. Under this case also the multifractality gets preserved. But the assessment shows an overall reduction in the multifractal property. The overall linear correlation between PM and MV is weak across the scales. Irrespective of the nature of the asymmetry of individual series, five out of eight spectra displayed a negative asymmetry.

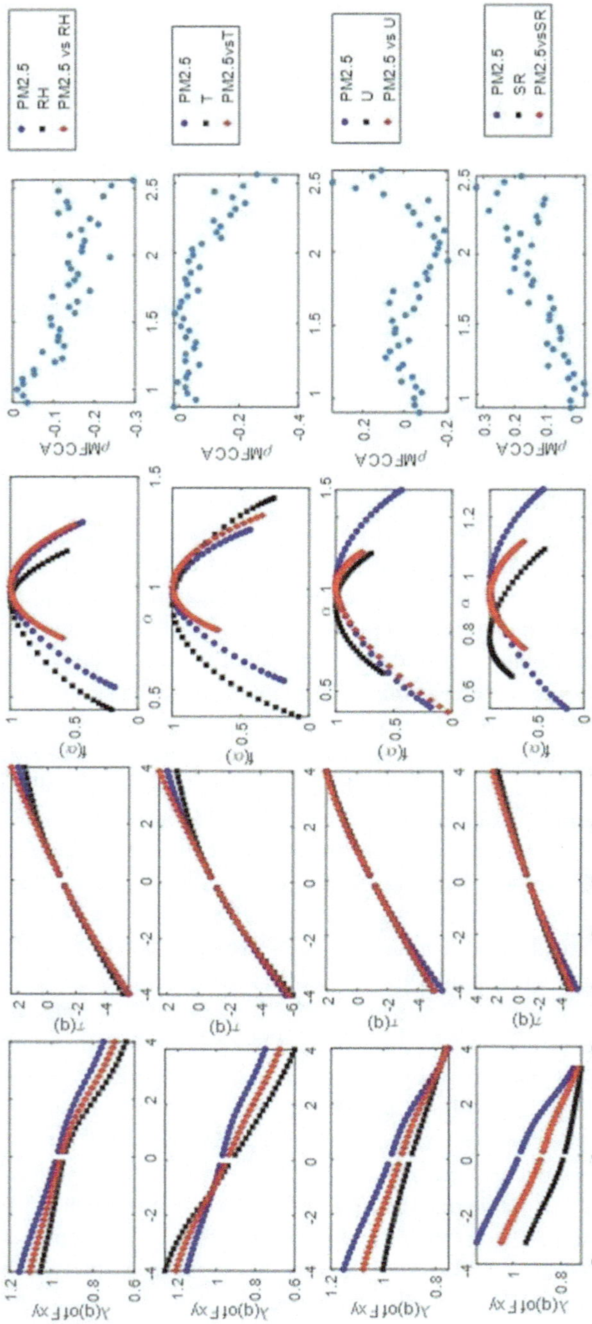

Fig. 5 MFCCA plots for $PM_{2.5}$

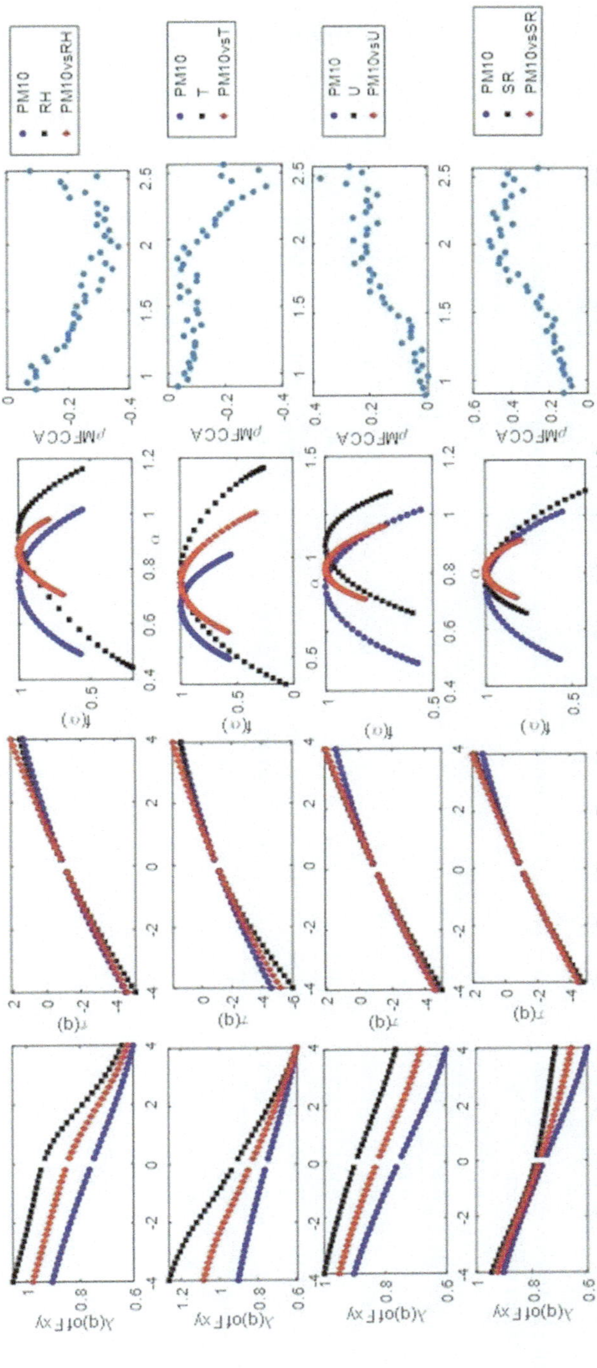

Fig. 6 MFCCA plots for PM_{10}

Table 2 Joint effect of multifractality of PM and meteorological variables (MVs), ρs-seasonal correlation; ρa-annual correlation; ρo-overall correlation; x-stands for the PM type, y-stands for MV

	H_x	H_y	H_{xy}	ρ_s	ρ_a	ρ_o	W_x	W_y	W_{xy}	R_x	R_y	R_{xy}
$PM_{2.5}$ and RH	0.881	0.791	0.837	−0.162	−0.291	0.027	0.749	0.721	0.519	0.160	0.394	−0.096
$PM_{2.5}$ and T	0.881	0.742	0.813	−0.036	−0.265	0.104	0.749	1.084	0.570	0.160	0.028	−0.268
$PM_{2.5}$ and U	0.881	0.826	0.853	−0.196	0.113	0.022	0.749	0.413	0.551	0.160	0.141	0.583
$PM_{2.5}$ and SR	0.881	0.746	0.813	0.166	0.182	0.074	0.749	0.434	0.366	0.160	0.426	0.012
PM_{10} and RH	0.669	0.791	0.734	−0.320	−0.095	−0.116	0.522	0.721	0.274	0.006	0.394	0.175
PM_{10} and T	0.669	0.742	0.705	−0.069	−0.206	0.240	0.522	1.084	0.596	0.006	0.028	−0.182
PM_{10} and U	0.669	0.826	0.746	0.210	0.265	0.145	0.522	0.413	0.249	0.006	0.141	−0.176
PM_{10} and SR	0.669	0.746	0.702	0.448	0.269	0.246	0.522	0.434	0.203	0.006	−0.426	−0.081

Figure 5 shows that the GHE values of PM are consistently more than the MVs for all q-order for the $PM_{2.5}$-U and PM-SR relationships. For $PM_{2.5}$-RH and $PM_{2.5}$-T association, such a behavior is noted only for positive moment orders. Interestingly the meteorological variables bear higher persistence than PM_{10}, for all moment orders. In general, joint multifractal spectra are smaller in base width when compared with that of individual spectra. This indicates a lower multifractality when considering the joint effect of PM-MV pairs in the multifractal analysis. Correlations are negative in PM-RH and PM-T relationships across all the scales. But the correlations are positive in PM-SR relationship over the scales. On considering the wind velocity, there is a fluctuation in the nature of relationships in $PM_{2.5}$-U relation, whereas relationship is consistently positive in PM_{10}- U relationship. In general, the correlation patterns showed an increasing trend for PM-U and PM-SR relations, whereas there is a reducing trend of correlations across scales in the PM-RH and PM-T relationships. Such dynamic nature of correlations may be because of the dominant group of gaseous pollutants as controlling factors, which also need to be accounted for in performing MF cross-correlation studies. Also, the interplay between the different meteorological factors needs to be untangled in order to decipher the physical reasoning behind the dynamic multifractal relationships of PM concentrations. This study being one of the initial attempts on multifractal cross-correlations of PM, a lot of improvisations need to be done. In order to arrive at strong conclusions, similar studies are to be performed for different cities of diverse geographical locations, varying data length, considering different temporal resolutions (like hourly and daily), etc. Moreover, this study has not accounted for the effect of rainfall in the form of wet and dry spells and has not considered the role of gaseous pollutants like CO, NOx, etc., but such studies are in progress. Understanding fractality is valuable for enhancing prediction models and informing policy decisions. These findings can contribute to a deeper comprehension of the factors governing the dynamics of the PM series and lead to more effective management of early-warning systems.

4 Conclusions

The research delved into the multifractality and cross-relationships among PM and MVs in Thiruvananthapuram City, utilizing daily data from the years 2018 to 2021. The MFDFA analysis indicated that both types of PM ($PM_{2.5}$ and PM_{10}) and meteorological factors (RH, T, U, and SR) displayed robust, long-term persistence and multifractal properties. Higher multifractality was found to be for $PM_{2.5}$ over $PM_{10,}$ and temperature series shows the highest multifractality among different MVs. All the series possessed multifractal behavior on considering pair-wise analysis using MFCCA, but there is a reduction of multifractality over individual cases, during PM-MV paired analysis. PM-SR relations showed weak to moderate correlation across diverse scale ranges. The correlations showed an increasing pattern for PM-U and

PM-SR relations, whereas there is a reduction of correlation in the PM-RH and PM-T relationships. The dynamics in the correlations are to be tracked, and it will be helpful for developing potential multiscale models which are believed to capture the highly complex fluctuations of PM concentrations.

References

1. Sankaran A, Plocoste T, Nourani V, Vahab S, Salim A (2023) Assessment of multifractal fingerprints of reference evapotranspiration based on multivariate empirical mode decomposition. Atmosphere 14(8):1219
2. Blessy A, Paul JJ, Gautam S, Shany JV, Sreenath M (2023) IoT-based air quality monitoring in hair salons: screening of hazardous air pollutants based on personal exposure and health risk assessment. Water Air Soil Pollut 234:336
3. Chelani AB (2016) Long memory in air pollutant concentrations. Atmos Res 171:1–4
4. Chelani AB, Gautam S (2022) Lockdown during COVID-19 pandemic: a case study from Indian cities shows insignificant effects on persistent property of urban air quality. Geosci Front 13:101284
5. Cleveland RB, Cleveland WS (1990) STL: a seasonal-trend decomposition procedure based on loess. J Off Stat 6:3–33
6. Dong Q, Wang Y, Peizhi L (2017) Multifractal behavior of an air pollutant time series and the relevance to the predictability. Environ Pollut 222:444–457
7. Drożdż S, Minati L, Oświęcimka P, Stanuszek M, Watorek M (2019) Signatures of the crypto-currency market decoupling from the Forex. Future Internet 11(7):154
8. Gao X, Wang X, Shi H (2019) Multifractal cascade analysis on the nature of air pollutants concentration time series over China. Aerosol Air Quality Res 19:2100–2114
9. He HD, Pan W, Lu WZ, Xue Y, Peng GH (2016) Multifractal property and long-range cross-correlation behavior of particulate matters at urban traffic intersection in Shanghai. Stoch Env Res Risk Assess 30:1515–1523
10. Ihlen EAFE (2012) Introduction to multifractal detrended fluctuation analysis in MATLAB. Front Physiol 3:141
11. Kantelhardt JW, Zschiegner SA, Bunde EK, Havlin S, Bunde A, Stanley HE (2002) Multifractal detrended fluctuation analysis of nonstationary time series. Physica A 316:87–114
12. Kwapień J, Oświęcimka P, Drożdż S (2015) Detrended fluctuation analysis made flexible to detect range of cross-correlated fluctuations. Phys Rev E 92(5):052815
13. Lee CK (2002) Multifractal characteristics in air pollutant concentration time series. Water Air Soil Pollut 135:389–409
14. Lee CK, Juang LC, Wang CC, Liao YY, Yu CC, Liu YC, Ho DS (2006) Scaling characteristics in ozone concentration time series (OCTS). Chemosphere 62:934–946
15. Li X (2021) On the multifractal analysis of air quality index time series before and during COVID-19 partial lockdown: a case study of Shanghai, China. Physica A 565:125551
16. Liu Z, Wang L, Zhu H (2015) A time-scaling property of air pollution indices: a case study of Shanghai, China. Atmos Pollut Res 6:886–892
17. Mandelbrot BB (1989) Multifractal measures, especially for the geophysicist (pp 5–42). Birkhäuser Basel
18. Manimaran P, Narayana AC (2018) Multifractal detrended cross-correlation analysis on air pollutants of University of Hyderabad Campus. India Phys A Stat Mech Appl 502:228–235
19. Masseran N (2022a) Multifractal characteristics on temporal maximum of air pollution series. Math 10:3910. https://doi.org/10.3390/math10203910
20. Masseran N (2022) Power-law behaviors of the severity levels of unhealthy air pollution events. Nat Hazards 112:1749–1766

21. Nikolopoulos D, Moustris K, Petraki E, Cantzos D (2021) Long-memory traces in PM10 time series in Athens, Greece: investigation through DFA and R/S analysis. Meteorol Atmos Phys 133:261–279
22. Oświęcimka P et al (2014) Detrended cross-correlation analysis consistently extended to multifractality. Phys Rev E 89(2):023305
23. Plocoste T, Calif R, Jacoby-Koaly S (2017) Temporal multiscaling characteristics of particulate matter PM10 and ground-level ozone O_3 concentrations in Caribbean region. Atmos Environ 169:22–35
24. Podobnik B, Stanley HE (2008) Detrended cross-correlation analysis: a new method for analyzing two nonstationary time series. Phys Rev Lett 100(8):084102
25. Rubin DB (1986) Statistical matching using file concatenation with adjusted weights and multiple imputations. J Bus Econ Stat 4(1):87–94
26. Shen C, Huang Y, Yan Y (2016) An analysis of multifractal characteristics of API time series in Nanjing, China. Physica A 451:171–179
27. Wang Q, Zhao T, Wang R, Zhang L (2022) Backward trajectory and multifractal analysis of air pollution in Zhengzhou region of China. Math Probl Eng 2022:1–17
28. Wątorek M, Drożdż S, Oświęcimka P, Stanuszek M (2019) Multifractal cross-correlations between the world oil and other financial markets in 2012–2017. Energy Eco 81:874–885
29. Xue Y, Pan W, Lu W-Z, He H-D (2015) Multifractal nature of particulate matters (PMs) in Hong Kong urban air. Sci Total Environ 532:744–751
30. Yuval D, Broday M (2010) Studying the time scale dependence of environmental variables predictability using fractal analysis. Environ Sci Technol 44:4629–4634
31. Zhou WX (2008) Multifractal detrended cross-correlation analysis for two non-stationary signals. Phys Rev E 77:066211

Comparative Analysis of Machine Learning Classification Techniques for Forecasting the $PM_{2.5}$ Category

Sajeed I. Ghanchi, Dishant M. Pandya, and Manan Shah

Abstract $PM_{2.5}$ is a hazardous form of air pollutant because of its diminutive size. Many Indian cities have experienced extremely high levels of $PM_{2.5}$ in recent years. This study aims to examine the accuracy of various algorithms for forecasting $PM_{2.5}$ category for the next day. Authorities and individuals alike might benefit from air quality forecasts that allow them to prepare for days with high pollution levels. Seven machine learning and one deep learning classification algorithms were employed to forecast the $PM_{2.5}$ category under Indian NAAQS for three diverse regions in Gujarat, India. According to the results, the support vector machine (SVM) outperformed the deep learning algorithm and other machine learning algorithms with an average accuracy of 71.33%. Furthermore, the result suggests that more accurate outcomes can be obtained by reducing the number of categories used for categorization, such as two-category classification, good or bad.

Keywords $PM_{2.5}$ · Category forecast · Machine learning · Classification algorithm

S. I. Ghanchi · D. M. Pandya (✉)
Department of Mathematics, Pandit Deendayal Energy University, Gandhinagar 382426, Gujarat, India
e-mail: dishantpandya777@gmail.com

S. I. Ghanchi
e-mail: ghanchisajeed786@gmail.com

M. Shah
Department of Chemical Engineering, Pandit Deendayal Energy University, Gandhinagar 382426, Gujarat, India
e-mail: Manan.Shah@spt.pdpu.ac.in

S. M. Shiva Nagendra et al. (eds.), *Select Proceedings of the 8th Indian International Conference on Air Quality Management (IICAQM 2023)*, Lecture Notes in Civil Engineering 582, https://doi.org/10.1007/978-981-96-2359-4_3

1 Introduction

Particulate matter (PM) refers to a heterogeneous combination of solid particles (e.g., dust, dirt, soot, smoke, etc.) and liquid droplets that are present in the atmosphere. In accordance with their aerodynamic diameter, particulate matter is commonly classified into two primary categories: (i) coarse particulate matter (PM_{10}), which refers to particles with an aerodynamic diameter of 10 μm or smaller, and (ii) fine particulate matter ($PM_{2.5}$), which refers to particles with an aerodynamic diameter of 2.5 μm or smaller [1]. Because of its diminutive size, $PM_{2.5}$ can enter the bloodstream and spread throughout the body. Therefore, $PM_{2.5}$ poses a greater health risk than larger particles like PM_{10} [1, 2, 7, 10, 11].

According to the IQAir study, 39 cities in India were among the 50 most polluted cities in the world, based on average annual $PM_{2.5}$ levels in 2022. Also, $PM_{2.5}$ levels were at least seven times higher than the yearly WHO limit in more than 60% of the Indian cities studied [4].

The Indian government has implemented various initiatives to minimize air pollution. Over time, these actions and precautions would lead to better air quality. However, it could take some time for the effects of these measures to become visible, so there is a lag between efforts and a noticeable decrease in air pollution in the real world. To bridge this gap and adapt to the current level of air pollution, novel approaches are required. One essential component of the system would be an advanced Real-Time Air Quality Forecasting (RT-AQF) tool [6].

Using data from 2004 – 2007, [9] developed artificial neural network models to forecast the highest PM_{10} class for the following day in Lodz, Poland. The results were better for the scenario with fewer prototypes and fewer air quality classes than for the situation with more prototypes and more classes. For the area of Campo de Gibraltar, Spain, [3] predicted the class (peak/no-peak) of the following day's PM_{10}. According to the results, support vector machines outperformed k-nearest neighbors, Bayesian classifiers, and backpropagation multilayer neural networks. Using data from 2009 – 2018 for the city of Cheonan, Korea, [5] examined the accuracy of logistic regression, decision tree, support vector machine, and ensemble models for predicting the PM_{10} class for the following day, as determined by the AQI. The f-score results suggest that SVM is superior to the other models. Shaziayani et al. [8] divided PM_{10} concentration into two classes to make a day-ahead categorical forecast. When compared to the other two methods, decision tree and boosted regression tree, random forest performed better.

This study aims to recommend the best model among several machine learning classification models for forecasting the next day's $PM_{2.5}$ category for nearby regions with diverse characteristics. Authorities and individuals alike need this kind of information in order to take appropriate precautions on days when pollution levels are high.

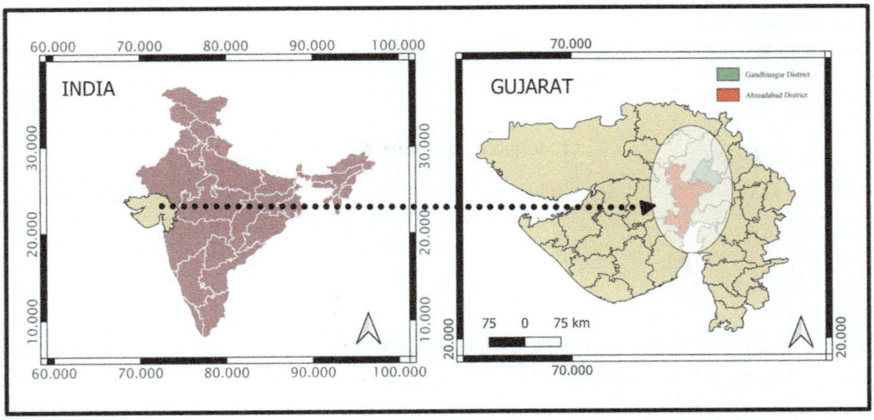

Fig. 1 Study area

2 Dataset and Methodology

2.1 *Study Area*

For this study, Sector 10 (latitude 23.22171 and longitude 72.65433) in the Gandhinagar district, together with Maninagar (latitude 23.002657 and longitude 72.591912) and Vatva (latitude 22.969611 and longitude 72.6435) in the Ahmedabad District, Gujarat, India, were selected as the areas of focus. While Gandhinagar serves as Gujarat's state capital, Ahmedabad is the first Indian city to be inducted into the UNESCO World Heritage City list (Fig. 1).

Maninagar, located in Ahmedabad, is characterized by its vibrant mixed-use environment and a significantly high population density. On the other hand, Vatva, also situated in Ahmedabad, is renowned as an industrial hub that offers ample manufacturing facilities and employment prospects. In contrast, Sector 10 is primarily a residential locality emphasizing well-organized urban development and incorporating environmentally friendly places. All three of these regions are situated within a 15-kilometer radius.

2.2 *Dataset*

The data of daily mean concentrations of $PM_{2.5}$, PM_{10}, nitrogen monoxide (NO), nitrogen dioxide (NO_2), nitrogen oxides (NO_x), ammonia (NH_3), sulfur dioxide (SO_2), carbon monoxide (CO), ozone (O_3), benzene, toluene, ethylbenzene, MP-Xylene, xylene, relative humidity (RH), wind speed (WS), wind direction (WD), and solar radiation (SR) were acquired from the official website of the Central Pollution

Table 1 $PM_{2.5}$ categorization

Category	Concentration range (in $\mu g/m^3$)
Good	$0 - 30$
Satisfactory	$30 - 60$
Moderately polluted	$60 - 90$
Poor	$90 - 120$
Very poor	$120 - 250$
Severe	250+

Control Board (CPCB), a government agency of India. The dataset covers a period from February 2019 to July 2023.

Classification of $PM_{2.5}$ was done in accordance with the Air Quality Index (AQI) and Indian National Ambient Air Quality Standards (NAAQS). The categorization of $PM_{2.5}$ is presented in Table 1.

2.3 Methodology

The analysis and model development processes are conducted utilizing the Python programming language. The method of linear interpolation was utilized to fill the missing values. The linear interpolation formula is as follows:

$$y = y_1 + \frac{y_2 - y_1}{x_2 - x_1}(x - x_1) \tag{1}$$

where (x_1, y_1) and (x_2, y_2) are coordinates and y is interpolated value at x (where x lies between x_1 and x_2). The box plot is used to detect outliers, whereas correlation is utilized to select parameters for model development.

Seven machine learning algorithms, including decision trees (DTs), random forests (RFs), K-nearest neighbor (KNN), support vector machines (SVMs), logistic regression (LR), Naive Bayes classifiers (NBCs), gradient boosted decision trees (GBDTs), and one deep learning approach, multilayer perceptron (MLP), are used to forecast the category of $PM_{2.5}$ for the following day. Figure 2 illustrates and explains the algorithm's flow.

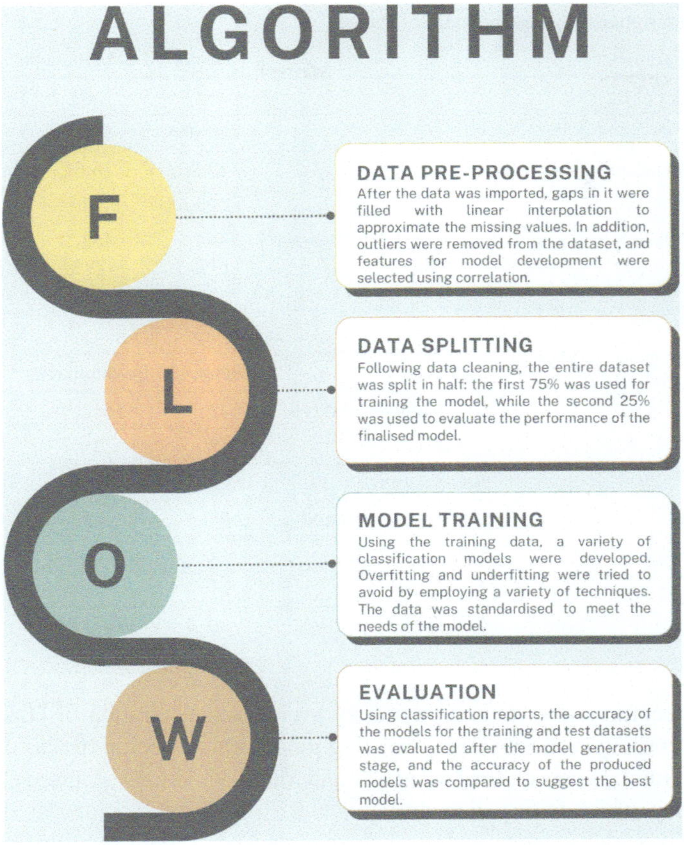

Fig. 2 Flow of algorithm

3 Results and Discussion

3.1 Descriptive Statistics

The daily average concentration of $PM_{2.5}$ for all three regions is statistically analyzed using descriptive statistics. Table 2 provides a summary of the descriptive statistics findings. Sector 10 and Maninagar received 1642 observations, whereas Vatva received 1639 observations. Compared to Maninagar and Vatva's daily average $PM_{2.5}$ concentration, Sector 10 has a noticeably lower mean concentration. The standard deviation of $PM_{2.5}$ average concentration was observed more than twice at Maninagar and Vatva than at Sector 10. The highest $PM_{2.5}$ concentrations were recorded in sectors 10 and Maninagar during Diwali. However, no similar events were recorded for Vatva, suggesting that the highest concentration is more likely to occur on regular days, which is a significant concern. The National Ambient Air Quality Standards

Table 2 Descriptive statistics summary of $PM_{2.5}$

Statistical Parameter	Sector 10	Maninagar	Vatva
Total number of days	1642	1642	1639
Mean	33.045	50.408	56.218
Standard deviation	13.893	26.613	28.633
Maximum	126.68	463.68	327.08
Number of days NAAQS guidelines violation $(60\mu g/m^3)$	56 (3%)	468 (28%)	576 (35%)

Table 3 Categorical $PM_{2.5}$ count

Category	Sector 10	Maninagar	Vatva
Total number of days	1642	1642	1639
Good	778	325	235
Satisfactory	808	849	828
Moderately polluted	53	365	390
Poor	2	80	129
Very poor	1	22	56
Severe	0	1	1

for India recommended a maximum daily average concentration of $60\mu g/m^3$ for $PM_{2.5}$. According to the findings, violating guidelines in Sector 10 was exceptionally uncommon. However, for Maninagar and Vatva, violations occurred every third or fourth day, which is very worrying.

Table 3 provides a categorical breakdown of the $PM_{2.5}$ count. Results indicate that Sector 10 did well, but Maninagar and Vatava did very poorly, with disproportionately high numbers of counts falling into the poor and very poor categories.

3.2 *Outliers Treatment and Feature Selection*

The presence of outliers can reduce the accuracy of a model. Therefore, box plot is employed as a means to identify extreme values. As depicted in Fig. 3, a limited number of $PM_{2.5}$ observations in Maninagar and Vatva are significantly away from the cluster of remaining observations. Data points with values greater than $200\mu g/\text{m}^3$, which exhibit significant deviation from the rest of the dataset, are omitted from the analysis.

Feature selection is a crucial step in the process of developing a model. The utilization of a proper subset of features has the potential to enhance the accuracy of a model. Features that exhibit a correlation coefficient greater than 0.25 with

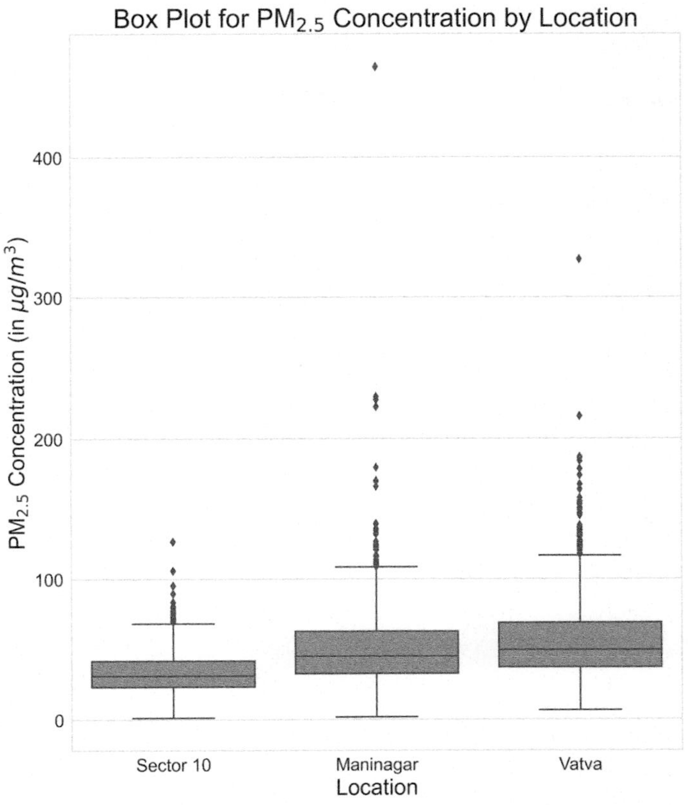

Fig. 3 Box plot

$PM_{2.5}$ were considered in the model development process. In this manner, uncorrelated variables were eliminated from the dataset. Table 4 provides a summary of the correlation between $PM_{2.5}$ and the other parameters.

3.3 Forecasting Analysis

The decision trees are constructed using two approaches: Gini impurity and Entropy. A parameter grid is utilized to mitigate the problem of overfitting and choose the optimal hyperparameters. The parameter grid consists of the following values for the 'max_depth' hyperparameter: 5, 10, 15, 20, and none, and the 'min_samples_leaf' hyperparameter is evaluated with 1, 2, 3, 4, and 5 values. The average accuracy for the test data was similar for both the Gini (0.680) and Entropy (0.673) approaches. For all the regions, the maximum depth of decision trees is found to be 5, while the minimum sample leaf was seen to be either 4 or 5. Furthermore, it was determined

Table 4 Correlation coefficient with $PM_{2.5}$

Parameter	Sector 10	Maninagar	Vatva
PM_{10}	0.786	–	0.847
NO	0.253	0.447	0.513
NO_2	0.386	0.423	0.344
NO_x	0.377	0.454	0.506
NH_3	0.185	–	0.216
SO_2	0.148	0.251	0.325
CO	0.475	0.200	0.520
O_3	0.054	0.011	0.058
Benzene	0.204	0.080	0.249
Toluene	0.075	0.123	–
Eth-Benzene	0.229	–	–
MP-Xylene	0.199	–	–
RH	−0.321	−0.284	–
WS	−0.416	−0.416	−0.275
WD	−0.356	−0.328	−0.323
SR	−0.057	0.117	–
Xylene	0.219	0.135	–

that $PM_{2.5}$ exhibited the highest level of dependability as a metric for forecasting the $PM_{2.5}$ category for the following day.

Random forests, like decision trees, are constructed using the Gini impurity and entropy approaches. In addition to the hyperparameter 'max_depth', the values of 'n_estimators' within the range of [20, 40, 60, 80, 100] are considered. The optimal model was achieved with a maximum depth of 5, while the number of estimators varied from 20 to 100. The random forest models utilizing the Gini index (0.680) and the Entropy measure (0.686) exhibited comparable performance. Furthermore, with the $PM_{2.5}$ concentration, the PM_{10} and relative humidity (RH) were also observed as significant predictors of the $PM_{2.5}$ category for the next day.

The K-nearest neighbor (KNN) algorithm utilizes the concept of distance between data points. Datasets were standardized during the model construction process to mitigate the potential impact of a single parameter on the model. The ideal values of k for Sector 10, Maninagar, and Vatva, selected from a range of 1–25, were determined to be 14, 18, and 24, respectively. The K-nearest neighbor (KNN) algorithm achieved an overall accuracy of 0.646.

Support vector machines (SVMs) are constructed using two kernels, namely linear and radial basis functions (RBF). The hyperparameters 'c' and 'gamma' are chosen as [0.1, 1, 10] and [0.1, 1, scale], respectively, for the construction of the model. The average accuracy achieved for the linear model was 0.713, whereas for the RBF model, it was 0.700. The SVM with a linear kernel demonstrated superior performance compared to the other models.

Logistic regression models were built using hyperparameter 'c' $= [0.001, 0.01, 0.1, 1, 10, 100]$ and 'penalty' $=$ ['l1', 'l2']. Optimal accuracy for all the regions was obtained using $c = 0.001$ and penalty $=$'$l2$' . Average accuracy obtained using this model is 0.690. It's worth noting that the model performed similarly well on both training (0.690) and test data (0.690).

The construction of Naive Bayes classifier models involved utilizing a Gaussian method due to the continuous nature of the input data. The model had an overall accuracy of 0.596, indicating its somewhat worse performance when compared to alternative machine learning techniques.

Gradient boosted decision trees (GBDTs) are generated using a 'learning_rate' of 0.01 and 300 'estimators'. The model demonstrates a noteworthy overall accuracy of 0.696, surpassing the performance of all the other tree-based models. Similar to decision tree and random forest of all the criteria under consideration, $PM_{2.5}$ emerged as the most significant factor in the development of the model.

The construction of the multilayer perceptron, a deep learning model, involved using three different optimizers: adam, sgd, and lbfgs. The hyperparameter 'learning_rate' was selected from the range of $[0.0001, 0.001, 0.01, 0.1]$, while the 'max_iter' was set to 300 and the 'alpha' value was set to 1. The lbfgs approach has the lowest accuracy (0.593) among the three approaches and also inferior to the accuracy attained by all machine learning models. The adam optimizer achieved the most favorable outcomes among the three approaches, with an accuracy of 0.706. Subsequently, the sgd approach produced similar results with an accuracy of 0.703.

The results of the developed models on both the test and train datasets are summarized in Table 5. The issue of overfitting is moreover solved by proper hyperparameter tuning. Sector 10, Maninagar, and Vatva had average accuracy of

Table 5 Result summary for developed models

Sr. No	Model	Approach	Sector 10		Maninagar		Vatva	
			Train	Test	Train	Test	Train	Test
1	DT	Gini	0.83	0.83	0.71	0.60	0.70	0.61
		Entropy	0.83	0.81	0.70	0.61	0.68	0.60
2	RF	Gini	0.85	0.82	0.73	0.61	0.72	0.61
		Entropy	0.84	0.83	0.73	0.63	0.72	0.60
3	KNN	–	0.81	0.82	0.68	0.60	0.65	0.52
4	SVM	Linear	0.81	0.82	0.66	0.67	0.65	0.65
		RBF	0.81	0.83	0.70	0.65	0.68	0.62
5	LR	–	0.80	0.82	0.65	0.66	0.62	0.59
6	NBC	–	0.76	0.76	0.55	0.51	0.59	0.52
7	GBDT	–	0.86	0.82	0.79	0.66	0.78	0.61
8	MLP	adam	0.81	0.82	0.65	0.67	0.66	0.63
		sgd	0.81	0.82	0.68	0.66	0.66	0.63
		lbfgs	1.00	0.74	0.98	0.54	0.98	0.50

0.810, 0.620, and 0.591, respectively. Table 6 summarizes the categorical classification of the $PM_{2.5}$ test data. This classification shows that for Sector 10, $PM_{2.5}$ has classified into mostly two groups, while the number of groups in Maninagar and Vatva is high. This might be the main cause of the Maninagar and Vatva region's decreased accuracy. This suggests that, if the categorization is primarily divided into two principal groups, good or bad, more accurate results can be obtained. Additionally, compared to Sector 10, Maninagar and Vatva's mean and standard deviation of $PM_{2.5}$ were greater. Figure 4 depicts a comparison of the developed model's accuracy using a bar graph. The accuracy can be depicted in the following sequence: SVM > MLP > GBDT > LR > RF > DT > KNN > NBC.

Table 6 Categorical $PM_{2.5}$ count of test data

Category	Sector 10	Maninagar	Vatva
Total number of days	411	410	410
Good	219	201	210
Satisfactory	179	134	96
Moderately polluted	11	61	76
Poor	1	10	22
Very poor	1	4	6
Severe	0	0	0

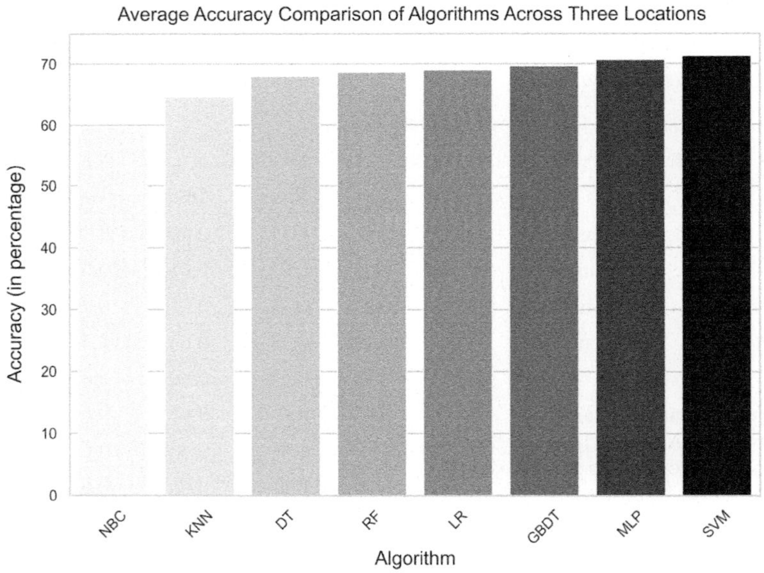

Fig. 4 Accuracy comparison for the developed models

4 Conclusions

In this paper, the accuracy of seven different classification machine learning and one deep learning algorithm models were compared in order to forecast $PM_{2.5}$ category for three regions in Gujarat, India, with diverse characteristics. The outcome indicates that support vector machine outperformed the other machine learning algorithms and a deep learning algorithm. The results obtained were highly accurate for Sector 10, because $PM_{2.5}$ was moreover placed in two categories for Sector 10. This suggests that more accurate results can be obtained if the classification is in fewer categories.

Nomenclature-Acronyms

CPCB: Central Pollution Control Board
DT: Decision Trees
GBDT: Gradient Boosted Decision Trees
KNN: k-Nearest Neighbour
LR: Logistic Regression
MLP: Multilayer Perceptron
NAAQS:National Ambient Air Quality Standards
NBC: Naive Bayes Classifiers
RF: Random Forests
RH:Relative Humidity
SR:Solar Radiation
SVM: Support Vector Machines
WS:Wind Speed
WD:Wind Direction

Ackknowledgements Mr. Sajeed I. Ghanchi is grateful for the financial assistance provided by Pandit Deendayal Energy University, Gandhinagar, Gujarat, India, in the form of a fellowship. The authors are also grateful to the Central Pollution Control Board (CPCB) for providing the data on air quality.

References

1. Particulate matter (PM) Basics|US EPA. https://www.epa.gov/pm-pollution/particulate-matter-pm-basics. Accessed 14 Nov 2023
2. Feng S, Gao D, Liao F, Zhou F, Wang X (2016) The health effects of ambient $PM_{2.5}$ and potential mechanisms. Ecotoxicol Environ Saf 128:67–74. https://doi.org/10.1016/j.ecoenv.2016.01.030
3. García EM, Rodríguez MLM, Jiménez-Come MJ, Espinosa FT, Domínguez IT (2011) Prediction of peak concentrations of PM_{10} in the area of campo de Gibraltar (Spain) using classification models. In Springer eBooks, pp 203–212. https://doi.org/10.1007/978-3-642-19644-7_22
4. IQAIR (2022) First in air quality (online), Available: https://www.iqair.com/world-air-quality-report. Accessed 6 Sept 2023

5. Jung YJ, Cho KW, Lee JS, Oh CH (2021) Comparative analysis of multiple classification models to improve PM_{10} prediction performance. Int J Electr Comput Eng (IJECE) 11(3):2500–2507
6. Marinov E, Petrova-Antonova D, Malinov S (2022) Time series forecasting of air quality: a case study of Sofia city. Atmosphere 13(5):788. https://doi.org/10.3390/atmos13050788
7. Pinkerton KE, Green FH, Saiki C, Vallyathan V, Plopper CG, Gopal V, ..., Schenker MB (2000) Distribution of particulate matter and tissue remodeling in the human lung. Environ Health Perspect 108(11):1063–1069. https://doi.org/10.1289/ehp.001081063
8. Shaziayani WN, Ul-Saufie AZ, Mutalib S, Noor NM, Zainordin NS (2022) Classification prediction of PM_{10} concentration using a tree-based machine learning approach. Atmosphere 13(4):538. https://doi.org/10.3390/atmos13040538
9. Skrzypski J, Kami K, Ski Jach-Szakiel E, Kami W (2009) Application of artificial neural networks for classification and prediction of air quality classes. WIT Trans Ecol Environ 127:219–228
10. Wang G, Zhen L, Lü P, Jiang R, Song W (2013) Effects of ozone and fine particulate matter ($PM_{2.5}$) on rat cardiac autonomic nervous system and systemic inflammation. J Hyg Res 42(4):554–560
11. Xu D, Huang N, Wang Q, Liu H (2008) Study of ambient $PM_{2.5}$ on the influence of the inflammation injury and the immune function of sub chronic exposure rats. J Hyg Res 37(4):423–428

Image-Based Air Quality Estimation Using DINOv2

Chandra Mohan Bhuma

Abstract Air pollution is a major threat to mankind nowadays. Major constituents in the polluted air, i.e., $PM_{2.5}$ (Particulate Matter 2.5) and PM_{10} (Particulate Matter 10) can be estimated using sensors. However, in this work, an algorithm is proposed for estimating the Air Quality Index, $PM_{2.5,}$ and PM_{10} from the outdoor images captured by a webcam. This problem is considered as a regression problem. Features from the images are computed by using a state-of-the-art self-supervised learning model DinoV2. The extracted features are given to a regressor. Various single output regressor and multiple output regressors are explored. The proposed algorithm is tested on the images collected from the cities in India and Nepal (12,240 images). Size of the feature vector is reduced by using genetic algorithm. For a tenfold cross-validation, the value of R^2 (Coefficient of determination) for AQI prediction is well above 0.98 for K-neighbourhood regressor.

Keywords AQI estimation · DINOv2 · Self-supervised model · Feature extraction

1 Introduction

We the human beings need air to breathe. Oxygen is a vital component in the air that when we breathe, it gives energy. In addition to oxygen, air also contains other gases. The quality of air affects the daily life of the people. Air quality is getting worse day by day in the major cities and towns due to industrialization, usage of vehicles, and many other reasons. Quality of the air is indicated by Air Quality Index (AQI). AQI is an index consisting of six pollutants, i.e., $PM_{2.5}$, PM_{10}, O_3 (Ozone), NO_2 (Nitrogen Dioxide), SO_2 (Sulphur Dioxide), and CO (Carbon Monoxide). Instead of considering one parameter, AQI takes care of all the six pollutants in a unified way.

An AQI of 0–50 is good, 51–100 satisfactory, 101–200 moderately polluted, 201–300 poor, very poor 301–400 and is severe if it lies between 401–500. Average

C. M. Bhuma (✉)
Department of ECE, Bapatla Engineering College, Bapatla, India
e-mail: chandramohan.bhuma@becbapatla.ac.in

S. M. Shiva Nagendra et al. (eds.), *Select Proceedings of the 8th Indian International Conference on Air Quality Management (IICAQM 2023)*, Lecture Notes in Civil Engineering 582, https://doi.org/10.1007/978-981-96-2359-4_4

permissible limits for PM $_{2.5}$ and PM_{10} are 40 μg/m^3 and 60 μg/m^3 in India and are 5 μg/m^3 and 15 μg/m^3 respectively by the World Health Organization. A recent scientific report by 2023 by the University of Chicago's Energy Policy Institute has revealed a fact that the life span of a New Delhi resident is decreasing by 11.9 years due to severe air pollution in the city. India faces the greatest health burden occurring due to air pollution in the metropolitan and urban areas. In 2020, the annual average $PM_{2.5}$ level in Delhi was 107 μg/m^3, and in 2021, it was 126.5 μg/m^3, a dramatic increase of over 20%.

Monitoring air quality in a timely manner is useful for the public health. Suitable warnings can be given, and preventive measures can be taken if AQI crosses certain limits. In most of the countries, the air quality is estimated from the sensor data collected from the monitoring stations located at prominent places in the city. Setup cost and maintenance restrict its application. AQI estimation based on the images [1, 4, 5, 9, 10] is captured by a mobile phone, or a web camera is a viable solution. As the technology in the machine learning and deep learning is maturing day by day, researchers have started using these approaches for the estimation of $PM_{2.5}$ or AQI from the images. Large deeper convolutional neural network architectures are available now.

An algorithm for estimating air pollution from the photos using CNN was proposed by Zhang et al. [12]. They have introduced a CNN-based network called as Estimation of Photo Air Pollution Level Network (EPAPLN). By introducing a negative log–log ordinal classifier, they have applied their algorithm on the images collected in Beijing City, China. Both $PM_{2.5}$ and PM_{10} were used in their work. Six categories were used. Cumulative gain and coincidence matrix were used as the performance metrics.

$PM_{2.5}$ prediction from the images can be improved [6] if suitable features of the images are considered. Six features were computed and applied on the images collected from Beijing (327 photos), Shanghai (1954 photos), and Phoenix (4306 photos). At first, Region Of Interest (ROI) was selected and watermarks were removed. The method was assessed using clear and cloudy photos taken at three fix locations in Beijing, Shanghai, and Phoenix with manually marked reference regions with different depths. The proposed method [6] achieves good performance for the images from Beijing and Shanghai but fails for the images from Phoenix due to the narrow region of $PM_{2.5}$ values.

Air pollution deteriorates the visual quality of the image. $PM_{2.5}$ estimation based on the image quality was done by Yang and Chen [11]. Transmission map was used as a feature based on a dark channel prior, and a linear regressor was used for predicting the $PM_{2.5}$ from smart phones. A haze model was used for this purpose. It is justified that the corrected transmission map, and $PM_{2.5}$ values bear a linear relationship. The algorithm was tested on four datasets, and recordings were taken from one month data. Mean Absolute Error (MAE) was taken as the performance metric.

Depth of a scene can be estimated by estimating the $PM_{2.5}$ value from a single image Jian [7]. By extracting the features of a single image by passing it through a CNN, $PM_{2.5}$ was estimated. A mapping is done with the features and support vector regression (SVR). 1545 images were captured using an iPhone under various weather

Table 1 DINOv2 models details

Model name	Model size	No. of parameters (millions)	Embedding dimension
ViT-S	84.2 MB	21	384
ViT-L	330 MB	86	768
ViT-B	1.13 GB	300	1024
ViT-g	4.23 GB	1100	1536

conditions. Corresponding $PM_{2.5}$ values were measured using a sensor. Features were extracted from a Visual Geometry Group (VGG) network. Three kernels, i.e., linear, Radial Basis Function (RBF), and Polynomial kernels, were used in the SVR. Out of 1545 images, only 30 images were used for testing. Rigorous cross-validations are not carried out in this work. In this work, the air quality parameters (AQI, $PM_{2.5}$, and PM_{10}) are estimated using the images. Features of the images are obtained by using DINOv2 models Maxime [8]. By using appropriate regressors, AQI, $PM_{2.5}$, and PM_{10} are estimated. Feature vector size is reduced using genetic algorithm John [3].

2 DINOv2

DINOv2 model is a self-supervised transformer model by Meta AI. It is an acronym for Distillation of knowledge with No labels and Vision Transformers 2. Training of this model does not need any labels. Models learn information from the images by themselves. It is useful for estimating the depth, instance segmentation, classification, and instance retrieval. DINOv2 models were trained on 142,109,386 (142 millions) curated images collected from diversified resources. A large Vision Transformer (ViT) model was trained initially and by using distillation approach, several small models were trained. DINOv2 is basically a foundation model with no textual guidance. As the model was trained on very large number of images, robust and diverse features of the images were learnt by the models. Hence, these models can be used as good feature extractors. There are four models released by Meta AI. A summary of the models is given in Table 1.

3 Dataset

The selected image dataset known as Air Pollution Image Dataset from India and Nepal Rouniyar et al [2] comprises the images collected in India and Nepal after visiting the sites physically and collecting the videos. Videos are converted into images and appropriate cleaning, and pre-processing steps were applied.

1. Collection of air pollution data is initiated at first by visiting physically the sites from where the images are to be collected. Images and videos of the areas are captured then and there itself.
2. There are many air pollution parameters to be noted. $PM_{2.5}$, PM_{10}, NO_2, SO_2, CO, etc., are some parameters that are included. By using the data from websites of Central Pollution Control Board (CPCB) website for India. For Nepal, the data is collected from a website tomorrow.io. Two websites are listed below:

   a. https://app.cpcbccr.com/AQI_India/
   b. https://www.tomorrow.io/weather/NP/4/Biratnagar/079711/hourly/

3. After capturing images, and videos, they are processed. Processing steps include removal of blur and overexposure. Poor-quality images were removed.
4. Both videos and images are captured at the site. Frames are obtained from the videos. Poor and improper videos are discarded.
5. All the data collected during the site visit, images, videos, and air pollution data from the website are integrated into a structured format useful for machine learning and deep learning applications.

There are 12,240 images in total and six categories of the images known under good (1541), moderate (1573), un-healthy for sensitive groups (2863), un-healthy (2622), very un-healthy (2194), and severe (1447). All the images are of 224 × 224 resolution. Sample images of the dataset are shown in Fig. 1 (Table 2).

4 Regression Metrics

Performance of a regressor is assessed by using several metrics. The list includes R^2 (Coefficient of Determination), Mean Absolute Error (MAE) and Mean Square Error (MSE). They are defined as follows. The metric R^2 gives an indication of goodness of the fit. Goodness of prediction of unseen samples is represented by this coefficient of determination by using proportion of variance. Ideal value is 1. MAE is calculated by summing the absolute difference between the actual value and the predicted value of each sample over all the samples and dividing by the number of samples. The ideal value is zero. MSE is computed by calculating the sum of squared difference between the actual and predicted value and dividing it by the total number of samples. If the model is error-free, then MSE is zero.

5 Methodology

All the images in the dataset are colour images and are 224 × 224 size. DINOv2 also requires the images to be of 224 × 224 size, and hence no resizing is required. However, all the images are to be normalized with appropriate mean and standard

Bengaluru		

Good AQI:41 $PM_{2.5}$:20 PM_{10}:36	Moderate AQI:136 $PM_{2.5}$:76 PM_{10}:137	Un-healthy for sensitive AQI:56 $PM_{2.5}$:27 PM_{10}:53

Delhi		

Unhealthy AQI:179 $PM_{2.5}$:287 PM_{10}:180	Very Un-healthy AQI:281 $PM_{2.5}$:174 PM_{10}:132	Severe AQI:449 $PM_{2.5}$:337 PM_{10}:198

Dimapur		Faridabad

Moderate AQI:63 $PM_{2.5}$:31 PM_{10}:52	Un-healthy for sensitive AQI:103 $PM_{2.5}$:53 PM_{10}:84	Very Un-healthy AQI:230 $PM_{2.5}$:185 PM_{10}:199

Greater Noida		

Moderate AQI:76 $PM_{2.5}$:68 PM_{10}:76	Un-healthy for sensitive AQI:133 $PM_{2.5}$:34 PM_{10}:199	Unhealthy AQI:199 $PM_{2.5}$:35 PM_{10}:345

Fig. 1 Sample images from the dataset and their AQI, $PM_{2.5}$, and PM_{10} values

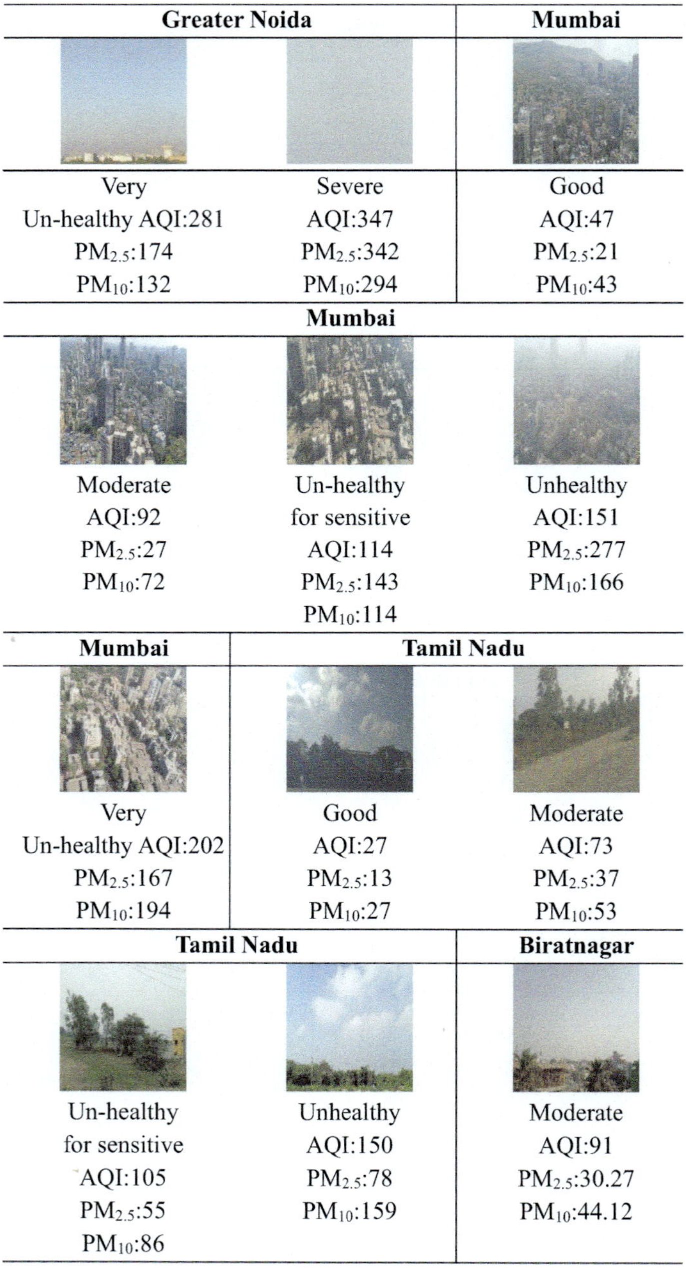

Fig. 1 (continued)

Biratnagar		
 Un-healthy for sensitive AQI:101 $PM_{2.5}$:33.18 PM_{10}:48.66	 Unhealthy AQI:154 $PM_{2.5}$:65.89 PM_{10}:95.23	Lowest AQI:15 Highest AQI:450 Lowest $PM_{2.5}$:4 Highest $PM_{2.5}$:500 Lowest PM_{10}:7 Highest PM_{10}:480

Fig. 1 (continued)

Table 2 Dataset statistics

City	No of classes	No. of images
ITO Delhi	3	2168
Dimapur, Nagaland	2	350
Spice Garden, Bengaluru	3	1926
Knowledge Park III, Greater Noida	5	1241
New Ind Town, Faridabad	1	641
Borivali East, Mumbai	5	2203
Oragadam, Tamil Nadu	4	1540
Biratnagar, Nepal	3	2171

deviation. Features are extracted by passing the processed images through DINOv2 models. As there are four models of varying sizes, features are extracted from all the four models. Features are given to the selected regressors. By applying various cross-validations, suitable DINOv2 model and appropriate regressor are selected. Selection of the model and the regressor plays a crucial role in the overall estimation of AQI. To improve the estimation further, a feature selection algorithm is applied, and a subset of features is selected. For all the images, AQI, $PM_{2.5}$, and PM_{10} are given. However, for some of the images, O_3, CO, NO_2, and SO_2 are not supplied. Hence, in this work, estimation of AQI, $PM_{2.5,}$ and PM_{10} are considered. Steps of the proposed algorithm are as follows.

6 Algorithm

1. All the images are normalized with a mean of 0.5 and standard deviation of 0.5.

2. Features are extracted by passing all the images through DINOv2 models.
3. As there are four models available, there are four feature arrays generated.
4. Pre-processing steps of feature array include mean imputation and a standard scalar operation.
5. Initially, DINOv2 ViT-L is chosen and is tested with several regressors to find an optimal regressor which offers higher R^2 for AQI estimation.
6. With the chosen regressor, all the four models of DINOv2 are tested with various cross-validations.
7. Simultaneous estimation of AQI, $PM_{2.5}$, and PM_{10} is done with multi-output regressors.
8. Feature vector size is reduced by using genetic algorithm, and the R^2 value is computed for a tenfold cross-validation.

7 Results

All the simulations are carried in Google Colab environment. Essential pre-processing steps, i.e., imputation and scaling, are carried on all the image features. As there are many regressors available, initially, DINOv2_ViTl14 is chosen, and it is tested with various regressors (38) available with a 80% train and 20% test strategy. A detailed comparison of estimation of AQI with R^2 as a regression metric is given Table 3. Out of 38 regressors, K-neighbour regressor (KNR) performance is higher with a R^2 value of 0.9857 and it took just 1 s. Hence, KNR regressor is chosen as the suitable regressor.

As there are four models available in DINOv2, a comparison of all the four models with KNR regressor is given for various K values ($K = 10, 5,2$) in K fold cross-validation. R^2 obtained with DINOv2_ViTl14 for a tenfold cross-validation is higher. But, for a twofold validation, DINOv2_ViTs14 is higher. A summary of four models with three cross-validations is given in Table 4. Both in Tables 3 and 4, the estimation is AQI estimation. Simultaneous estimation of AQI, $PM_{2.5}$, and PM_{10} can be done with multi-output regressors. KNR regressor, linear regressor, and decision tree regressor (DT) are used in this work, and a summary is given in Table 5. As shown in Table 5, KNR regressor is superior in simultaneous prediction of AQI, $PM_{2.5}$, and PM_{10}.

In addition to R^2, various other metrics are also used in the regression domain. AQI estimation using DINOv2_ViTl14 model, with KNR regressor, and other metrics for a tenfold cross-validation, are given in Table 6.

Even for a 90% test and 10% train case, the estimated AQI values are closer to the actual values as shown in Fig. 2. Figures. 3 and 4, $PM_{2.5}$ and PM_{10} estimations vs actual values are shown. Even though there are substantial number of images 12,240, certain images have strong correlation in each set, and hence redundancy exists in the dataset. This is due to the generation of images from videos. As the video is captured continuously, there is a minor variation in the visual content due to pollution which is insignificant. AQI value also doesn't change much for every 10 s or so. This implies

Table 3 Various regressors with a 80% train and 20% test split

Regressor	R^2
K neighbours regressor	0.9857
MLP regressor	0.9723
Extra trees regressor	0.9222
Hist gradient boosting regressor	0.9212
LGBM regressor	0.9213
Bayesian ridge	0.9051
Lasso CV	0.9045
Ridge CV	0.9037
Ridge regression	0.9034
Linear regression	0.9033
Transformed target regressor	0.9033
Elastic net CV	0.8972
Huber regressor	0.8971
SGD regressor	0.8945
Linear SVR	0.8918
Random forest regressor	0.8904
XGB regressor	0.8885
Lasso lars CV	0.8701
Lasso lars IC	0.8701
Poisson regressor	0.8701
Bagging regressor	0.8630
Lasso regression	0.8307
Lasso lars	0.8307
Gradient boosting regressor	0.8299
Elastic net	0.8199
Tweedie regressor	0.8170
Passive aggressive regressor	0.8111
Orthogonal matching pursuit	0.8106
Orthogonal matching pursuit CV	0.8106
Gamma regressor	0.7836
Decision tree regressor	0.6611
Extra Tree regressor	0.6568
Ada Boost regressor	0.6358
Lars CV	0.5734
SVR	0.5272
Nu SVR	0.5010

Table 4 Performance of DINOv2 models(R^2)

Model name	10 Fold	5 Fold	2 Fold
DINOv2_ViTs14	0.987	0.986	0.980
DINOv2_ViTb14	0.986	0.985	0.976
DINOv2_ViTl14	0.988	0.986	0.978
DINOv2_ViTg14	0.987	0.986	0.979

Table 5 Simultaneous estimation of AQI, $PM_{2.5}$, and PM_{10} with multi-output regressors (tenfold cross-validation and R^2)

Model name	KNR	Linear	DT
DINOv2_ViTs14	0.977	0.806	0.647
DINOv2_ViTb14	0.975	0.871	0.634
DINOv2_ViTl14	0.976	0.887	0.633
DINOv2_ViTg14	0.976	0.913	0.711

Table 6 Various metrics in AQI estimation

Metric	Value
R^2	0.988
Explained variance	0.988
Maximum error	143.940
Mean absolute error (MAE)	3
Mean square error (MSE)	128.492
Median absolute error	0
Mean squared log error	0.008
Mean absolute percentage error	0.029

that some test images are already seen by the model in the training process. This can lead to higher R^2 values, and it cannot reflect the model generalization capabilities. One way to overcome this problem is to remove the redundant images from the dataset.

Compared to AQI and $PM_{2.5}$, PM_{10} estimation is more accurate as can be seen from Fig. 4. Very few samples are far from the actual values.

Hence, even for a 10% train set, the predicted AQI, $PM_{2.5,}$ and PM_{10} values are very close. Extensive cross-validations may be carried out to assess the generalization of the model. Another way to tackle this problem is to take the images for every half an hour or so, hoping that there is a significant variation of AQI for every half an hour. Quality of the dataset plays a key role in the AQI estimation from the images. To the knowledge of the author, this is the first work on this dataset. Hence, comparison of the proposed work with the existing works is not possible.

Feature vector size of the DINOv2_ViTl14 is 1024 and it is reduced to 505 by using genetic algorithm. With a tenfold cross-validation, the R^2 value for AQI estimation is 0.987. Variation of fitness function over the generations is given in Fig. 5. Parameters of the genetic algorithm used in the simulations are given in Table 7.

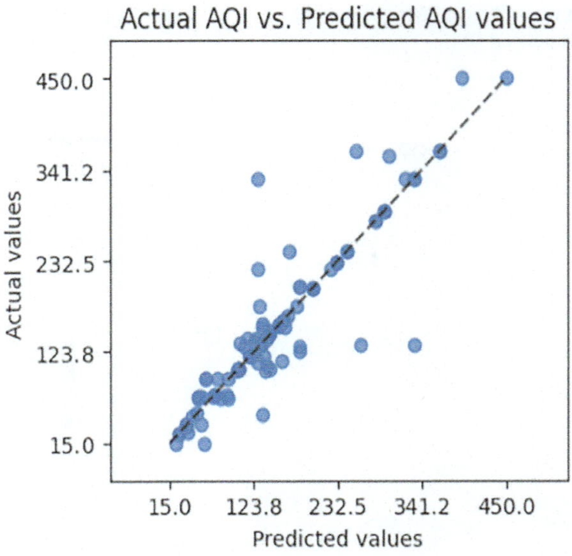

Fig. 2 Actual AQI versus predicted AQI

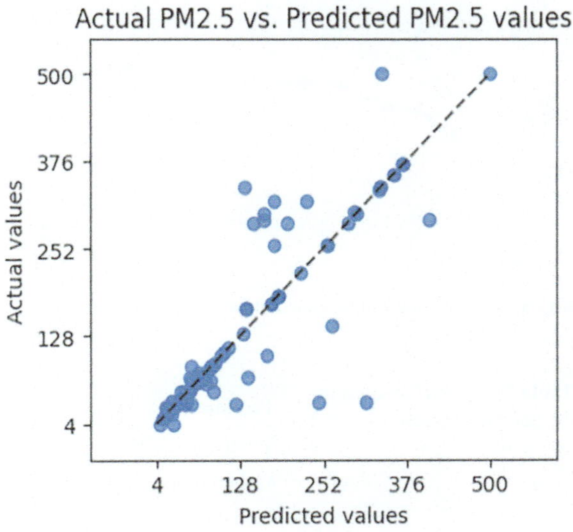

Fig. 3 Actual $PM_{2.5}$ and Predicted $PM_{2.5}$

8 Conclusions

In this work, an algorithm is proposed for estimating the Air Quality Index (AQI) from the images, using self-supervised DINOv2 models. DINOv2 is the state-of-the-art model that is publicly available and is a large foundation model trained on millions of images. In addition to AQI, $PM_{2.5}$ and PM_{10} are also estimated by using various regressors. Features are extracted from the DINOv2 models and are given

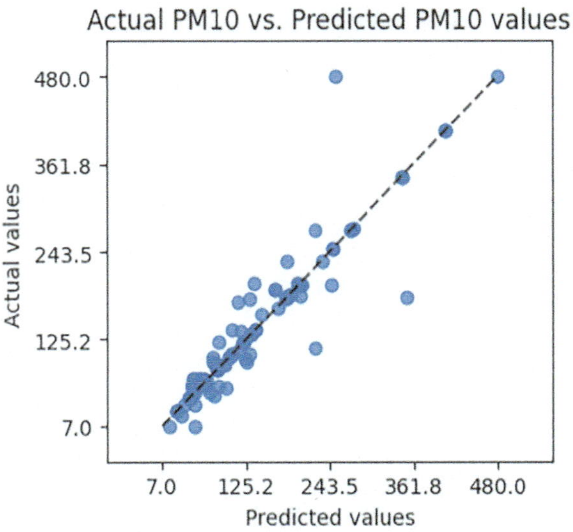

Fig. 4 Actual PM_{10} and Predicted PM_{10}

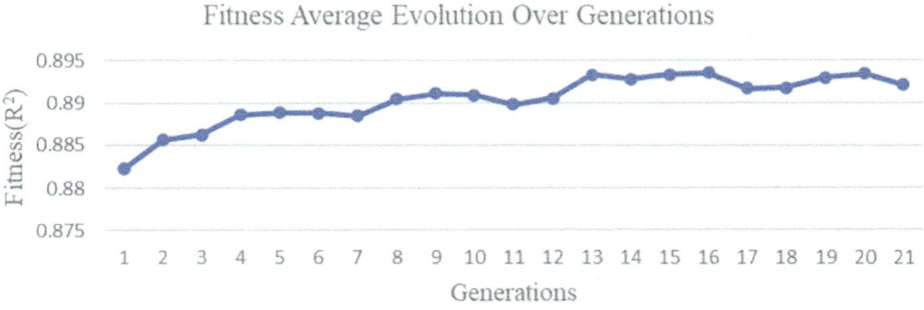

Fig. 5 Fitness function (R^2) evolution vs generations

Table 7 Parameters used in the genetic algorithm

Parameter	Remarks
Population Size	30
Generations	20
Elitism	True
Cross-Over probability	0.2
Mutation probability	0.8
Tournament size	3

to regressors. Various cross-validations are used to assess the generalization of the models and regressors in estimating the AQI, $PM_{2.5,}$ and PM_{10}. Further, simultaneous estimation of all the three parameters is also carried out by using multi-output regressors. R^2 is calculated for various folds and the proposed approach is justified with extensive simulations. DINOv2 models are computationally expensive ones and hence difficult to deploy on resource-constrained devices, i.e., mobile phones and tablets. However, by suitable knowledge distillation methods the knowledge can be transferred to lightweight models and then can be deployed. This would be the future work.

Acknowledgements Author would like to thank the Principal and the Management of Bapatla Engineering College, Bapatla, for their support.

References

1. Ahmed M, Xiao Z, Shen Y (2022) Estimation of ground $PM_{2.5}$ concentrations in Pakistan using convolutional neural network and multi-pollutant satellite images. Remote Sens 14:1735
2. Rouniyar A, Utomo S, John A, Hsiung P-A (2023) Air pollution image dataset from India and Nepal . Kaggle. https://doi.org/10.34740/KAGGLE/DS/3152196
3. Holand JH (1992) Genetic algorithms. Sci Am 267(1):66–73
4. Kamble A, Champrasert P (2022) Using photo images with deep residual network for $PM_{2.5}$ value estimation. In Barolli L. (eds) Advances on P2P, parallel, grid, cloud and internet computing. 3PGCIC 2021. Lecture Notes in Networks and Systems, vol 343. Springer, Cham. https://doi.org/10.1007/978-3-030-89899-1_14
5. Liaw JJ, Huang YF, Hsieh CH, Lin DC, Luo CH (2020) $PM_{2.5}$ concentration estimation based on image processing schemes and simple linear regression. Sensors (Basel) 20(8):2423
6. Liu C, Tsow F, Zou Y, Tao N (2016) Particle pollution estimation based on image analysis e0145955. Plos One.4
7. Ma J, Li K, Han Y, Du P, Yang J (2018) Image-based $PM_{2.5}$ estimation and its application on depth estimation. In: 2018 IEEE international conference on acoustics, speech and signal processing (ICASSP)
8. Oquab M et al. (2023) DINOv2: learning robust visual features without supervision, arxiv: 2304.07193
9. Pudasaini B, Kanaparthi M, Scrimgeour J, Banerjee N, Mondal S, Skufca J, Dhaniyala S (2020) Estimating $PM_{2.5}$ from photographs. Atmospheric Environment: X 5:100063
10. Sun K et al (2022) A photo-based quality assessment model for the estimation of $PM_{2.5}$ concentrations. IET Image Process 16:1008–1016
11. Yang B, Chen Q (2017) $PM_{2.5}$ concentration estimation based on image quality assessment. In: 2017 4th IAPR Asian conference on pattern recognition (ACPR)
12. Zhang C, Yan J, Li C, Rui X, Liu L, Bie R (2016) On estimating air pollution from photos using convolutional neural network. In: Proceedings of the 2016 ACM on multimedia conference–MM' 16

The Role of Precursor Gases and Meteorological Parameters on Surface O_3 Over Bangalore City, Southern Part of India

H. N. Sowmya, B. T. Shivendra, R. Prajwal, R. Shruthi, G. P. Shivashankara, H. K. Ramaraju, and B. S. Shilpa

Abstract In the urban area of Bangalore, from January 2017 to March 2018, comprehensive assessments of surface-level O_3, CO, NOx, CH_4, NMHC, SO_2, and NH_3 were conducted across five distinct zones. Hourly average concentrations, accompanied by standard deviations, revealed the following values: O_3 (31.8 ± 17.5 μg/m^3), CO (0.7 ± 0.2 μg/m^3), NOx (34.6 ± 9.6 μg/m^3), CH_4 (412.8 ± 259 μg/m^3), NMHC (1.1 ± 1.1 μg/m^3), SO_2 (4.8 ± 1.6 μg/m^3), and NH_3 (28.8 ± 15.7 μg/m^3). A noteworthy observation was the rapid surface ozone production (peaking at + 8.01 ppb/h) occurring between 8:00 am and 12:00 pm LT. This surge in ozone levels was attributed to the influx of freshly emitted precursors and photochemical reactions. Conversely, during the evening hours (between 17:00 and 19:00 h LT), a sharp decline in surface ozone (− 7.82 ppb/h) was evident, primarily due to reduced production of nitrogen oxides. When considering seasonal variations, the most significant rates of ozone increase (09.43 ppb/h) were noted in the winter season, followed by the post-monsoon (09.42 ppb/h), summer (07.78 ppb/h), and monsoon (05.38 ppb/h) periods. Conversely, the rapid destruction of surface ozone followed a similar seasonal order of winter (− 09.67 ppb/h), post-monsoon (− 08.97 ppb/h), summer (− 07.11 ppb/h), and monsoon (− 05.54 ppb/h) during morning hours (between 17:00 and 19:00 h LT). The elevated surface ozone concentrations in Bangalore City raise concerns about potential impacts on human health. In light of this, it is imperative to devise and implement measures aimed at reducing the emissions of ozone precursors in the study area.

H. N. Sowmya (✉) · B. T. Shivendra · R. Prajwal · G. P. Shivashankara · H. K. Ramaraju
Department of Civil Engineering, Dayananda Sagar College of Engineering, Bangalore, India
e-mail: soumya-cvl@dayanandasagar.edu

R. Shruthi
Department of Civil Engineering, BGS Institute of Technology, Adichunchanagiri University, Mandya, India

B. S. Shilpa
Department of Civil Engineering, Vidya Vardhaka College of Engineering, Mysore, India

S. M. Shiva Nagendra et al. (eds.), *Select Proceedings of the 8th Indian International Conference on Air Quality Management (IICAQM 2023)*, Lecture Notes in Civil Engineering 582, https://doi.org/10.1007/978-981-96-2359-4_5

Keywords Ozone · Photochemical reaction · Precursors · Boundary layer height · Relative humidity

1 Introduction

Air pollution poses a critical challenge to the well-being of humans, animals, and the environment. In developing countries, particularly in urban areas, the issue of air pollution has worsened over the last few decades due to a surge in human activities, a substantial increase in population density, and emissions from vehicular traffic. These factors have collectively led to deteriorating air quality and environmental threats in and around urban environments. One of the primary concerns in urban settings is the presence of trace gases in the atmosphere generated by human activities [26]. The exponential growth in population, vehicles, and industrial facilities within urban areas has significantly degraded regional air quality [2, 12, 18]. Wang et al. [23] reported that vehicular emissions contribute to approximately 70–80% of air pollution in developing megacities. Moreover, many megacities continue to use older vehicles with subpar maintenance, low-quality fuel mixtures, and deficient transport infrastructure [8]. Previous studies have highlighted that the combustion of fossil fuels in industrial processes and urban transportation has led to a notable increase in various atmospheric gaseous pollutants [3]. Additionally, the burning of agricultural residues during different seasons poses a significant threat, contributing to both particulate and gaseous pollutants in Asian countries [4]. Kandlikar [12] pointed out that atmospheric pollutants are unevenly distributed in the environment without respecting geographical boundaries. Furthermore, meteorological factors and local topography play a significant role in determining pollutant concentrations, leading to daily, diurnal, temporal, and spatial variations [9].

Ozone, a secondary atmospheric pollutant not directly emitted, is formed through complex reactions involving nitric oxide and nitrogen dioxide. The combustion of fossil fuels in human activities releases a range of pollutants such as volatile organic compounds (VOCs), carbon monoxide (CO), and nitrogen oxides (NOx), all of which have direct implications for both climate and human health [21]. CO, a potential health hazard, primarily originates from vehicular emissions, industrial processes, and other sources. The emission of CO also indirectly contributes to global warming, as suggested by Crutzen et al. [5]. Sulfur dioxide (SO_2) in the lower atmosphere is directly linked to adverse health effects.

2 Methodology

2.1 Study Area & Data Collection

Bengaluru (located at 12.9716° N, 77.5946° E, and 900 m above sea level) is situated approximately 450 km away from the Bay of Bengal and 400 km from the Arabian Sea. This city covers an area of 741 km^2 (286 square miles) and serves as the capital of Karnataka and an industrial hub [6]. Bengaluru boasts a population of over ten million, making it the fifth-most populous urban agglomeration in India (*source* World Population Review).

Data on GPs, including NO_x (measured with an Ecotech NOx analyzer model Serinus 40 T), CO (measured with a Serinus 30 from Ecotech, Australia), NH_3 (measured with an analyzer from Ecotech Pvt Ltd Australia, "Serinus 44 analyzer"), SO_2 (measured with an analyzer from Ecotech Pvt Ltd Australia, "Serinus 50 T trace sulfur dioxide (SO_2)"), O_3 (measured with the Serinus series 10 from ACOEM Ecotech Industry Pvt Ltd, Australia), CH_4 (measured by Ecotech Pvt Ltd Australia), and BTX (measured with the Syntech Spectras GC955 series 600 BTX analyzer) [11, 17], were collected simultaneously with Meteorological Parameters (MPs).

These data were gathered from five monitoring stations run by the Karnataka State Pollution Control Board (KSPCB) and the Central Pollution Control Board (CPCB) during the years 2017 and 2018, at a height of 5 m from the ground. Surface ozone and its precursors data were obtained from online monitoring of GPs conducted by the central and Karnataka Pollution Control Boards. Online GPs samplers were deployed at five distinct locations with various climatic conditions, including:

(i) BTM Layout, Bengaluru (KSP)—(12.97° N, 77.59° E), an industrial area with mixed residential and rural sources. (ii) Peenya, Bengaluru (PB)—KSPCB—(13.03° N, 77.51° E), another industrial area. (iii) BWS Kadabesanahalli, Bengaluru—KSPCB—(12.93° N, 77.69° E), featuring mixed pollution sources. (iv) Sanegurava Halli Bengaluru (SHB)—KSPCB—(12.99° N, 77.54° E), a mixed source of atmospheric pollution from residential and urban areas. (v) City Railway Station (CRS)—Bengaluru KSPCB—(12.97° N, 77.57° E), located in the heart of Bangalore city. GPs data were collected at one-hour intervals from January 2017 to March 2018 (Fig. 1).

3 Results and Discussion

3.1 Gaseous Pollutants in the Study Area

Gaseous pollutant concentrations, including O_3, CH_4, NMHC, CO, NH_3, NO, NO_2, and SO_2, were monitored in Bengaluru from January 1, 2017, to March 20, 2018. The daily mean concentrations and 30-day moving averages were plotted (Fig. 2). The

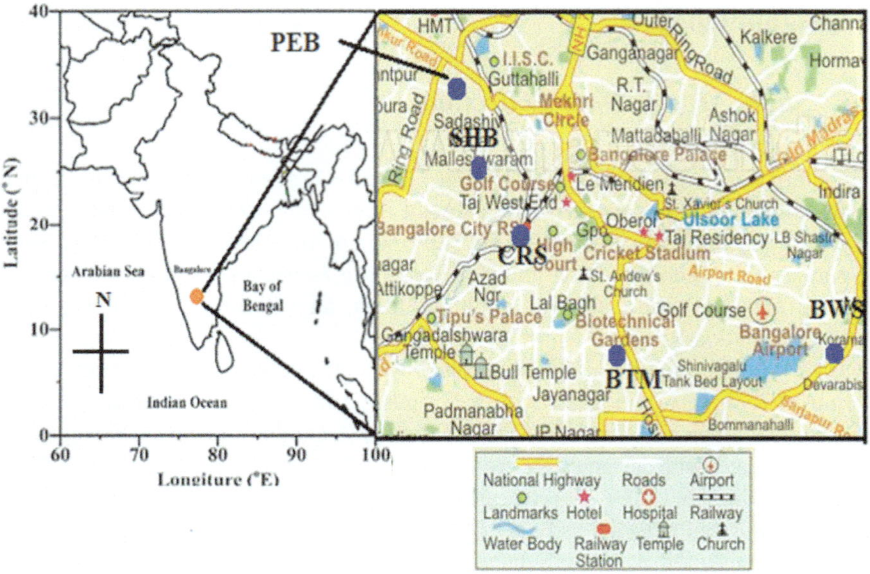

Fig. 1 Sampling Station over Bangalore City

daily mean concentrations were as follows: O_3 (ppb)—35.00 ± 17.50, CH_4 (ppm)—412.80 ± 259, NMHC (ppb)—1.10 ± 1.10, CO (ppm)—0.70 ± 3.10, NH_3 (ppb)—28.80 ± 15.70, NO (ppb)—7.20 ± 3.10, O_2 (ppb)—27.50 ± 8.40, and SO_2 (ppb)—4.80 ± 1.60. The surface ozone concentrations in Bengaluru are lower than the World Health Organization (WHO) standard, which is set at 100 $\mu g/m^3$ or 50 ppb for an 8-h maximum exposure, and also lower than the national standard stipulated by the Central Pollution Control Board (100 $\mu g/m^3$). Gaseous pollutant concentration data were further categorized on a monthly and seasonal basis in Bengaluru. Seasonally, the mean O_3 concentration was highest during the winter period (mean = 49.7 ppb), followed by the summer (33.2 ppb), post-monsoon (25.8 ppb), and the southwest summer monsoon season (23.7 ppb) (Fig. 3). The lowest concentration of O_3 and its precursor gases during the monsoon period can be attributed to the "washing out" effect. Conversely, the higher concentration during the winter period is linked to increased emissions of precursor gases, elevated photochemical reactions due to higher solar radiation, and increased fossil fuel consumption [7, 22, 25]. Additionally, the higher concentrations observed during the winter and post-monsoon seasons may be attributed to excessive biomass burning [13]. Station-wise mean values of NO_x ($NO + NO_2$) concentrations across Bengaluru were found to be 17.3 (BTM), 30.5 (BWS), 56.9 (CRS), and 28.4 (SHB) ppb.

The diurnal variations of SO_2 concentration over Bengaluru are represented in Fig. 4. Seasonally, it was in the order of winter season (37.90 ppb highest) followed by the Post-monsoon (26.00-ppb), summer (24.50-ppb), and monsoon (23.30-ppb) (Fig. 5).

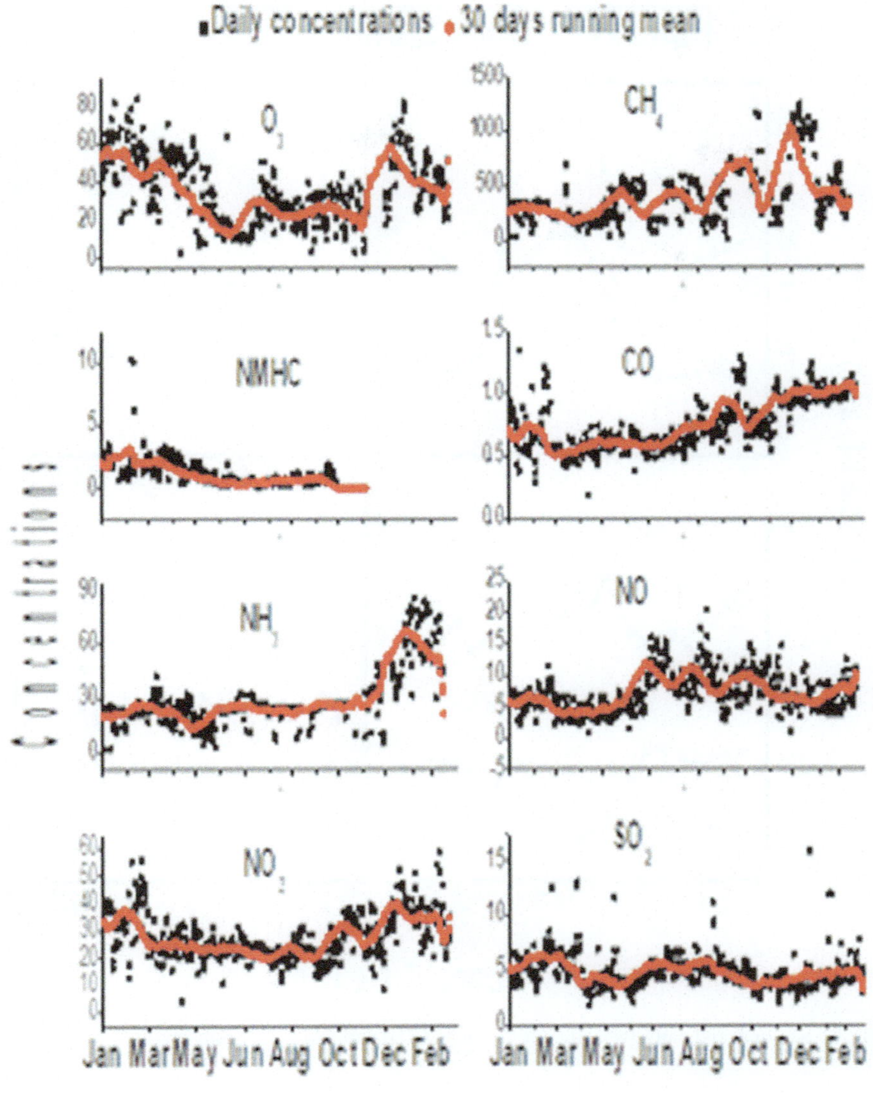

Fig. 2 Daily values of gaseous pollutants from 1/01/2017 to 20/3/2018 in study area

3.2 Diurnal Variations of Surface-Ozone

The diurnal pattern of O_3 concentrations and its precursors across Bengaluru city is depicted in Fig. 4. Surface ozone concentrations exhibit distinct diurnal variations, characterized by higher levels during the daytime hours, typically from 10:00 to 18:00 local time (LT), varying from 59.30 to 35.70 ppb. Conversely, lower concentrations, around 20.00 ppb, are observed during the late evening (18:00 to 06:00

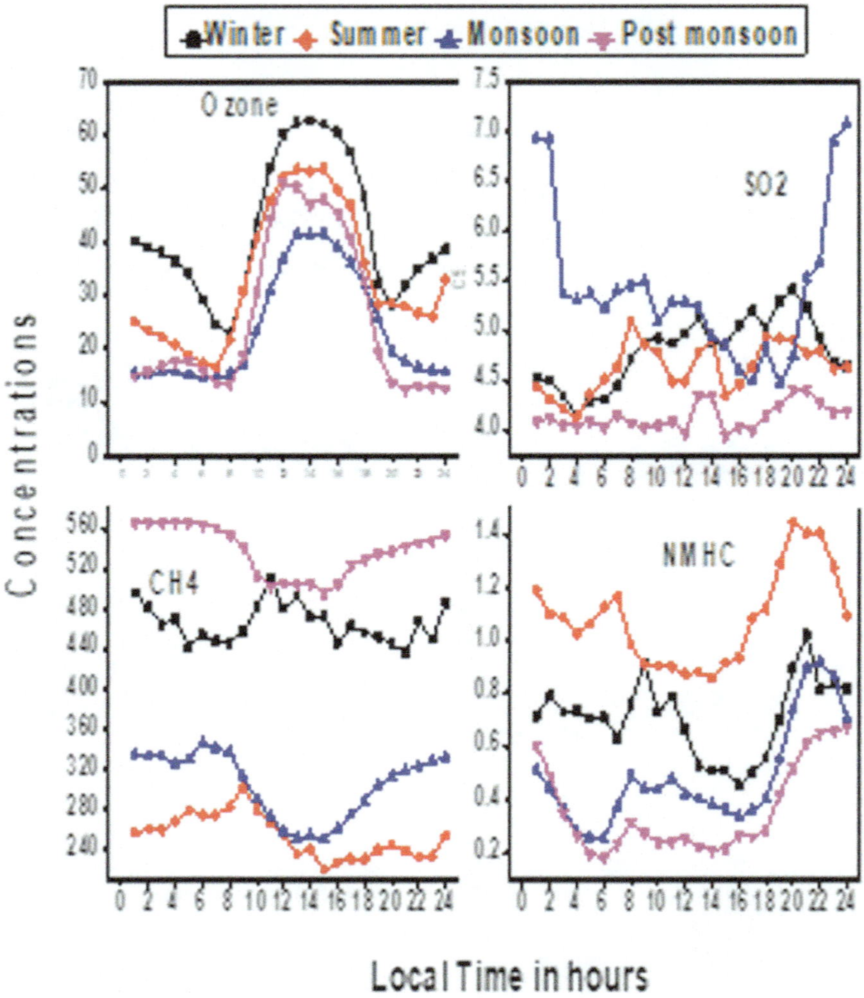

Fig. 3 Seasonal and daily variability of O_3, SO_2, CH_4, and NMHC in study area

LT) and morning hours (07:00 to 09:00 LT). O_3 concentrations gradually increase after 09:00 LT, following sunrise and coinciding with the rising solar radiation (SR). The highest concentrations, reaching 51.70 ppb, occur around 14:00 LT, after which they start to decrease and maintain lower values throughout the night due to the absence of sunlight [10]. Seasonally, there are slight variations, with higher night-time concentrations during the winter and summer seasons compared to the monsoon and post-monsoon periods.

It was observed that as NO_2 levels decrease during the day, O_3 begins to increase due to its role as a precursor, playing a crucial part in NO_x formation. A similar trend is observed for CO, which follows the NO pattern and exhibits a reverse diurnal

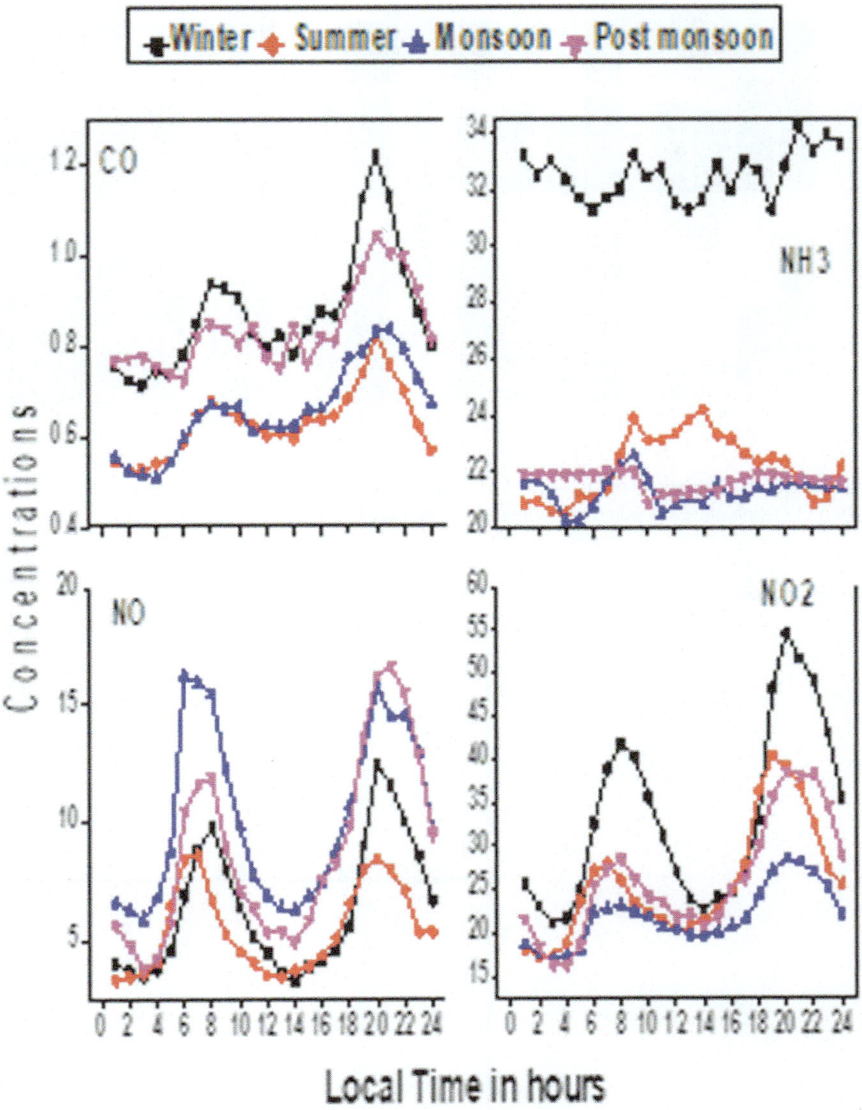

Fig. 4 Seasonal and diurnal variability of CO, NH_3, NO, and NO_2 in the study area

variation in comparison to O_3, with the lowest concentrations of NO and CO occurring during the afternoon hours when O_3 peaks. During the evening hours, as NO dominates NO_x, ozone concentrations decrease. The variation in NO clearly indicates the dominance of local NO sources. However, once surface ozone is depleted, NO_2 concentrations decrease steadily from their maximum values. In the monsoon season, the rates of change of O_3 are lower due to lower precursor concentrations and the washout effect from rain (Fig. 6).

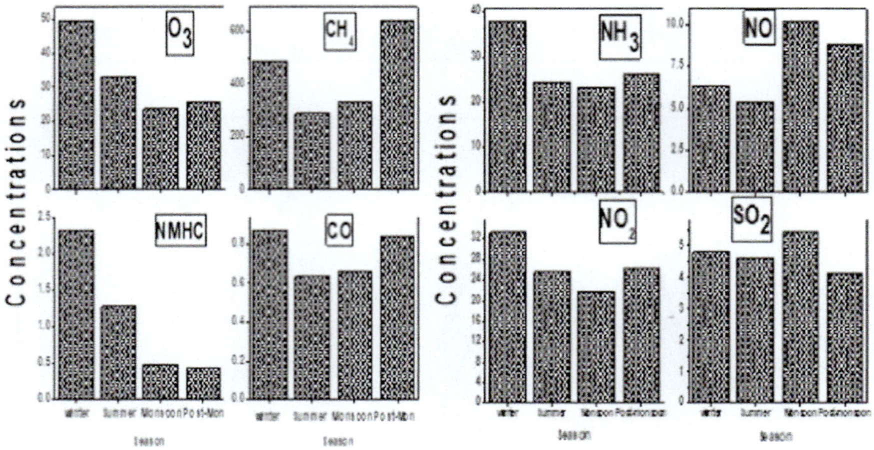

Fig. 5 Seasonal variations of O_3—ppb, CH_4—ppm, NMHC—ppb, CO—ppm, NH_3—ppb, NO—ppb, NO_2—ppb and SO_2 ppb over Bengaluru Urban area

Fig. 6 Rate of change of O_3 in different seasons

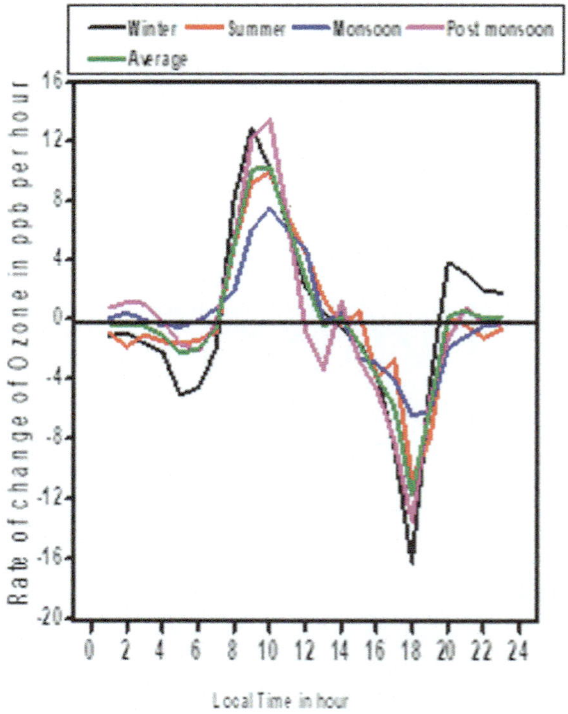

3.3 Impact of Ventilation Coefficient on Surface O_3

The Ventilation Coefficient (VC) is a pivotal parameter that offers insights into the dispersion and accumulation of atmospheric pollutants in the lower atmosphere. It is determined by the interplay between boundary layer heights (BLH) and surface wind speed. Seibert et al. [16] highlighted the role of BLH in regulating atmospheric pollutants, as it generates turbulence within the boundary layer and influences land surface processes. Additionally, Krishnan and Kunhikrishnan [14] suggested that BLH alone is insufficient to explain atmospheric pollutants and introduced the ventilation coefficient as a superior tool for assessing lower atmospheric pollutants. In essence, a lower VC indicates reduced dispersion of atmospheric pollutants, while a higher VC signifies greater dispersion in the lower atmosphere.

Considering its significance, the diurnal variations of VCs were examined in four different seasons in the Bengaluru region, as depicted in Fig. 7. Hourly VC values exhibited a consistent trend across all seasons, with the highest VC (1935 m^2/s) observed during the summer and the lowest (209 m^2/s) in the post-monsoon season. Throughout the study period, the VC was highest in the summer season (518 m^2/s: annual mean), followed by the monsoon (435 m^2/s), winter (407 m^2/s), and post-monsoon (306 m^2/s).

During the study, the ventilation coefficient showed a consistent increase from 0800 h LT (137 m^2/s: annual mean) and peaked at 1500 h LT (1019 m^2/s). VC values

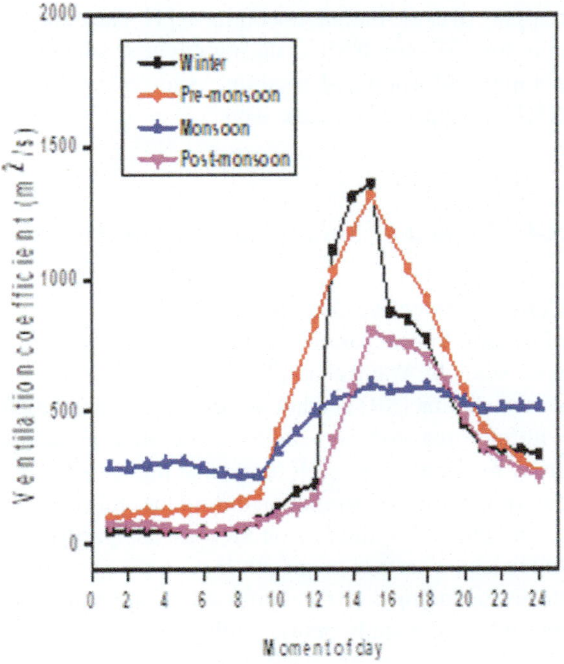

Fig. 7 Hourly variations of ventilation coefficient in different seasons

were significantly higher during daytime (587 m^2/s) compared to nighttime (246 m^2/s), representing a difference of approximately one and a half orders of magnitude. However, the variation in VC during day and night was more pronounced seasonally, with differences in winter (2.4 times higher during the day), summer (2.4), monsoon (0.3), and post-monsoon (1.4). This variation is attributed to local meteorological conditions. Generally, higher VC values indicate lower pollution due to convection and pollutant mixing, while lower VC values are associated with increased pollution due to accumulation.

Interestingly, in the case of surface ozone, the pattern is reversed. Ozone concentrations are higher during the daytime due to photochemical processes facilitated by sunlight and higher solar radiation, particularly evident in the summer with the highest VC (annual mean—802 W/m^2). Conversely, ozone precursor concentrations (CO, NO, NO_2, CH_4, and NMHC) are lower during the daytime, influenced by higher VC in all seasons. This pattern is most pronounced in the pre-monsoon period between 11:00 and 17:00 h LT when pollution dispersion into the atmosphere is favored, though this pattern is not observed in the winter and monsoon seasons.

Correlations analyses between BLH and measured pollutants were conducted, revealing a positive correlation (0.75) with O_3, illustrating the influence of BLH and VC in the study region. Praveena and Kunhikrishnan (2004) reported that the Atmospheric Boundary Layer height is lower during the winter period over the Indian subcontinent. In the monsoon season, strong prevailing winds facilitate the rapid dispersion of atmospheric pollutants, which are subsequently washed out by precipitation. In contrast, during the winter, low winds and shallow BLH lead to pollutant trapping over the Indian subcontinent [1]. The impact of BLH was particularly noticeable for CH_4 and NMHC during daytime when wind speed was highest, favoring pollutant diffusion and resulting in lower concentrations of atmospheric CH_4 and NMHC, especially on clear days.

3.4 *Impact of Solar Radiation on Gaseous Pollutants*

Currently, extensive research worldwide has been conducted to investigate the impact of meteorological parameters on measured surface ozone concentrations [15, 20]. A prior study by Wang et al. [24] suggested that heavy emissions, coupled with higher solar radiation (SR) and hot weather, create a photochemically reactive environment with elevated hydroxyl radical concentrations and abundant surface O_3. Another study by Singla et al. [19] found that higher SR and lower-level temperatures act as catalysts for ozone formation during the summer season.

In light of this, the data on SR were analyzed annually and seasonally to assess its influence on surface ozone production via photochemical reactions. Solar radiation fluxes, including direct and diffused radiation in clear sky conditions, were measured throughout the study period, with an annual mean value of 169.09 $\pm$ 55.89 W/m^2. Seasonally, the order of SR was highest in the pre-monsoon (summer: 181.6 W/

m^2), followed by winter (175.5 W/m^2), monsoon (152.4 W/m^2), and post-monsoon (142.9 W/m^2) seasons (see Figs. 3 and 4).

Seasonally, the mean O_3 concentration was highest during the winter period (mean = 49.7 ppb). This was attributed to the combined impact of solar radiation and the higher presence of aerosols from precursor pollutants. Biomass burning and a lower boundary layer, along with photochemical smog reactions, also contributed to this elevated concentration. In the summer season, O_3 concentrations were the second-highest (33.2 ppb) and were primarily influenced by solar radiation.

As reported by Soni et al. (2015), atmospheric aerosols can either absorb or scatter solar radiation, thus reducing the solar radiation flux at the Earth's surface. Stanhill and Cohen (2001) have suggested that variations in the optical properties of man-made atmospheric aerosols are responsible for fluctuations in solar fluxes.

Furthermore, the consequences of increased greenhouse gases have led to a rise in mean surface temperatures. In the twentieth century, the mean surface temperature increased by about 0.8 °C, and if emissions continue at the same rate, it is projected to increase by approximately 1.1–6.4 °C in the twenty-first century.

4 Conclusions

In this study, the ambient concentrations of gaseous pollutants (O_3, CH_4, NMHC, CO, NH_3, NO, NO_2, and SO_2) were measured from urban sites, in Bengaluru, India from 1st January 2017 to 20th March 2018. The data reporting over the complete study phase was remarkably good quality, with more than 90% of the data accessible. The obtained data of gaseous pollutants were performed by statistical analysis to characterize their daily, monthly, yearly, and diurnal pattern jointly with meteorological parameters to see the influence on measured trace gases. The hourly averaged mean concentrations along with standard deviation of O_3 (31.80 ± 17.50 μg/m^3), CO (0.7 ± 0.2 μg/m^3), NOx (34.6 ± 9.6 μg/m^3), CH_4 (412.8 ± 259 μg/m^3), NMHC (1.1 ± 1.1 μg/m^3), SO_2, (4.8 ± 1.6 μg/m^3), and NH_3 (28.8 ± 15.7 μg/m^3) were observed. The highest concentrations of surface O_3 were in the winter (48.8 μg/m^3) season due to biomass burning and lower boundary layer along with photochemical smog reaction, however, the second-highest was during the summer (33.2 μg/m^3) period due to the impact of solar radiation through photochemistry. However, such long-term unbroken measurements of gas pollutants and meteorological parameters are very important to better understand the description of atmospheric pollutants at diverse sites, including the urban environment that is at high health in developing nations.

Acknowledgements The authors are thankful to the IITM-Pune, CPCB, and KSPCB for allowing to utilize the air quality data for Bangalore city.

References

1. Alappattu DP, Kunhikrishnan PK, Aloysius M, Mohan M (2009) A case study of atmospheric boundary layer features during winter over a tropical inland station—Kharagpur (22.32°N, 87.32°E). J Earth Syst Sci 118(4):281–293. https://doi.org/10.1007/s12040-009-0028-3
2. Amato F, Cassee FR, Denier van der Gon HAC, Gehrig R, Gustafsson M, Hafner W, Harrison RM, Jozwicka M, Kelly FJ, Moreno T, Prevot ASH, Schaap M, Sunyer J, Querol X (2014) Urban air quality: the challenge of traffic non-exhaust emissions. J Hazard Mater 275:31–36. https://doi.org/10.1016/j.jhazmat.2014.04.053
3. Andreae MO, Merlet P (2001) Emission of trace gases and aerosols from biomass burning. Global Biogeochem Cycles 15(4):955–966. https://doi.org/10.1029/2000GB001382
4. Awasthi A, Agarwal R, Mittal SK, Singh N, Singh K, Gupta PK (2011) Study of size and mass distribution of particulate matter due to crop residue burning with seasonal variation in rural area of Punjab, India. J Environ Monit 13(4):1073–1081. https://doi.org/10.1039/c1em10019j
5. Crutzen PJ, Lawrence MG, Pöschl U (1999) On the background photochemistry of tropospheric ozone. Tellus B Chem Phys Meteorol 51(1):123. https://doi.org/10.3402/tellusb.v51i1.16264
6. Devaraj S, Tiwari S, Ramaraju HK, Dumka UC, Sateesh M, Parmita P, Shivashankara GP (2019) Spatial and temporal variation of atmospheric particulate matter in Bangalore: a technology-intensive region in India. Arch Environ Contam Toxicol 77(2):214–222. https://doi.org/10.1007/s00244-019-00643-8
7. Sowmya HN, Wollur C, Shivashankara GP, Ramaraju HK (n.d.) Identifying source apportionment of atmospheric particulate matter and gaseous pollutants using receptor models: a case study of Bengaluru, India. MAUSAM 75(1):1–16. https://doi.org/10.54302/mausam.v75i1.6080
8. Gaur A, Tripathi SN, Kanawade VP, Tare V, Shukla SP (2014) Four-year measurements of trace gases (SO_2, NOx, CO, and O_3) at an urban location, Kanpur, Northern India. J Atmos Chem 71(4):283–301. https://doi.org/10.1007/s10874-014-9295-8
9. Gokhale S, Khare M (2007) Vehicle wake factor for heterogeneous traffic in urban environments. Int J Environ Pollut 30(1):97–105. https://doi.org/10.1504/IJEP.2007.014505
10. Sowmya HN, Shivashankara GP, Ramaraju HK (2021) Diurnal and seasonal behavior of gaseous pollutants in Bengaluru urban area: a silicon valley of India. J Earth Sci Climatic Change 12(6)
11. Sowmya HN, Dinamani M, Shivashankara GP, Ramaraju HK, Surendra BS, Vijaya V, Shanbhag RP (2024) A novel diurnal and seasonal variation analysis of BTEX in Bengaluru urban area" research in cold and arid regions. Res Cold Arid Reg. https://doi.org/10.1016/j.rcar.2024.03.005
12. Kandlikar M (2007) Air pollution at a hotspot location in Delhi: detecting trends, seasonal cycles and oscillations. Atmos Environ 41(28):5934–5947. https://doi.org/10.1016/j.atmosenv.2007.03.044
13. Kaskaoutis DG, Kumar S, Sharma D, Singh RP, Kharol SK, Sharma M, Singh AK, Singh S, Singh A, Singh D (2014) Effects of crop residue burning on aerosol properties, plume characteristics, and long-range transport over northern India. J Geophys Res 119(9):5424–5444. https://doi.org/10.1002/2013JD021357
14. Krishnan P, Kunhikrishnan PK (2004) Temporal variations of ventilation coefficient at a tropical Indian station using UHF wind profiler. Curr Sci 86(3):447–451
15. Olszyna KJ, Luria M, Meagher JF (1997) The correlation of temperature and rural ozone levels in southeastern U.S.A. Atmos Environ 31(18):3011–3022. https://doi.org/10.1016/S1352-2310(97)00097-6
16. Seibert P, Feldmann H, Neininger B, Bäumle M, Trickl T (2000) South foehn and ozone in the Eastern Alps—case study and climatological aspects. Atmos Environ 34(9):1379–1394. https://doi.org/10.1016/S1352-2310(99)00439-2
17. Sheoran R, Dumka UC, Sowmya HN, Bisht DS, Srivastava AK, Tiwari S, Attri SD, Hopke PK (2022) Changes in inorganic chemical species in fog water over Delhi. Asian J Atmos Environ. https://doi.org/10.5572/ajae.2021.092

18. Shivashankara GP, Adarsh S, Soumya HN (2018) Urbanization and air quality—Bengaluru, India. Int J Innov Sci Res Technol
19. Singla V, Satsangi A, Pachauri T, Lakhani A, Kumari KM (2011) Ozone formation and destruction at a sub-urban site in North Central region of India. Atmos Res. https://doi.org/10.1016/j.atmosres.2011.03.011
20. Tarasova OA, Karpetchko AY (2003) Accounting for local meteorological effects in the ozone time-series of Lovozero (Kola Peninsula). Atmos Chem Phys 3(4):941–949. https://doi.org/10.5194/acp-3-941-2003
21. Verma RL, Sahu LK, Kondo Y, Takegawa N, Han S, Jung JS, Kim YJ, Fan S, Sugimoto N, Shammaa MH, Zhang YH, Zhao Y (2010) Temporal variations of black carbon in Guangzhou, China, in Summer 2006. Atmos Chem Phys 10(14):6471–6485. https://doi.org/10.5194/acp-10-6471-2010
22. Verma SR, Pervez S, Chow JC, Watson JG, Andrabi SM, Mandal P, Khan NA, Tiwari S, Chandra Dumka U, Chakrabarty RK, Verma M, Pervez YF, Mishra A, Tamrakar A, Sowmya HN, Deb MK, Ghosh KK, Jain VK, Karbhal I, Satnami ML (2023) Optical properties of fine mode aerosols over high-altitude Himalayan glacier regions. ACS Earth and Space Chemistry. https://doi.org/10.1021/acsearthspacechem.3c00088
23. Wang H, Fu L, Zhou Y, Du X, Ge W (2010) Trends in vehicular emissions in China's mega cities from 1995 to 2005. Environ Pollut 158(2):394–400. https://doi.org/10.1016/j.envpol.2009.09.002
24. Wang M, Penner JE (2009) Aerosol indirect forcing in a global model with particle nucleation. Atmos Chem Phys 9(1):239–260. https://doi.org/10.5194/acp-9-239-2009
25. Wang S, Hao J (2012) Air quality management in China: issues, challenges, and options. J Environ Sci 24(1):2–13. https://doi.org/10.1016/S1001-0742(11)60724-9
26. Yang J, Fu Q, Guo X, Chu B, Yao Y, Teng Y, Wang Y (2015) Concentrations and seasonal variation of ambient PM2.5 and associated metals at a typical residential area in Beijing, China. Bull Environ Contam Toxicol 94(2):232–239. https://doi.org/10.1007/s00128-014-1443-y

Multigas Monitoring Using Hollow Core Waveguide

Vivek Bargate, Ramya Selvaraj, R. N. Patel, K. Saran Kumar, and Nilesh J. Vasa

Abstract Natural gas leakage, mainly ethane, and methane, are potent greenhouse gases and significant contributors to climate change. This study proposes utilizing a broadband supercontinuum laser (SCL) to measure ethane (C_2H_6) and methane (CH_4) absorption characteristics in the near-infrared range. The SCL is guided through hollow core waveguides (HCW). The approach utilizes Direct Absorption Spectroscopy, which involves measuring the absorption of specific wavelengths of light by the target gas. A hollow core waveguide of length 10 m is used for measurements. The system provides a minimum detection limit (MDL) of 30 and 23 ppm for C_2H_6 and CH_4 in the 2.3 and 1.65 μm wavelength range, respectively, with a signal-to-noise ratio (SNR) of 3 dB. The suggested method permits real-time monitoring of gas pipelines, allowing prompt maintenance to prevent catastrophic failures.

Keywords Natural gas leakage · Supercontinuum laser · Hollow core waveguide · Absorption spectroscopy · Multigas sensing

1 Introduction

Natural gas, composed primarily of methane with trace amounts of other hydrocarbons such as ethane, has long been regarded as a cleaner alternative to conventional fossil fuels. However, the unplanned release of CH_4 and C_2H_6 during extraction, transportation, and distribution processes poses a significant environmental challenge. These gases, infamous for their potency as greenhouse gases, are key contributors to the rising climate change crisis [1]. Understanding the implications of natural gas leakage becomes paramount as the world grapples with the imperative to reduce

V. Bargate · R. Selvaraj (✉) · R. N. Patel
Department of Electrical Engineering, National Institute of Technology Raipur, Raipur, India
e-mail: rselvaraj.ee@nitrr.ac.in

K. Saran Kumar · N. J. Vasa
Department of Electrical Engineering, Indian Institute of Technology Madras, Chennai, India
e-mail: njvasa@iitm.ac.in

S. M. Shiva Nagendra et al. (eds.), *Select Proceedings of the 8th Indian International Conference on Air Quality Management (IICAQM 2023)*, Lecture Notes in Civil Engineering 582, https://doi.org/10.1007/978-981-96-2359-4_6

greenhouse gas emissions. This discussion discovers the deep impact of ethane and methane leakage on the Earth's climate, investigates their roles as global warming catalysts, and clarifies the far-reaching consequences on ecosystems, air quality, and human health. In light of these challenges, it becomes evident that addressing natural gas leakage is crucial for mitigating climate change and safeguarding the delicate balance of our planet's environmental equilibrium [3].

For CH_4 to ignite and sustain combustion, the concentration in the air must be within the range of 5–17% by volume. Below 5%, the mixture is too "lean" to support combustion, and above 17%, it is too "rich." Like CH_4, C_2H_6 requires a concentration within the range of 3–12.5% by volume in air to form an explosive mixture.

There are a variety of methods that can detect natural gas pipeline leaks. They are classified into optical and non-optical methods. Typical non-optical methods include vapor sampling [16], acoustic monitoring [13], ultrasonic flow meters, soil monitoring, and software-based techniques like dynamic pressure wave modeling and statistical approaches [19].

On the other hand, optical methods include LIDAR [15], thermal imaging [17], distributed fiber sensing [2], and tunable diode absorption spectroscopy [11].

Several criteria are used to evaluate the performance of a leak detection system: the ability to identify the location and size of the leak, sensitivity, and the detection speed. When identifying the leak's location, distributed optical sensors can pinpoint it with high precision, while other techniques, like software-based methods, can only give an estimate. When it comes to detection speeds, ground-based optical techniques like LIDAR or handheld thermal imagers have the longest, and optical gas sensors mounted on a UAV are relatively faster. However, other techniques can measure the leak in real time. On the other hand, optical gas sensing technology has the highest sensitivity (usually in the ppm range).

With the advent of IR and MID IR lasers like Quantum cascade lasers, Interband cascade lasers, and Optical Parametric Oscillators (QCLs, ICLs, OPO, and detectors where many gas molecules exhibit strong rovibrational absorptions are present, several compact optical gas sensor technologies have been developed that can find uses in pipeline leak detection. Driven by the need for in-situ analysis, high accuracy, sensitivity, detection limits, and reasonable prices, laser spectroscopic-based techniques have drawn increasing attention in recent decades. MID-IR QCLs/OPOs offer wide tunability and pronounced output power, leading to molecular selectivity, sensitivity, and improved signal-to-noise ratio [4, 6].

Single-mode emission is essential in laser sources to achieve high selectivity in gas sensing. Tunability of the laser enables precise adjustment of the laser emission on an absorption line of the target species, which is necessary for high-sensitivity detection and selectivity where multiple target gases might have overlapping absorption lines.

Li et al. [10], have used the ICL-based TDLAS approach for ethane detection at 3.34 μm. A detection limit of 1.2 ppbv was achieved at 4 s data acquisition time.

Ghorbani and Schmidt [7], used an ICL-based TDLAS system for the identification of carbon monoxide (CO) at 4.69 μm using a multi-pass gas cell with a detection limit of 9 ± 5ppbv at 0.07 s acquisition time. On the other hand, photoacoustic spectroscopy (PAS) and quartz-enhanced photoacoustic spectroscopy (QEPAS) are trace gas detection techniques that use a conventional microphone and a quartz tuning fork (QTF) with a microresonator (mR) as an acoustic transducer for detecting weak photoacoustic signals in minimal sample volumes respectively [18]. This technique uses a low-cost, commercially available sound detector as an acoustic wave transducer with high selectivity and immunity to ambient noise.

Conventional spectroscopic methods like direct absorption spectroscopy are limited in sensitivity by the path length, as a longer path length improves the sensitivity of the measurement. In such cases, integrated cavity output spectroscopy (ICOS), cavity-enhanced absorption spectroscopy (CEAS), and CRDS allowing a sufficient path length of many kilometers with sensitivity as small as parts per billion or even parts per trillion, are used to improve the sensor system 12. [5, 14, 20]. Though such techniques offer high sensitivity and can be adapted for continuous real-time monitoring, such methods need excellent cavity length tuning. The optical cavity or multi-pass cell is also sensitive to mechanical vibration. Even the availability of either tunable light sources or high reflectance mirrors precisely for a wavelength of the target gas sample limits this technique from being widely used. Moreover, the need for high-reflectance mirrors makes it relatively more expensive when compared with the other optical methods.

On the contrary, optical fibers make it easier to transport the beam from the source to the sensing cell [21]. These arrangements are especially appealing in the near-infrared spectral region, where several gases, especially CH_4, and C_2H_6, can be detected [8]. The ability to use optical fibers as gas cells, namely hollow core waveguides (HCW), for molecular detection is another fascinating potential of optical fiber technology. In this direction, an innovative approach of utilizing a supercontinuum laser (SCL) in combination with an HCW for detecting ethane and methane gas concentration is proposed in this work. The proposed technique utilizes a broadband SCL to cover a wide spectral range [22], from 0.8 to 2.8 μm. This broad spectrum is achieved using nonlinear optical effects in a medium, often through self-phase modulation and subsequent spectral broadening. By tuning the laser to the target gas absorption bands, the transmitted light intensity changes can be used to quantify gas concentration. Fiber-based supercontinuum lasers can be integrated into existing optical setups easily. Depending on the pump laser and nonlinear processes, SCL can deliver relatively high power and energy levels.

The HCW is an optical waveguide that utilizes a core with a low refractive index region surrounded by a higher refractive index cladding. Unlike traditional solid-core waveguides, which confine light within a solid material, hollow-core waveguides confine light within a hollow region [12]. This unique design offers several advantages and applications in various fields. HCWs typically consist of three main layers: The core, the central region through which light propagates. This region is usually filled with air or a low-density gas, resulting in a lower refractive index than the cladding. Cladding is the material that surrounds the core. It has a higher refractive index and helps to confine the light within the core. Coating or Outer Cladding is an additional layer surrounding the cladding to protect the waveguide and minimize losses due to environmental factors. The core is often surrounded by a coating of metallic and dielectric layers, usually made of silver (Ag) and silver iodide (AgI).

HCWs offer advantages like broadband transmission, low dispersion, and the ability to guide gas and liquids. HCW can enhance the interaction between light and the target gas molecules, making them useful for trace gas sensing applications. However, the choice of technique depends on factors like sensitivity, accuracy, required detection limit, sample matrix, and the specific application. Conventional sensing methods face challenges in detecting low methane/ethane concentrations amidst other gases. When combined with HCW, Broadband SCL-based Sensing offers a novel solution to overcome these limitations.

2 Experimental Setup

Figure 1 illustrates the HCW absorption spectroscopy setup utilized in the laboratory. An infrared detector card was used to orient the setup and guarantee that the laser beam enters the cell through the existing window after passing through all the optics. This arrangement used a (Leukos 250-IR) low power, low energy supercontinuum light source. The measurement of several gas species contained in the gas analyte was explored using the broadband spectrum provided by the supercontinuum laser source, which spectral range from 0.8 to 2.5 μm. A Hollow core waveguide (HFP 1500 SWMW-SMA, Guiding Photonics) fiber with a hollow core diameter of 1500 μm and a length of 10 m was used. In order to send the metered gas sample through a gas mixing system, the HCW's two ends were attached to stainless steel gas cell blocks (Guiding Photonics). The light output from the supercontinuum source was directed via an appropriate filter that matched the target gas analyte's absorption wavelength (2.3 μm for C_2H_6 and 1.65 μm for CH_4 with a spectral bandwidth of approximately 50 nm) from the filter. The free-space optical system couples the output to the HCW. The output is connected in free space to the optical spectrum analyzer (OSA 203C, Thorlabs) to obtain the transmission characteristics.

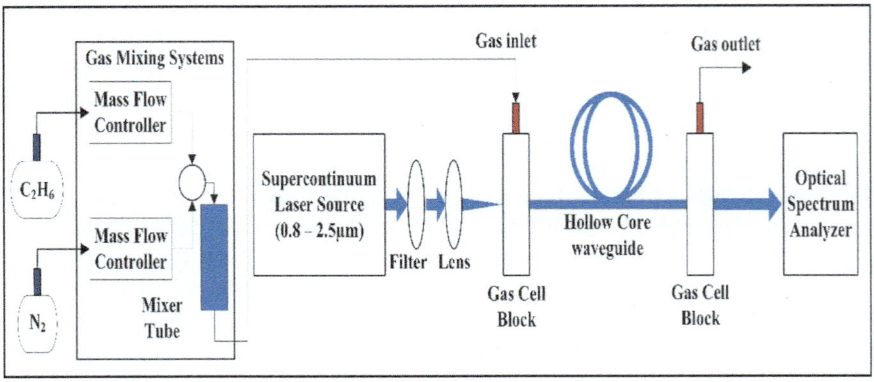

Fig. 1 Schematic of the hollow-core fiber absorption spectroscopy experimental setup [9]

3 Result and Discussion

The HCW absorption experiment aimed to determine the C_2H_6 and CH_4 concentrations. The transmission spectra of the supercontinuum source were obtained using the calibrated C_2H_6 (1% balanced with N_2) and CH_4 (each 10.6% balanced with N_2). The absorption spectra for C_2H_6 (2.3 μm) and CH_4 (1.65 μm) were compared with the PNNL molecular absorption and HITRAN (high-resolution transmission molecular absorption) database. The absorption cross-sections of C_2H_6 and CH_4 in the 2.3 μm and 1.65 μm bands are 10^{-21} cm^2/molecule.

A gas mixing system with a mass flow controller capable of providing gases at various concentrations was used for the experiment to obtain C_2H_6 and CH_4 in different concentrations.

Figure 2a displays the transmission spectrum of the SCL with the 2.3 μm filter obtained from the FTIR spectrum analyzer with an optical path length (OPL) of 10 m. Figure 2b shows the transmittance spectra of C_2H_6 at a concentration of 1%, and Fig. 2c shows the absorption spectra of C_2H_6 for concentrations varying from 1000 ppm to 1%. From Fig. 2a and b, the transmission spectra are seen with multiple etalon interferences that could result from forming any two-plane optical surfaces like the inlet and outlet gas cell blocks fitted with BaF windows.

Figure 3a displays the CH_4 transmission spectrum of the SCL with the 1.65 μm filter obtained from the FTIR spectrum analyzer with an optical path length (OPL) of 10 m. The etalon peaks seen in Fig. 3a contribute to the background noise, thereby severely limiting the detection limit of the sensor system. Nevertheless, normalization and signal processing techniques were followed to remove the background noise. Figure 3b shows the transmittance spectra of CH_4 at a concentration of 10.6%, and Fig. 3c shows the absorption spectra of CH_4 for concentrations varying from 500 ppm to 10.6%.

The minimum detection limit of C_2H_6 and CH_4 was determined by considering the absorbance values as a function of concentration.

The linear fit was obtained with a regression coefficient (R^2) of 0.938 and 0.973, respectively, as shown in Fig. 4a and b. A 10 m path length and a 3 dB signal-to-noise ratio were used to determine the C_2H_6 and CH_4 detection limits of 30 and 23 ppm. This technology can be further developed by extending the HCW's path length to achieve considerably lower detection limits. The major advantage of this technique is that considerably longer path lengths can be achieved with a compact footprint, allowing this sensor system to be mounted on aerial vehicles for continuous monitoring of natural gas pipelines.

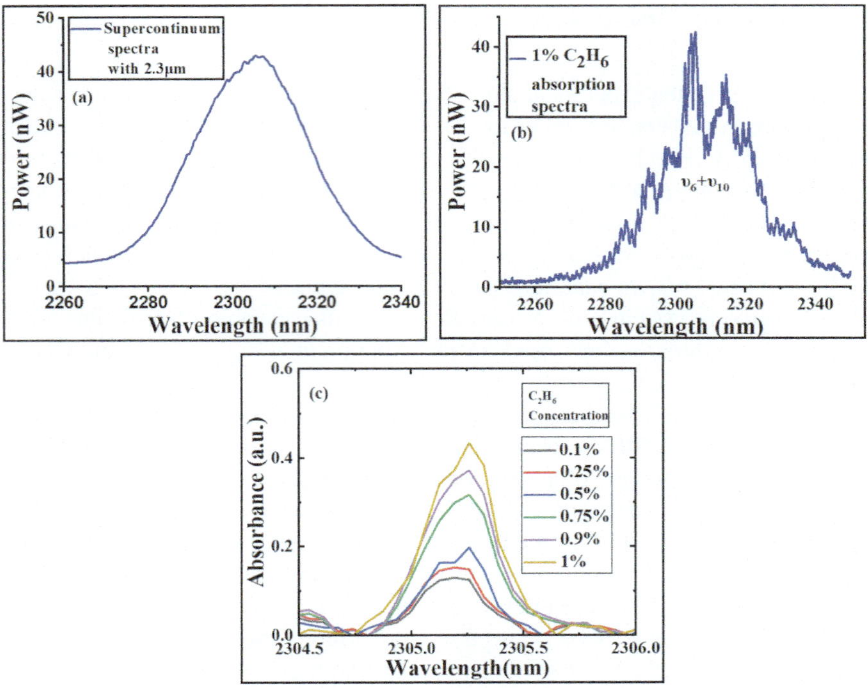

Fig. 2 Spectrum of **a** 2.3 μm nm filter, **b** C_2H_6 absorption, **c** C_2H_6 absorbance for various concentrations

4 Conclusions

Broadband direct absorption spectroscopy for multigas sensing was demonstrated using a hollow core waveguide. Target gas molecules like ethane and methane, the primary natural gas hydrocarbon, must be monitored to avoid disasters due to pipeline leakage. C_2H_6 concentration was varied from 0.1 to 1%, with the remaining N_2 gas as a buffer using a Mass Flow Controller system. Similarly, CH_4 concentration varied from 0.05 to 1% inside 10 m HCW. The minimum detection limits achieved for SNR of 3 dB are 30 ppm and 23 ppm, respectively, for ethane and methane. The method may include more gases such as CO, H_2S, H_2O vapor, NH_3, and CO_2. The sensor can be a complete multigas sensing unit for various environmental applications.

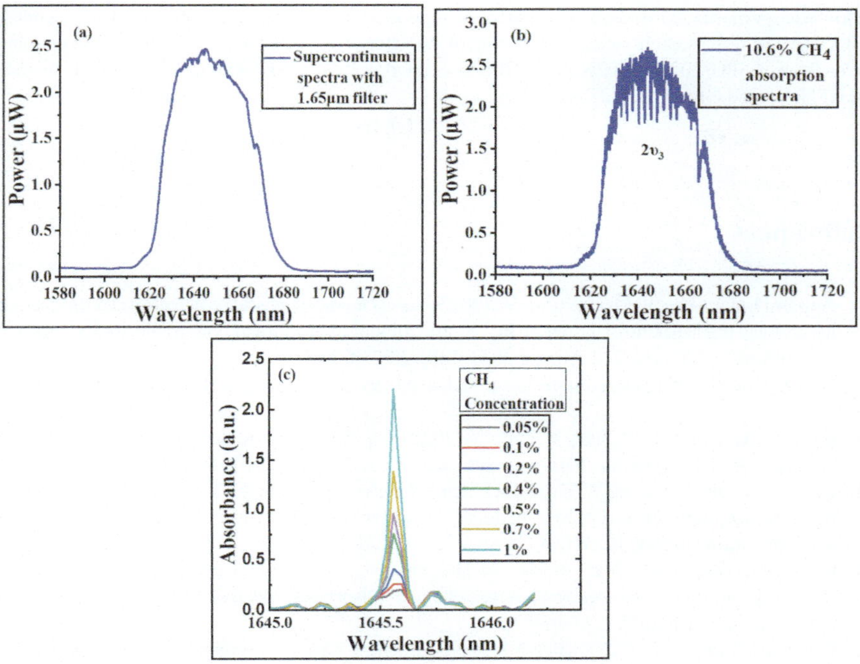

Fig. 3 **a** Spectrum of **a** 1.65 μm filter, **b** CH_4, absorption, **c** CH_4 absorbance for various concentrations

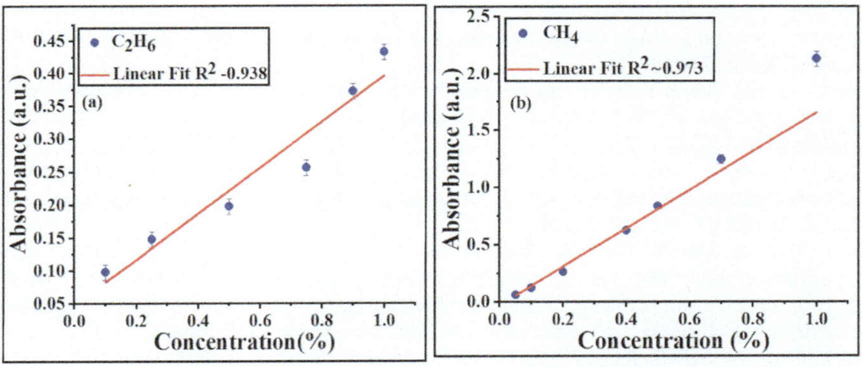

Fig. 4 **a** Absorbance as a function of the concentration of **a** C_2H_6 in 2.3 μm band, **b** CH_4 in 1.65 μm band

Acknowledgements This work was partially supported by a Research Grant from the Science and Engineering Research Board—Department of Science and Technology (SERB-DST), India, under Project SPG/2021/003626, and partially supported by the NIT Raipur SEED grant, project no: NITRR/Seed Grant/2021-22/35.

References

1. Appah D, Aimikhe V, Okologume W (2023) Assessment of gas leak detection techniques in natural gas infrastructure: a review. In: Nigeria annual international conference and exhibition held in Lagos, Nigeria, 2–4 August 2021, SPE-208236-MS
2. Bao X, Chen L (2012) Recent progress in distributed fiber optic sensors. Sensors 12(7):8601–8639
3. Cai P, Li M, Liu Z, Li P, Zhao Y, Zhou Y (2022) Experimental and numerical study of natural gas leakage and explosion characteristics. Am Chem Soc 7(29):25278–25290
4. Cao Y, Sanchez NP, Jiang W, Griffin RJ, Xie F, Hughes LC, Zah CE, Tittel FK (2015) Simultaneous atmospheric nitrous oxide, methane and water vapor detection with a single continuous wave quantum cascade laser. Opt Express 23(3):2121–2132
5. Ciaffoni L, Hancock G, Harrison JJ, van Helden JPH, Langley CE, Peverall R, Wood S (2013) Demonstration of a mid-infrared cavity enhanced absorption spectrometer for breath acetone detection. Anal Chem 85(2):846–850
6. Dong L, Tittel FK, Li C, Sanchez NP, Wu H, Zheng C, Yu Y, Sampaolo A, Griffin RJ (2016) Compact TDLAS based sensor design using interband cascade lasers for mid-IR trace gas sensing. Opt Express 24(6):528–535
7. Ghorbani R, Schmidt FM (2017) ICL-based TDLAS sensor for real-time breath gas analysis of carbon monoxide isotopes. Opt Express 25(11):12743–12752
8. Jiang J, Wang Z, Han X, Zhang C, Ma G, Li C, Luo Y (2019) Multigas detection in power transformer oil based on tunable diode laser absorption spectrum. IEEE Trans Dielectr Electr Insul 26(1):153–161
9. Kumar KS, Vidhya YEB, Selvaraj R, Seshadri S, Nagendra SS, Vasa NJ (2023) Dual absorption broadband photoacoustic technique to eliminate interference in gas mixtures. IEEE Sens J 23(6):5703–5712
10. Li C, Dong L, Zheng C, Tittel FK (2016) Compact TDLAS based optical sensor for ppb-level ethane detection by use of a 3.34 μm room-temperature CW interband cascade laser. Sens Actuators B Chem 232:188–194
11. Li J, Yu B, Zhao W, Chen W (2014) A review of signal enhancement and noise reduction techniques for tunable diode laser absorption spectroscopy. Appl Spectrosc Rev 49(8):666–691
12. Li J, Yan H, Dang H, Meng F (2021) Structure design and application of hollow core micro structured optical fiber gas sensor: a review. Opt Laser Technol 135:106658
13. Loth JL, Morris GJ, Palmer GM, Guiler R, Mehra D (2003) Portable Acoustic Monitoring Package (PAMP). U.S. Department of Energy. https://doi.org/10.2172/822766
14. McCurdy MR, Bakhirkin Y, Wysocki G, Tittel FK (2007) Performance of an exhaled nitric oxide and carbon dioxide sensor using quantum cascade laser-based integrated cavity output spectroscopy. J Biomed Opt 12(3):034034–034034
15. Mishra RK, Zhang Y (2012) A review of optical imagery and airborne LiDAR data registration methods. Open Remote Sens J 5:54–63
16. Ren ZY et al (2011) Characterization of microbial fuel cells at microbially and electrochemically meaningful time scales. Environ Sci Technol 45(6):2435–2441. https://doi.org/10.1021/es103115a
17. Ring F (2010) Thermal imaging today and its relevance to diabetes. J Diabetes Sci Technol 4(4):857–862

18. Shi C, Wang D, Wang Z, Ma L, Wang Q, Xu K, Ren W (2017) A mid-infrared fiber-coupled QEPAS nitric oxide sensor for real-time engine exhaust monitoring. IEEE Sens J 17(22):7418–7424
19. Souza de Joode A, Hoffman A (2011) Pipeline leak detection and theft detection using rarefaction waves. Presented at the Pipeline Technology Conference. Retrieved from Atmos International
20. Wojtas J (2015) Application of cavity enhanced absorption spectroscopy to the detection of nitric oxide, carbonyl sulphide, and ethane—breath biomarkers of serious diseases. Sensors 15(6):14356–14369
21. Woodward RI, Majewski MR, Hudson DD, Jackson SD (2019) Swept-wavelength mid-infrared fiber laser for real-time ammonia gas sensing. APL Photonics 4(2)
22. Yamazoe N (2005) Toward innovations of gas sensor technology. Sens Actuators B Chem 108(1–2):2–14

Laser-Assisted Techniques for Trace Gas Sensing and Measurements

K. Saran Kumar, Ramya Selvaraj, S. Satyanarayanan, S. M. Shiva Nagendra, and Nilesh J. Vasa

Abstract Various state-of-the-art optical techniques used for trace gas sensing are presented. Hollow-core fiber (HCF) based absorption spectroscopy and Photoacoustic Spectroscopy (PAS) low-volume measurement techniques are demonstrated for ammonia (NH_3), methane (CH_4), carbon dioxide (CO_2), carbon monoxide (CO) and acetone (C_3H_6O), which are important for environmental, other industrial monitoring, and biomedical monitoring applications. A supercontinuum laser (SCL) source (from 0.9 to 2.5 μm) and a quantum cascade laser (QCL) source (from 7.9 to 8.4 μm) are used. The HCF-based measurements using a 10 m long fiber measured concentrations of NH_3, CH_4, CO_2 and C_3H_6O in the order of 13, 15, 154 and 30 parts-per-million (ppm), respectively. While, in the PAS, the detection limits in sub-ppm levels for NH_3 as 0.36 ppm, CH_4 as 0.48 ppm and CO as 2.59 ppm are attained. The interference of various species in the gas mixture is eliminated by adopting the dual-wavelength measurement approach.

Keywords Photoacoustic spectroscopy · Hollow-core fiber · Broadband source · Atmospheric monitoring · Multi-gas sensing

---

K. Saran Kumar · N. J. Vasa (✉)
Department of Engineering Design, Indian Institute of Technology Madras, Chennai, India
e-mail: njvasa@iitm.ac.in; ed19d755@smail.iitm.ac.in

R. Selvaraj
Department of Electrical Engineering, National Institute of Technology, Raipur, India

S. Satyanarayanan
Department of Applied Mechanics, Indian Institute of Technology Madras, Chennai, India

S. M. Shiva Nagendra
Department of Civil Engineering, Indian Institute of Technology Madras, Chennai, India

S. M. Shiva Nagendra et al. (eds.), *Select Proceedings of the 8th Indian International Conference on Air Quality Management (IICAQM 2023)*, Lecture Notes in Civil Engineering 582, https://doi.org/10.1007/978-981-96-2359-4_7

1 Introduction

Laser-assisted optical gas sensing is used frequently in various applications, including industrial exhaust, clinical and environmental monitoring. Laser-based gas sensing techniques offer high selectivity and sensitivity to the measurements when compared to the other non-optical methods. The quick response time makes it a suitable candidate for online and real-time monitoring. Hodgkinson and Tatam [10] reported several optical techniques that are used for gas sensing, however, most of them are application-specific which governs the measurement such as concentration, resolution, sensitivity and operating environment. The reason for the widespread use of laser-assisted gas sensing is due to the fact that most of the gases from industrial exhausts, environment and clinical have strong absorption characteristics in the infrared (IR) wavelengths. Popa and Udrea [20], the mid-IR region is of particular interest due to the presence of characteristics vibration and rotation spectra with narrow and non-overlapping bands. Other non-optical gas sensing techniques such as electrochemical (EC), metal oxide semiconductor (MOS), carbon nanotubes (CNTs) and gas chromatography (GC) are also used. Khan et al. [13] extensively reviewed the EC-based gas sensors, they are specific to gases and have very high sensitivity in the order of ppm to parts-per-billion (ppb), but they have shorter operational lifetimes. Dey [3] described in detail the working and uses of MOS for sensitive gas measurements however, these sensors show drift and cross-sensitivity responses based on the operating environmental conditions. On the other hand, Han et al. [8] commented on CNTs for gas sensing applications and their sensitivity to a few gases like CH_4 at room temperature. The major setback in using CNTs as gas sensors lies in the fact that the synthesis of CNT is a complex process and also, it is difficult to maintain the characteristics of the designed prototype. GC is a broadly used method for analyzing volatile compounds. The system is combined with several detectors to improve the range of its selectivity and sensitivity. Aslani and Armstrong [1] explained in detail about the various detectors used in GC, such as mass spectrometer, flame ionization detector, thermal conductivity detector, electron capture detector, flame photometric and photoionization detector. Though GC is a mature technique, it is bulky and laboratory-based.

However, on the other hand, optical sensors involve a direct method of measurement based on the physical properties of the gas molecule. Several laser-assisted techniques such as direct absorption spectroscopy (DAS), hollow-core fiber-based direct absorption spectroscopy (HCF-DAS), cavity ring-down spectroscopy (CRDS), photoacoustic spectroscopy (PAS), quartz enhanced photoacoustic spectroscopy (QEPAS), laser-induced fluorescence spectroscopy (LIFS), evanescent-wave spectroscopy (EWS), Raman spectroscopy (RS) and photo-thermal based spectroscopy (PTS) are available for application specific uses. DAS works on the principle that a wavelength (or frequency) of light is made to strike the gas molecule, if the transition mode frequency matches the input radiation, then the gas molecule absorbs that portion of the light. Zhang et al. [30] emphasized improving the absorbance signal by reducing the noise using particle swarm optimization. Generally, absorption spectroscopy uses a single-line laser with a single detector ideally suited for a particular gas molecule. Kumar et al. [17] used broadband laser source to cover multiple lines/ absorption bands of several gases. Such broadband absorption is prone to the effect of cross-sensitivity, which can be eliminated by dual-wavelength measurements. Divya et al. [4] measured NH_3 in the presence of H_2O using a Superluminescent light-emitting diode (SLED) and obtained an MDL of 120 ppm. Tunable diode lasers (TDL) can be used to span across different wavelengths for investigating different gas molecules, Tm:$GdVO_4$ based tunable laser absorption spectroscopy (TDLAS) with continuous operation having tunability over 130 nm (1.86–1.99) μm was investigated by Sorokin et al. [26]. Dong et al. [5], showed an inter-band cascade laser (ICL) based multi-pass gas cell that was used for the detection of CH_4 and ethane (C_2H_6) in 3.2 μm and 3.3 μm, respectively in ppb order. Cao et al. [2], measured atmospheric moisture (H_2O), CH_4, and nitrous oxide (N_2O) using a QCL at 7.7 μm for CH_4 detection and an MDL of 8.5 ppb was achieved. Wavelength modulation DAS (WM-DAS) is used to combine the advantages of absorbance from DAS with improved sensitivity and enhanced noise rejection using wavelength modulation spectroscopy (WMS) was experimented by Du et al. [6]. CRDS effectively improves the path length from meters to a few kilometres which can improve the sensitivity of the system to as low as ppb or even parts-per-trillion (ppt) order. An absorption cavity with two highly reflective mirrors is used, and the system is inert to noise fluctuations as the absorbance is measured from the decay time as demonstrated by Maity et al. [18]. Techniques such as noise immune cavity-enhanced optical heterodyne molecular spectroscopy (NICE-OHMS) have been used by Schmidt et al. [23] to measure acetylene (C_2H_2) at 1.5 μm with a detection limit of 103 ppt. He et al. [9] reported on off-axis integrated cavity output spectroscopy (OA-ICOS) for the detection of CH_4 at 1.65 μm with a distributed feedback (DFB) laser. LIFS is widely used for flow diagnostics and combustion studies as reported by Giezendanner et al. [7]. Though this method can be used for determining multiple gases, the quantitative examination is complex and the method requires advanced detection systems. EWS works on the principle that when a light source is coupled between the core and cladding region to undergo total internal reflection, the reflected light (evanescent wave) penetrates the cladding region. Vlk et al. [27] used EWS sensor to measure C_2H_2 in 2.56 μm to an order of 7 ppm with a 2 cm long fiber, also CH_4 detection was reported with a microfiber of length 5.6 cm and 2.36 μm diameter, MDL of 160 ppb was measured in nitrogen (N_2) atmosphere. RS is an effective technique that works on the effect of scattering, when a laser source with frequency υ_o passes through the gas, the incident light excites the molecules to produce spherical divergent Raman scattering with frequency $\upsilon_o \pm \upsilon_R$ as stated by Qian et al. [21]. Each gas species has its specific υ_R called the Raman frequency shift, thus multi-gas measurements can be made simultaneously. However, the sensitivity is less due to the low Raman scattering cross-section. The sensitivity of the system can be enhanced when an optical resonator cavity is used as shown by Wang et al. [29]. Knebl et al. [14] reported another way to improve the signal strength by combining the RS with HCF or a capillary that improves the collection efficiency of spherical divergent Raman scattered light to about 360°. PTS is similar to the DAS that illuminates the gas sample with the laser source (pump laser), once the absorption takes place a part of the energy absorbed is converted to heat (due to non-radiative relaxation) which in turn causes pressure and density change in the gas molecules [15]. The change in density in turn changes the refractive index (RI) of the gas. However, this change is very small most of the PST system is combined with interferometers. By employing additional WMS can further enhance the signal. Hu et al. [11] reported a study on nitric oxide (NO) using a HCF of 35 cm length with QCL emitting 5.2 μm, a noise equivalent concentration of 50 ppb corresponding to a normalized noise equivalent absorption coefficient (NNEA) of 8.6×10^{-9} cm^{-1} $WHz^{-1/2}$ was achieved. Selvaraj et al. [25] explained PAS technique in detail, in this method light energy is converted to sound energy in series of steps. A laser light matching the absorption wavelength of the gas sample is modulated by means of an optical chopper that runs at the resonance frequency of the gas cell. The gases in the ground state get excited and during relaxation, heat is given out. By periodically modulating the light, a periodic local heating is produced in the gas which in turn gives rise to a periodic pressure change which is captured using an acoustic detector. PAS is ideally used for trace gas sensing as it has zero background noise and the photoacoustic signal is not getting scattered in the medium. Broadband photoacoustic (BPA) techniques are also reported for measuring multiple gases using a single broadband laser source [17, 25]. However tunable lasers are also used for multi-gas sensing, such as NH_3, CO, CO_2, CH_4, H_2S and H_2O. The detection limit in the order of sub-ppm level is reported by Wang and Wang [28]. QEPAS systems work with the same principle as that of PAS, the only difference being that a quarts tuning fork (QTF) is used as an acoustic detector instead of a microphone. Sampaolo et al. [22] reported on a custom-designed QTF, the quality factor of the system can be improved in the order of a few thousands, the unique acoustic quadrupole structure and narrow resonance make the QTF immune to environmental noises, and this design enhances the SNR of the sensor. Zheng et al. [31] measured nitrogen dioxide (N_2O) at 450 nm a MDL of 0.6 ppb was obtained with an integration time (τ) of 1 s. HCFs are specially designed optical fibers that allow gas and guide the light to travel through it. HCFs are specially designed to enhance the path length of light–matter interaction with low propagation and bend losses [19]. Johnson et al. [12], used a hollow-core photonic bandgap fibers (HC-PBGFs) with optical parametric oscillator (OPO) source for the measurement of CH_4 with a detection limit of 50 ppm. A QCL-based system operating at 4.52 μm was used to measure N_2O in the order of 5.4 ppb with a 3.2 m long HCF. The other major advantage of using a HCF is the sample volume, as the HCF is thin and long the gas sample required is only in the order of few microliters (μL) which makes it ideal for breath gas analysis for clinical diagnosis.

The major motivation to carry out this work is to shed light on various types of laser-assisted techniques available for gas sensing applications. Particularly in this paper, we have explained two systems namely HCF-DAS and BPAS with experimental demonstrations. An HCF, 10 m long along with a QCL was used to measure NH_3, CH_4, and C_3H_6O and An SCL based BPAS for NH_3, CH_4, and CO. The SCL

Table 1 Gas concentrations

S. No	Gas	Concentration (%)
1	Ammonia (NH_3)	5
2	Carbon dioxide (CO_2)	5
3	Carbon monoxide (CO)	0.2
4	Methane (CH_4)	10
5	Acetone (C_3H_6O)	1

source was also used with HCF to measure NH_3. Atmospheric CO_2 was measured with both HCF and PAS techniques using the SCL source.

2 Experimental Setup

Calibrated gases with concentrations safe for laboratory usage were used for the experimental study. Table 1 shows the concentrations of gases used during the experiment.

The calibrated gases are balanced with the homo nuclear diatomic molecule, nitrogen (N_2) in this case. They have no dipole moment and are thus IR inactive. N_2 was further used to dilute the gases to achieve different concentrations. The gas concentrations were varied using a dedicated gas mixing system with three mass flow controllers (MFCs), the system offers the ability to mix three gases at the same time. The MFC library has the list of most commonly used gases that are pre-calibrated and stored, the gases can be easily changed based on their use. The required concentration is fed to the MFC using system-specific software (Flow Vision MX). A digital pressure gauge is used to read out the outlet gas pressure.

3 Hollow-Core Fiber Absorption Spectroscopy

HCF-based absorption spectroscopy (HCF-AS), using QCL is shown in Fig. 1. A broadband QCL (NL-8100-7023-00373) from Sacher Lasertechnik was used as the laser source. The QCL offers a spectrum ranging from (7.9 to 8.4) μm, covering strong absorption lines of C_3H_6O, NH_3 and CH_4 gases with absorption cross-sections in the order of (~ 10^{-19}, ~ 10^{-20} and ~ 10^{-19}) cm^2/molecule, respectively. An HCF (HFP 1500 SWMW SMA) from Guiding Photonics was used for the study. Two 5 m long HCF were coupled using a sub-miniaturized assembly (SMA) coupler to obtain a 10 m long HCF whose ends are connected to the gas cell block. The HCF core is made of silver iodide (AgI) coating as a dielectric layer over silver (Ag) on the glass fiber. The gas cell block includes three parts, a fiber port, a gas port and an optical window. The ends of the HCF are connected to the fiber port and the gas sample is fed to the system using the gas port located perpendicular to the fiber port. The optical

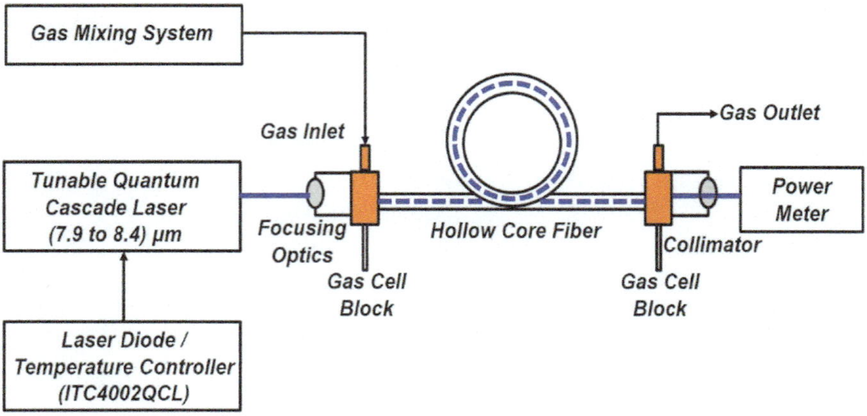

Fig. 1 Laboratory setup of HCF-based absorption spectroscopy

window is made of Barium fluoride (BaF_2) having transmission from (0.2–11) μm to facilitate the laser to pass through the HCF containing the gas.

The proposed system can be used to measure trace gases by increasing the HCF path lengths. The QCL wavelengths were tuned across the absorption lines of the gases using dedicated software (motor-motion controller V3.2). The output of the system is collimated and connected in free space to a power meter (Gentec, MAESTRO) to obtain the transmission characteristics. High humidity levels in the atmosphere during the measurement helped in removing the NH_3 from the inner wall of HCFs.

4 Broadband Photoacoustic Spectroscopy

A schematic of a laboratory setup of a SCL-based broadband photoacoustic spectroscopy (BPAS) approach is shown in Fig. 2. A SCL (Leukos, SM-250-IR) was used as a source for the study, the SCL source consists of an optical fiber-based output with a spectrum ranging from (0.9 to 2.5) μm. The pulse width is approximately 1 ns with a pulse repetition rate of 250 kHz. An optical chopper (NF instruments, 5584A) was used to modulate the laser source at the resonance of the photoacoustic gas cell. Two focusing lenses are used, one is placed in front of the cell to focus the laser beam at the centre of the gas cell where the acoustic detector (electric microphone, Ono Sokki, MI 1235) is located for picking up the photoacoustic signal. One more lens is placed after the gas cell that collects the beam and focuses it onto a power meter (OPHIR NOVA 82495).

The signal from the detector is passed on to a preamplifier for enhancing the signal. A lock-in amplifier (Signal Recovery SR 7225) was used to extract the signals produced at the reference frequency (optical chopper). The lock-in amplifier is used

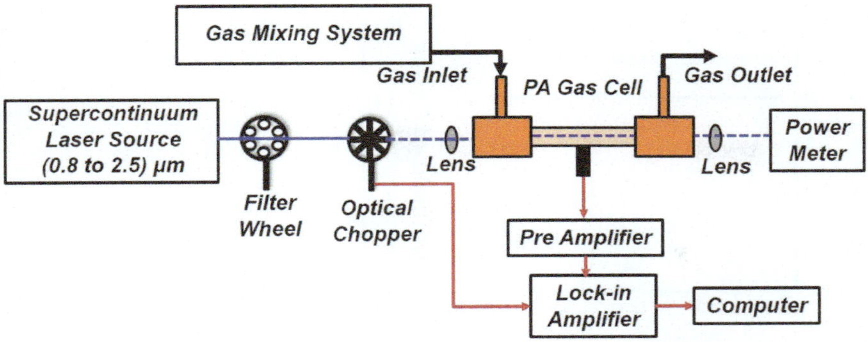

Fig. 2 Laboratory setup of broadband photoacoustic spectroscopy

to detect small signal from a highly noisy background. The device works on phase-sensitive detection for capturing the signal at a specific frequency and phase. Hence, any signals out of phase or at random frequencies are rejected.

5 Results and Discussion

Figure 3a–c show the transmission spectra of C_3H_6O, NH_3 and CH_4 from (7.9 to 8.4) μm band. The normalized power of N_2 is shown in a black line and for that of the gas is shown in red. The obtained absorption spectra were compared with the high-resolution transmission molecular absorption (HITRAN) database for verification. Absorption studies were done with different concentrations of C_3H_6O, NH_3 and CH_4 at 8.21, 8.24 and 8.16 μm, respectively.

NH_3 was also measured using the HCF with the SCL in the 2 and 2.3 μm bands, as shown in Fig. 4a and b.

The experiments were repeated 3 times to analyze the repeatability of the results, also the system was purged with atmospheric air and N_2 after every trial. A linear fit was obtained with regression coefficient (R^2) of 0.98, 0.99 and 0.99 for C_3H_6O, NH_3 and CH_4, respectively as shown in Fig. 5a–c.

With path length of 10 m, and SNR of 3 dB the MDL of 30, 13 and 15 ppm was obtained for C_3H_6O, NH_3 and CH_4, respectively.

For PAS experiments, the measurements were carried out using the SCL source. The calibration gases were mixed with N_2 to vary the concentration and then passed into the PA gas cell. The optical chopper was set to run at 982 Hz, which was calculated to be the resonance frequency of the gas cell. Figure 6a–c show a linear relationship between the PA signal and the varying concentrations.

A linear fit was obtained with regression coefficient (R^2) of 0.98, 0.98 and 0.99 for NH_3, CH_4 and CO, respectively, in the 2.3 μm band. Higher concentration measurements were also carried out and found that the PA signal had a linear dependency on the concentration. The MDL of the gases was determined by Allan deviation

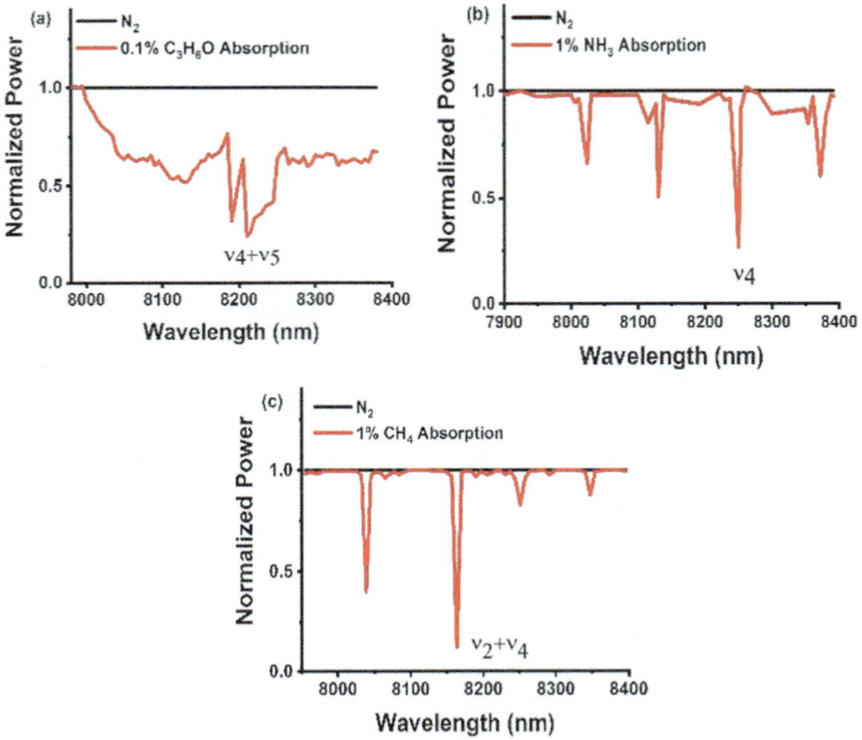

Fig. 3 Absorption spectrum of **a** C_3H_6O, **b** NH_3, **c** CH_4 from 7.9 to 8 μm band

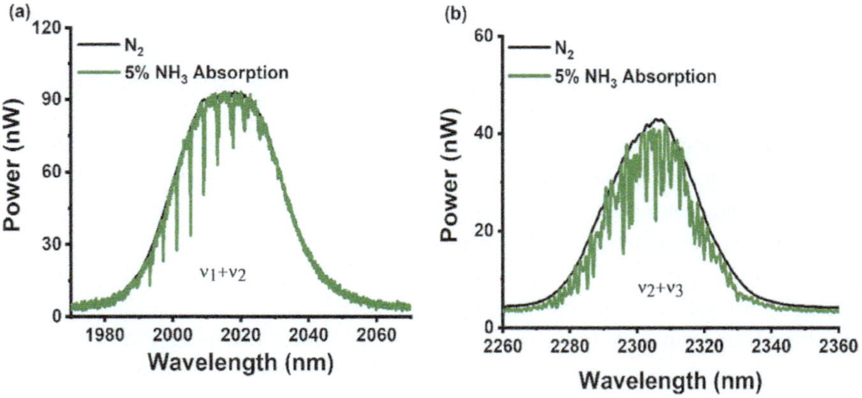

Fig. 4 Absorption spectrum of NH_3, **a** 2 μm, **b** 2.3 μm band

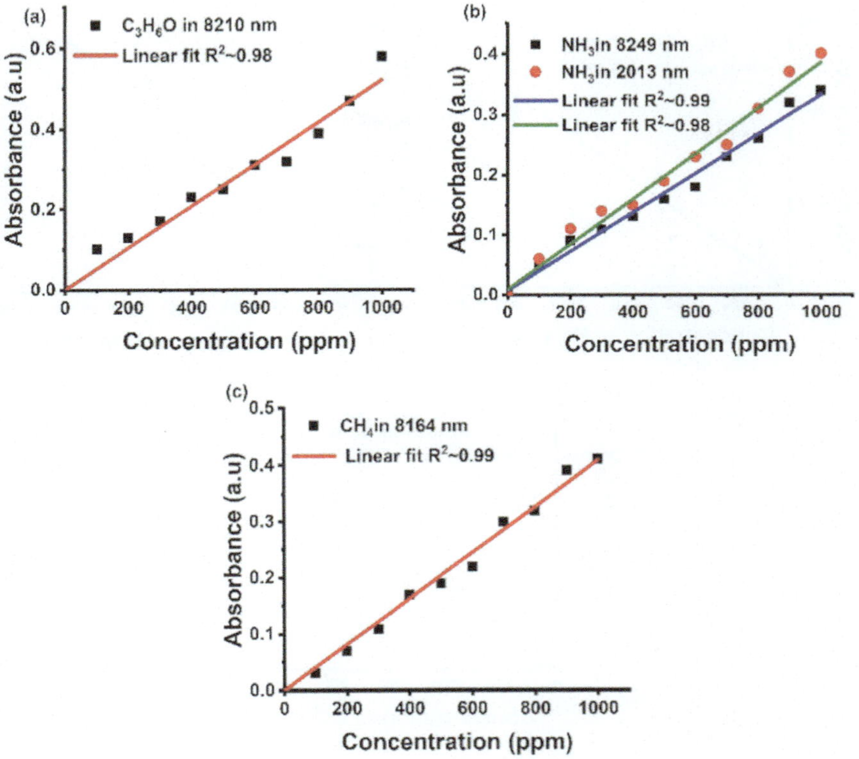

Fig. 5 Absorption as a function of **a** C_3H_6O, **b** NH_3 in 2 and 8 μm band, **c** CH_4

analysis experimentally. A 25-ppm concentration of NH_3, CH_4 and CO was used for the study. With τ of 100, 142 and 71 s for an MDL of 0.36, 0.48 and 2.59 ppm were obtained for NH_3, CH_4 and CO, respectively.

Atmospheric CO_2 measurement was carried out with both HCF and PAS techniques using the SCL source and a 2 μm bandpass filter targeting the $(2\upsilon_1 + 2\upsilon_2 + \upsilon_3)$ transition lines of CO_2. The measurements were made at two different timings of the day (daytime, 10:00 am and nighttime, 8:00 pm). Figures 7 and 8 show the CO_2 measurements made using HCF and PAS, respectively.

6 Conclusions

This paper gives a broad overview of the optical techniques that are available and widely used for gas sensing applications. In this paper, we have focused on HCF and PAS techniques for sensing gases such as NH_3, CH_4, CO, CO_2, and C_3H_6O. A QCL source was used along with HCF to measure NH_3, CH_4 and C_3H_6O in the 8 μm band

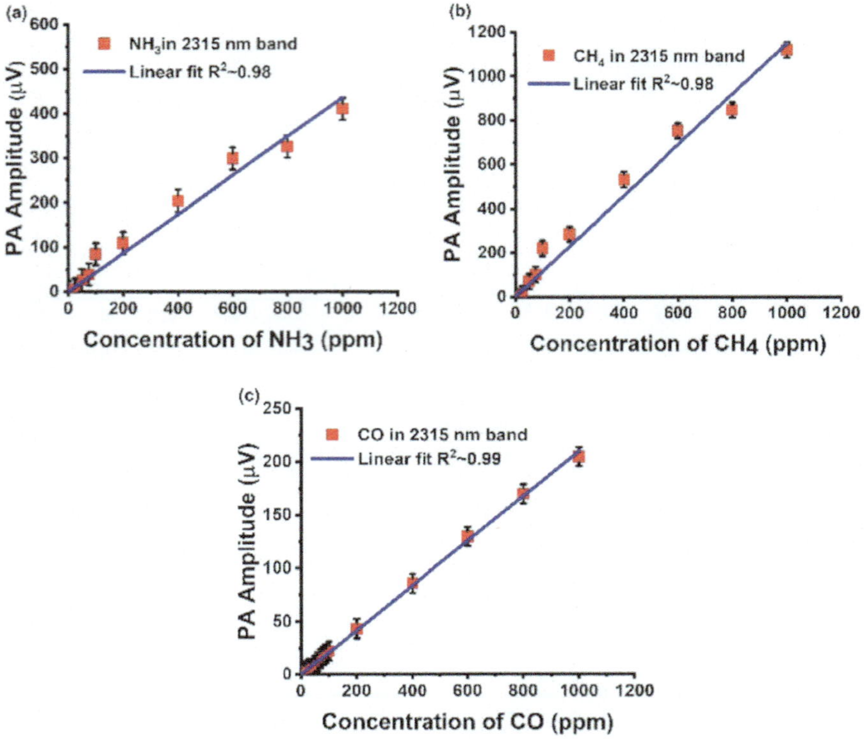

Fig. 6 Variation of PA signal amplitude as a function of concentration, **a** NH_3, **b** CH_4, **c** CO in 2.3 μm band

Fig. 7 Variation of absorbance as a function of CO_2 concentration showing atmospheric CO_2

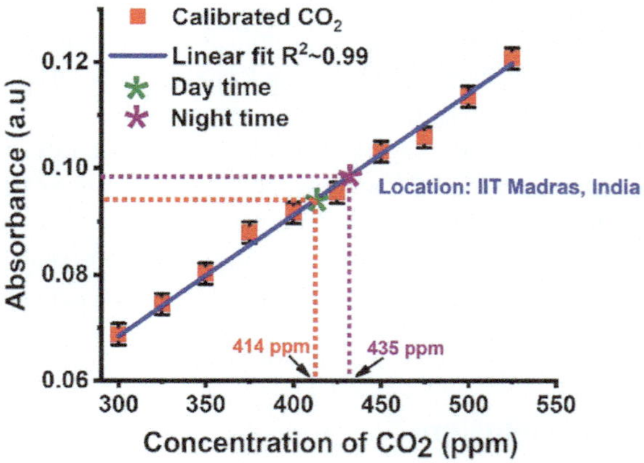

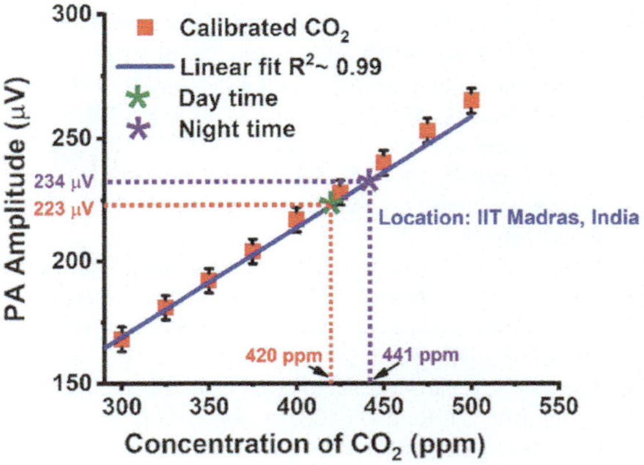

Fig. 8 Variation of PA amplitude as a function of CO_2 concentration showing atmospheric CO_2

and a SCL source with PAS to measure NH_3, CH_4, CO_2 and CO. Using a 10 m long HCF the least detectable concentrations of NH_3, CH_4, CO_2 and C_3H_6O were in the order of 13, 15, 154 and 30 ppm, respectively. While, in the PAS, the least detectable concentrations of NH_3, CH_4 and CO were in sub-ppm order, 0.36, 0.48 and 2.59 ppm, respectively. Atmospheric CO_2 (during the day and night) was also measured using both techniques. CO_2 estimate during the day and at night using HCF was 414 and 435 ppm, respectively, and using PAS was 420 and 441 ppm, respectively. To verify the results, the atmospheric CO_2 was simultaneously measured using an NDIR sensor (IAQ-Calc Indoor Air Quality Meter 7545). The measured value and the actual sensor values fall within an error percent of $\pm$ 1.

Acknowledgements This work was partly supported by a Research Grant from the Science and Engineering Research Board—Department of Science and Technology (SERB-DST), India, under Project CRG/2020/003918. The work of K. Saran Kumar was supported by the Prime Minister's Research Fellowship (PMRF), India.

References

1. Aslani S, Armstrong DW (2022) High information spectroscopic detection techniques for gas chromatography. J Chromatogr A 1676:1–23
2. Cao Y, Sanchez NP, Jiang W, Griffin RJ, Xie F, Hughes LC, Zah CE, Tittel FK (2015) Simultaneous atmospheric nitrous oxide, methane and water vapor detection with a single continuous wave quantum cascade laser. Opt Express 23(3):2121–2132
3. Dey A (2018) Semiconductor metal oxide gas sensors: a review. Mater Sci Eng B 229:206–217
4. Divya K, Sulochana K, Vasa NJ (2011) Superluminescent diode-based multiple-gas sensor for NH_3 and H_2O vapor monitoring. IEEE J Sel Top Quantum Electron 18(5):1540–1546
5. Dong L, Tittel FK, Li C, Sanchez NP, Wu H, Zheng C, Yu Y, Sampaolo A, Griffin RJ (2016) Compact TDLAS based sensor design using interband cascade lasers for mid-IR trace gas sensing. Opt Express 24(6):528–535
6. Du Y, Peng Z, Ding Y (2018) Wavelength modulation spectroscopy for recovering absolute absorbance. Opt Express 26(7):9263–9272

7. Giezendanner R, Keck O, Weigand P, Meier W, Meier U, Stricker W, Aigner M (2003) Periodic combustion instabilities in a swirl burner studied by phase-locked planar laser-induced fluorescence. Combust Sci Technol 175(4):721–741
8. Han T, Nag A, Mukhopadhyay SC, Xu Y (2019) Carbon nanotubes and its gas-sensing applications: a review. Sens Actuators A 291:107–143
9. He Q, Zheng C, Zheng K, Tittel FK (2021) Off-axis integrated cavity output spectroscopy for real-time methane measurements with an integrated wavelength-tunable light source. Infrared Phys Technol 115:1–7
10. Hodgkinson J, Tatam RP (2012) Optical gas sensing: a review. Meas Sci Technol 24(1):1–59
11. Hu M, Yao C, Ventura A, Hayashi JG, Poletti F, Ren W (2022) Mid-infrared photothermal gas sensor enabled by core-cladding mode interference in a hollow-core fiber. J Lightwave Technol 40(19):6568–6575
12. Johnson K, Castro-Marin P, Farrell C, Davidson IA, Fu Q, Jasion GT, Wheeler NV, Poletti F, Richardson DJ, Reid DT (2022) Hollow-core fiber delivery of broadband mid-infrared light for remote spectroscopy. Opt Express 30(5):7044–7052
13. Khan MAH, Rao MV, Li Q (2019) Recent advances in electrochemical sensors for detecting toxic gases: NO_2, SO_2 and H_2S. Sensors 19(4):1–39
14. Knebl A, Domes R, Wolf S, Domes C, Popp J, Frosch T (2020) Fiber-enhanced Raman gas spectroscopy for the study of microbial methanogenesis. Anal Chem 92(18):12564–12571
15. Krzempek K (2019) A review of photothermal detection techniques for gas sensing applications. Appl Sci 9(14):1–19
16. Kumar SK, Vidhya YEB, Selvaraj R, Seshadri S, Nagendra SMS, Vasa NJ (2023) Low volume gas sensing using hollow-core fiber for detection of methane and ammonia with supercontinuum source. CLEO, San Jose, USA, 7–12 May 2023, ATh4I.1
17. Kumar KS, Vidhya YEB, Selvaraj R, Seshadri S, Nagendra SS, Vasa NJ (2023) Dual absorption broadband photoacoustic technique to eliminate interference in gas mixtures. IEEE Sens J 23(6):5703–5712
18. Maity A, Maithani S, Pradhan M (2021) Cavity ring-down spectroscopy: recent technological advancements, techniques, and applications. Anal Chem 93(1):388–416
19. Ni W, Yang C, Luo Y, Xia R, Lu P, Hu DJJ, Danto S, Shum PP, Wei L (2021) Recent advancement of anti-resonant hollow-core fibers for sensing applications. Photonics 8(4):1–13
20. Popa D, Udrea F (2019) Towards integrated mid-infrared gas sensors. Sensors 19(9):1–15
21. Qian G, Wan F, Zhou F, Wang J, Kong W, Chen W (2022) Fiber-enhanced Raman spectroscopy for trace-gas sensing in the high-concentration gas background with an anti-resonant hollow core fiber. Front Phys 10:1–8
22. Sampaolo A, Patimisco P, Giglio M, Zifarelli A, Wu H, Dong L, Spagnolo V (2022) Quartz-enhanced photoacoustic spectroscopy for multi-gas detection: a review. Anal Chim Acta 1202:1–27
23. Schmidt FM, Foltynowicz A, Ma W, Axner O (2007) Fiber-laser-based noise-immune cavity-enhanced optical heterodyne molecular spectrometry for Doppler-broadened detection of C_2H_2 in the parts per trillion range. JOSA B 24(6):1392–1405
24. Selvaraj R, Vasa NJ, Shiva Nagendra SM (2019) Off-resonance photoacoustic spectroscopy technique for multi-gas sensing in biogas plants. Anal Chem 91(22):14239–14246
25. Selvaraj R, Vasa NJ, Nagendra SS (2019) Off-resonant photoacoustic spectroscopy for analysis of multicomponent gas mixtures at high concentrations using broadband vibrational overtones of individual gas species. Appl Opt 58(15):4118–4126
26. Sorokin E, Alpatiev AN, Sorokina IT, Zagumennyi AI, Shcherbakov IA (2002) Tunable efficient continuous-wave room-temperature Tm^{3+}:$GdVO_4$ laser. ASSL, Québec City Canada, 3–6 February 2002, WA2
27. Vlk M, Datta A, Alberti S, Yalle HD, Mittal V, Murugan GS, Jágerská J (2021) Extraordinary evanescent field confinement waveguide sensor for mid-infrared trace gas spectroscopy. Light Sci Appl 10(1):1–7
28. Wang J, Wang H (2014) Tunable fiber laser based photoacoustic gas sensor for early fire detection. Infrared Phys Technol 65:1–4

29. Wang P, Chen W, Wan F, Wang J, Hu J (2019) Cavity-enhanced Raman spectroscopy with optical feedback frequency-locking for gas sensing. Opt Express 27(23):33312–33325
30. Zhang L, Li Y, Wei Y, Wang Z, Zhang T, Gong W, Zhang Q (2022) SNR enhancement of direct absorption spectroscopy utilizing an improved particle swarm algorithm. Photonics 9(6):1–11
31. Zheng H, Dong L, Ma Y, Wu H, Liu X, Yin X, Zhang L, Ma W, Yin W, Xiao L, Jia S (2016) Scattered light modulation cancellation method for sub-ppb-level NO_2 detection in a LD-excited QEPAS system. Opt Express 24(10):752–761

Air Quality Measurement and Monitoring

Traffic-Induced Ultrafine Particle Concentration in Megacity Delhi

Kanagaraj Rajagopal, S. Ramachandran, and Rajeev Kumar Mishra

Abstract Urbanization is occurring at an alarming rate, causing severe pollution events in different developing cities. Delhi is one among them, and it experiences frequent air pollution episodes. Most of the time, the concentration of the pollutants is found beyond the prescribed limits. This study measured and analysed the traffic-induced nanoparticle concentration in the roadside micro-environment. Two periods were considered for the study: Period I, which had less emissions (April–June 2021), and Period II which experienced increased emissions (October–November 2021). During Period I, due to vehicle restrictions, the concentration measured was 31% less than that under normal conditions, and in Period II, due to the high emissions the concentrations were around 35% more than usual. During Period I, the mean concentration of the particles ranged from 10^3 to $10^5\,cm^{-3}$ particles, and during higher emissions, the concentration increased up to 10^4 to 10^5 cm^{-3}. The results obtained through different emission scenarios in the atmosphere will help us better understand the connection between vehicle emissions and contaminants, especially in the microenvironment around metropolitan roadway.

Keywords Ultrafine particles · Traffic emission · Particle number concentration · Nano pollutants

1 Introduction

Air pollution from traffic-related sources is contributing more to pollution in urban regions [11]. A city like Delhi experiences around 50% of the contribution from transportation sources [10]. The majorly known sources are particulate matter and

K. Rajagopal · R. K. Mishra (✉)
Department of Environmental Engineering, Delhi Technological University, New Delhi, India
e-mail: rajeevkumarmishra@dtu.ac.in

S. Ramachandran
Space and Atmospheric Sciences Division, Physical Research Laboratory, Ahmedabad, Gujarat, India

S. M. Shiva Nagendra et al. (eds.), *Select Proceedings of the 8th Indian International Conference on Air Quality Management (IICAQM 2023)*, Lecture Notes in Civil Engineering 582, https://doi.org/10.1007/978-981-96-2359-4_8

gaseous pollutants [12]. The transport-related emissions emit particles in the nano-size ranges as well [8]. These nanoparticles are present in the atmosphere due to natural and anthropogenic sources [10, 5]. Recent advancements in science and technology have made possible the measurement and analysis of these nanoparticles in the atmosphere, and also the estimation of health impacts due to them [4]. In the nanoparticle concentration, the size range classification or standard was not clearly defined, However various studies justified these nanoparticle size ranges into nucleation (10–30 nm), Aitken (30–100 nm), and accumulation mode (100–1000 nm) ranges [3]. The Aitken mode particles can be further classified into small Aitken (30–50 nm) and large Aitken mode (50–100 nm). The 1–100 nm particles are often referred as Ultrafine particles (UFP) [9]. The nanoparticles in roadside microenvironments are directly associated with the engine exhausts in urban regions. The majority of the engine exhaust emits particles in ultrafine particle range [2]. The size range of the particles varies based on the type of fuel and combustion technology used. The nanoparticles are measured using the number concentration due to their lower mass contribution [13]. The nanoparticle concentration in an urban region may range from 10^2 to 10^7 cm^{-3}. The diesel engine exhaust emits particles from 20 to 130 nm, whereas the gasoline engines can emit particles from 20 to 60 nm. The concentration of the nanoparticles was observed to be high near the emission source, and it became less when measured away from the source due to the various physiochemical processes that undergoes in the atmosphere [6]. The nanoparticles undergo processes like coagulation and condensation in the atmosphere depending upon the environmental conditions, such as temperature and relative humidity, and get converted into higher-size particles, such as fine and sub-fine particles [1]. The UFP alone can contribute around 80–90% of the total particle number concentration in a highly polluted urban environment [7]. The study's objective is to estimate the role of transportation sources in the urban environment, especially near the roadside region. Delhi, one of the highly polluted and populated urban cities, was selected for the study. The study was conducted from April to November 2021 where the emission scenarios was different. The study's outcomes will help to estimate the transportation source's contribution on nanoparticles emission in the urban roadside environments and helps to develop control measures and to initiate policies for abatement.

2 Materials and Methods

2.1 Description of the Monitoring Location

The study was conducted in the urban city of Delhi near a traffic source (Roadside) adjacent to Delhi Technological University (Fig. 1) campus. The monitoring station was maintained and operated by members of the Advanced Air and Acoustics Research Laboratory. The university is loacted in Delhi's North West district in Rohini region and is categorized as an typical urban region. The monitoring site

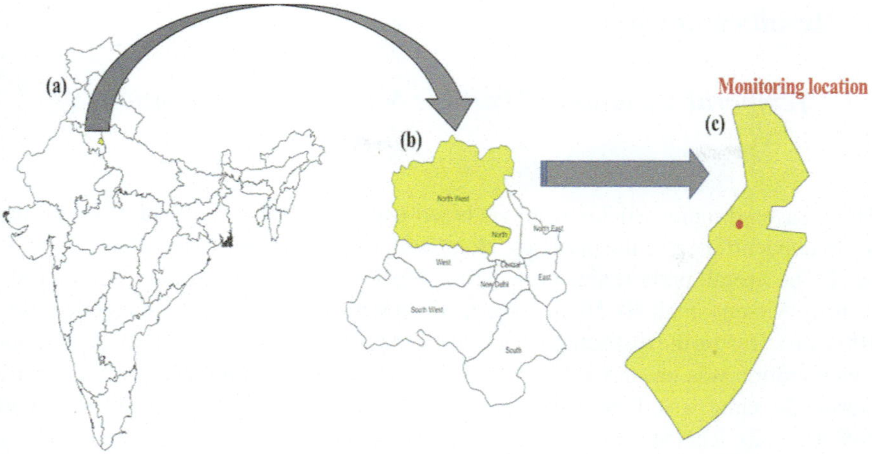

Fig. 1 Map showing the study location, **(a)** map of India, **(b)** Delhi city, and, **(c)** monitoring station

lies in 28°36′ N and 77°33′ E coordinates. The road adjacent to the monitoring location is a busy road with a high density of vehicle population throughout the year.

2.2 *Instrumentation, Data Collection and Analysis*

The scanning mobility particle sizer (GRIMM) was used to measure the traffic-induced ultrafine particles. The SMPS can detect particles in the size range from 10 to 1090 nm using a Large DMA unit. The study was conducted in 2021 when the study region experienced different emission scenarios. In the first phase of the study, lockdown conditions were implemented, leading to restricted traffic sources. In the second phase, there was an event of Diwali and related pollution episodes which experienced more pollution load. The collected data was classified into two periods and three different events in each period and then analyzed using MATLAB and R statistical tools. Meteorological data was collected using a automated weather station installed along with other instruments.

3 Results and Discussion

3.1 Temporal Variation of Particle Number Concentration (PNC)

The measured nanoparticles were analyzed using the box whisker plot, where the mean concentration of the particles, along with its median and standard deviation, is used. The overall particle size range (Total PNC) was classified into nucleation (10–30 nm), Aitken (small 30–50 nm), (large 50–100 nm), and accumulation mode (100–1000 nm). During the restriction phase (DR), the particles' mean total particle number concentration was around 1.7×10^3, 31% less than the normal emission scenario. During the same period, the vehicular density in the study area also reduced by around 45% (Fig. 2). This shows that the role of vehicular emissions in the study area is significant in determining the particle number concentration of the nanoparticles. Recent research studies in study region's indicates that there is a rise in baseline concentration of the pollutants due to the continuous emissions. [8].

The geometric mean diameter (GMD) of the particle is directly corelated with type of emission sources, through which the particle emission source can be identified. Based on the combustion process the GMD of the emitted particle concentration changes. During Period I, the deviation of GMD was less due to the limited emission sources, whereas, during Period II, it was quite more deviated due to the complex mixture of traffic and Diwali sources. The Period after the restriction (AR) phase in Period I recorded the highest concentration in Period I due to the resuming of vehicular activities. Throughout Period I, the concentration of the nanoparticles lies between 10^3 cm^{-3} and 10^5 cm^{-3}, whereas during the Diwali period, it was 10^4 to 10^5 cm^{-3} (Fig. 3). During Period II, there was no restriction for the anthropogenic sources in the study area. During Period II, the GMD also ranges between 20 and 200

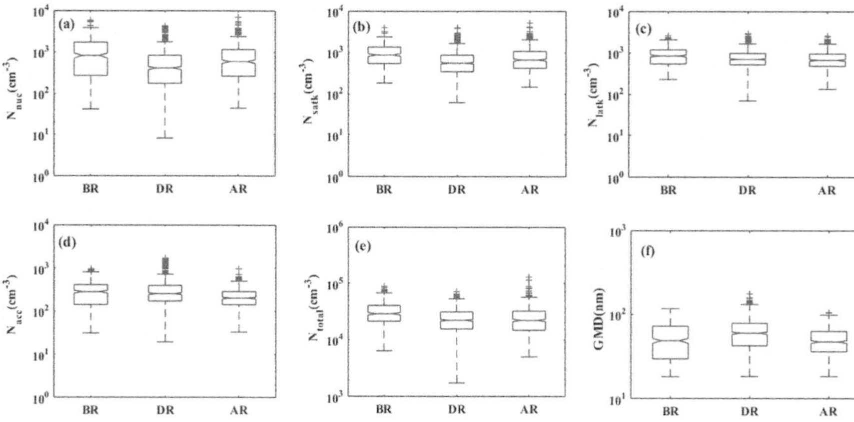

Fig. 2 Box whisker plot analysis for hourly concentration during Period I

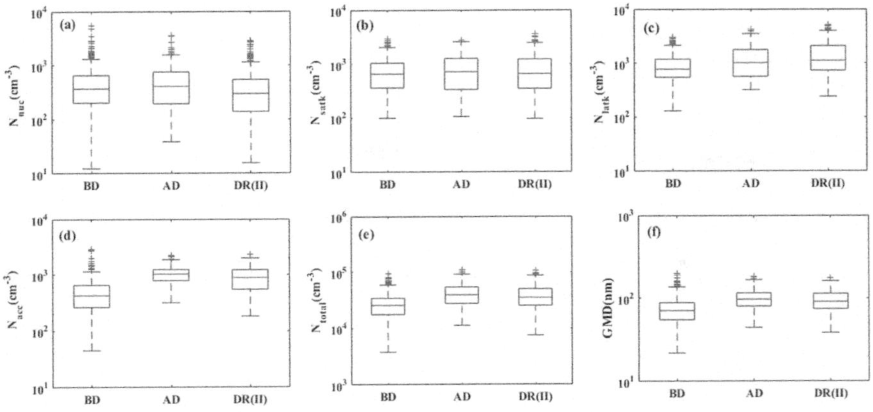

Fig. 3 Box whisker plot analysis for hourly concentration during period II

due to the complex emission of multiple sources. The Difference in GMD concentration during the different periods is due to changes in the emission sources. The emission measurement under different conditions proves that vehicular emission and density are the different diverse factors that help in estimating the concentration of particle number concentration of the nanoparticles and their behavior in the urban roadside microenvironment.

3.2 *Particle Number Concentration During Different Periods*

The time series measurement of different sized particles and their measured concentration are depicted in Figs. 4 and 5. The analysis showed that before restriction (BR), the concentration of the total PNC was found to be 3.2×10^4 cm^{-3} and reduced to around 2.4×10^4 cm^{-3}, 24% reduction from the previous phase due to the reduction of vehicular density (Fig. 4). The gap in the figure represents the data points missing. In Period II (Fig. 5), the higher particle numbers concentration received was after Diwali (AD) 4.2×10^4 cm^{-3}.

The concentration received during the AD phase was high compared to the other phases in the year 2021, a result of the Diwali emissions. This reduction was due to implementing phase-wise ban on firecrackers in the city. The concentration received during the AD phase was the highest mean concentration recorded throughout the study period, i.e., both Period I and II. During both Periods of monitoring, the study region experienced precipitation events, resulting in the wet removal of the particles which is not explored in the present study.

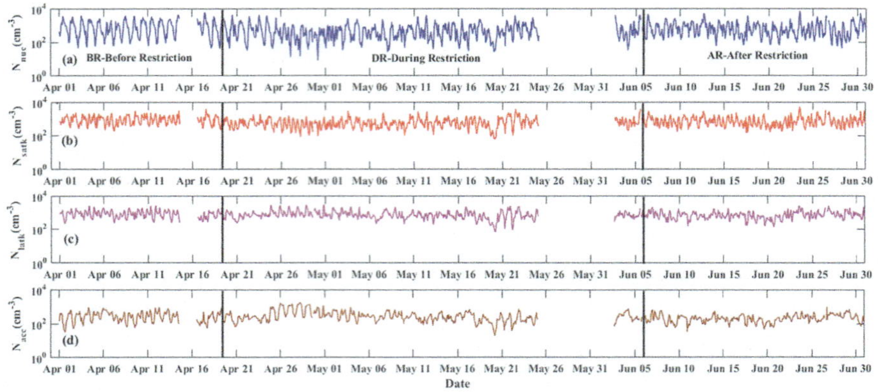

Fig. 4 Temporal distribution of different size range particles during Period I

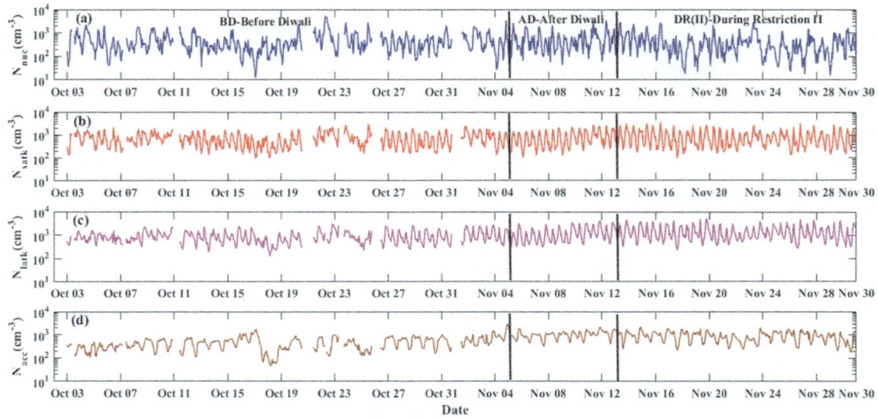

Fig. 5 Temporal distribution of different size range particles during Period II

4 Diurnal Behaviour

The diurnal behavior of the nanoparticles showed the evolution of the particle's concentration in a day with respect to time. The diurnal analysis in an urban region helps to identify the sources through their concentration, e.g., the vehicular sources show more concentration during the peak hours, and during the nonpeak hours, it will be less. In Period II (Fig. 7), the concentration of the nanoparticles was seen high during 9–12 hrs, especially the smaller sizes such as nucleation and Aitken mode particles, which was related to traffic emissions due to the revoke of restriction of vehicles during this Period. A similar trend was also observed during the evening peak hour. This pattern of increased concentration during the morning and evening peak hours is called the double hump model. When there was no vehicular emission, the concentration changes that occured during peak and nonpeak hours were not clearly visible. But, during Period I (Fig. 6), the increased concentration of the nanoparticles during the afternoon hours was due to the particles' UFP burst and photochemical formation under favorable temperature, solar radiation, and relative humidity.

During the restriction phase, nucleation contributed to around 31% of total PNC due to bursts of particles, and the accumulation mode contributed to around 28%. The Aitken mode contributed to around 42% of the total PNC. In general, nucleation contributed to around 30–40%, small Aitken 10–15% and large Aitken 25–30%, and accumulation 20–30% to the total PNC in both the period. During Period II, vehicle density was only 1% less compared to the previous year. During the AD phase accumulation mode, particles contributed around 46–50% to the total PNC of the particles. This was due to the contribution of Diwali emissions to the total PNC. The comparison of the concentration of the particles during the different emission scenarios showed that along with the PNC concentration, the different size fractions

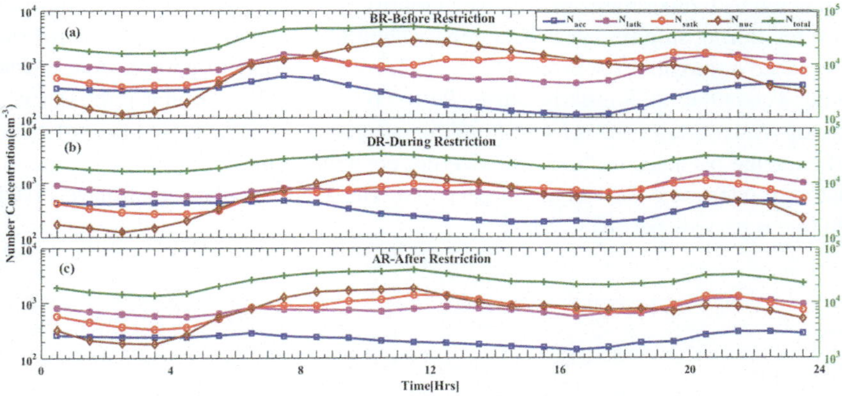

Fig. 6 Diurnal concentration of the nanoparticles during Period I

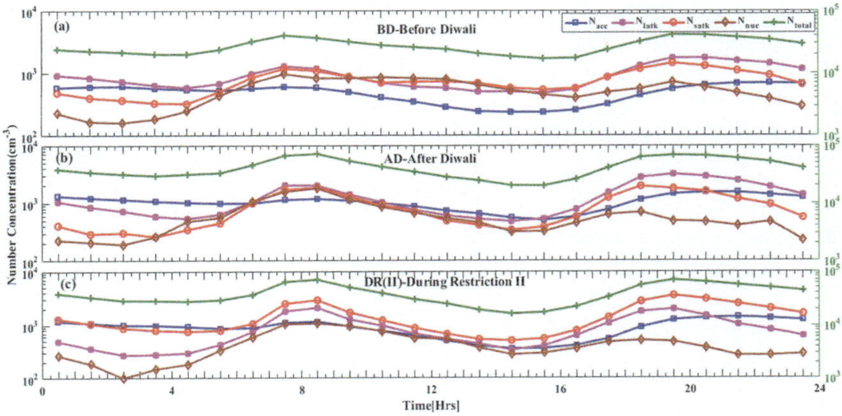

Fig. 7 Diurnal concentration of the nanoparticles during Period II

that contributed to the total PNC also varied based on the sources. When the sources are fewer, the contribution pattern will be different, and when the sources are more, the distribution will also vary based on their dominating size.

5 Impact of Meteorology on Particle Number Size Concentration

The particle number concentration of the locality is not only based on the source's intensity but also influenced by local meteorological factors such as wind speed, relative humidity, and solar radiation. The relative humidity helps particles to undergo coagulation, which results in increased particle size. The particles transform from nucleation to accumulation mode due to the particles' physical transformation, which was favored by temperature and relative humidity. Period I (Fig. 8) of the monitoring period fell during summer, where the relative humidity was lower, and the temperature reached a maximum of around 44.5 °C, favoring the gas-to-particle conversion instead of coagulation. Similarly, during Period II (Fig. 9), the season was the starting phase of winter where the relative humidity was high, which favored coagulation, and hence the particles in the lower size got coagulated into higher size i.e., accumulation mode was observed at a higher rate. The average solar radiation received during both periods was 468 Wm^{-2}.

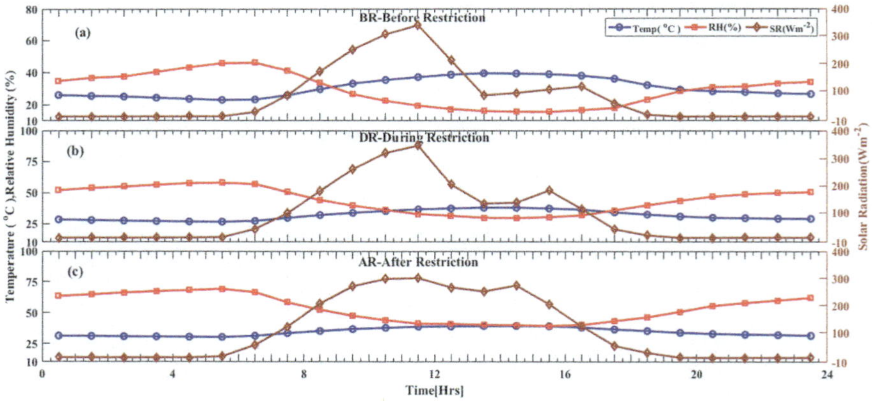

Fig. 8 Diurnal variation of meteorological parameters during Period I

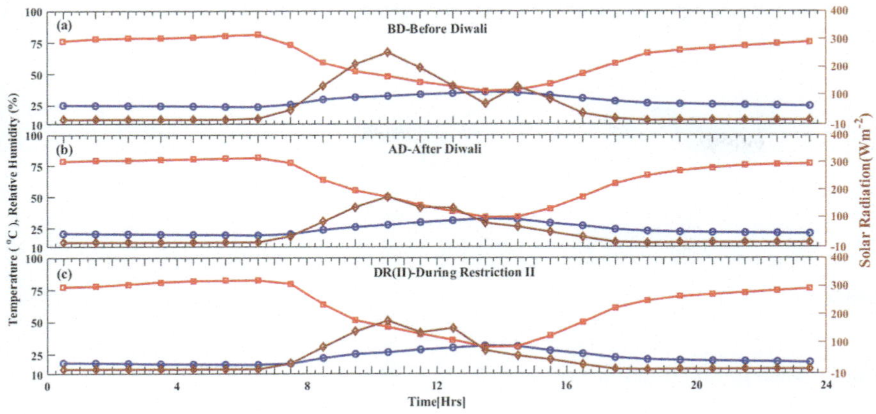

Fig. 9 Diurnal variation of meteorological parameters during Period II

6 Influence of Wind Speed on Measured Particle Number Concentration

The windspeed in the locality determines the dispersion of the particles. The dispersion of the particles transfers the roadside emission to the nearby regions. For a city like Delhi, where the residential areas are located adjacent to the road, where dispersion plays a major role in transferring the nanoparticles from the roadside to the residential areas.

Period I (Fig. 10) received a maximum wind speed of around 11 m/s; during Period II, it was 6.5 m/s (Fig. 11).

The bivariant polar plot showed the role of wind speed and direction on the dispersion of pollutants. In the study area, the dispersion of the pollutants was from the southwest towards the northeast direction, where the generation of the pollutants was more in the south direction of the study region. The south direction of the study

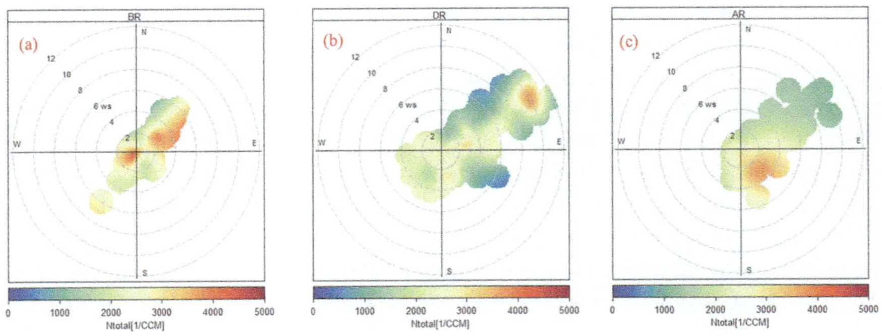

Fig. 10 Polar plots of total PNC during Period I

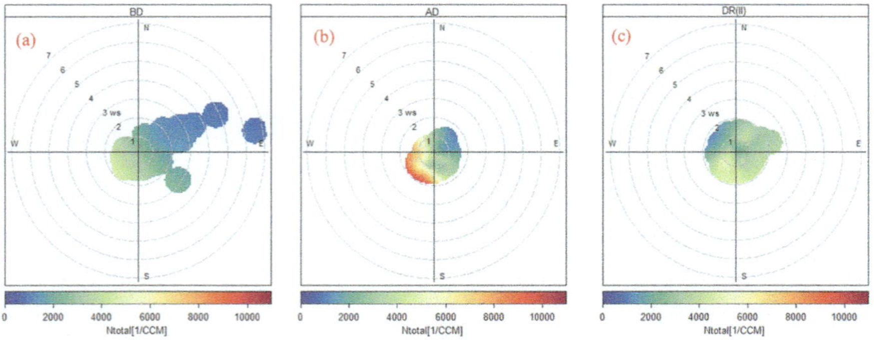

Fig. 11 Polar plots of total PNC during Period II

region was covered by an arterial road, which was the main source of nanoparticles in the study area. During Period I, the windspeed was high, causing more dispersion of the particles. Whereas during Period II, even though the concentrations were high, it was not dispersed to different regions and accumulated at the source region. The dispersion rate is based on the wind speed, such as high or low. The higher windspeed, i.e., > 6 or 7 m/s, favors more dispersion, and the lesser windspeed, < 2 m/s, causes accumulation of the particles in the same region.

7 Summary and conclusions

The study was conducted in 2021 under different emission conditions in Delhi to estimate the role of vehicular emissions in nanoparticle concentration in the study region. The study conducted in two different periods revealed that the vehicular sources play a crucial role in the nanoparticle concentration in the roadside environment. The reduction in nanoparticle emissions to 31% was achieved due to vehicular density reduction of 45%. When the restriction was imposed, the average total PNC concentration were found in the range of 10^3 to 10^5 cm^{-3}, whereas during multiple mixtures of emissions, the range was 10^4 to 10^5 cm^{-3}. During the complex emission period (Period II), the GMD of the particle also ranged from 20 to 200 nm. The concentration of the total PNC during Period I before the restriction was around 3.2×10^4 cm^{-3}, and it reduced to around 2.4×10^4 cm^{-3}, which is a reduction of around 24%. The highest concentration was observed around 4.2×10^4 cm^{-3} due to the emission from Diwali sources. The Diurnal behavior showed a double hump model concentration in which the concentration was higher during the peak hours of the vehicular flow. During the study period, the average contributions of the different-sized particles to the total PNC was 31% by nucleation, 42% by Aitken, and 28% by accumulation. Higher wind speed favored more dispersion, and lesser wind speed caused accumulation of the particles. The study findings clearly indicate that vehicular emission's role in urban regions is dominant for nanoparticle emission and signifies the importance of policy development and mitigation measures for reducing the impact of these nanoparticles on human health.

Acknowledgements The authors thank Advanced Air and Acoustics Research Lab, Environmental Engineering Department, Delhi Technological University for providing instrumentation facilities and support to conduct this study.

References

1. Apte JS, Kirchstetter TW, Reich AH, Deshpande SJ, Kaushik G, Chel A, Marshall JD, Nazaroff WW (2011) Concentrations of fine, ultrafine, and black carbon particles in auto-rickshaws in New Delhi, India. Atmos Environ 45(26):4470–4480. https://doi.org/10.1016/j.atmosenv.2011.05.028
2. Dwivedi D, Agarwal AK, Sharma M, Program EE (1994) Diesel exhaust particulates characterization for heavy metals. 1977:0–3
3. Kumar P, Morawska L, Birmili W, Paasonen P, Hu M, Kulmala M, Harrison RM, Norford L, Britter R (2014) Ultrafine particles in cities. Environ Int 66:1–10. https://doi.org/10.1016/j.envint.2014.01.013
4. Manojkumar N, Srimuruganandam B, Shiva Nagendra SM (2019) Application of multiple-path particle dosimetry model for quantifying age specified deposition of particulate matter in human airway. Ecotoxicol Environ Saf 168(July 2018):241–248. https://doi.org/10.1016/j.ecoenv.2018.10.091
5. Rajagopal K, Ramachandran S, Mishra RK (2024a) Traffic-induced nanoparticle emissions and associated respiratory risk analysis using measurements conducted in a roadside environment. Air Qual Atmos Health. https://doi.org/10.1007/s11869-024-01689-x
6. Rajagopal K, Ramachandran S, Mishra RK (2024b) Seasonal variation of particle number concentration in a busy urban street with exposure assessment and deposition in human respiratory tract. Chemosphere 366:143470. https://doi.org/10.1016/j.chemosphere.2024.143470
7. Rajagopal K, Mohan V, Mishra RK (2024c) Are Delhi residents exposed to lesser particle number concentration due to the firework ban in the city? Air Qual Atmos Health 17:1617–1627. https://doi.org/10.1007/s11869-024-01532-3
8. Ramachandran S, Rupakheti M (2022) Trends in the types and absorption characteristics of ambient aerosols over the Indo-Gangetic Plain and North China Plain in last two decades. Sci Total Environ 831:154867. https://doi.org/10.1016/j.scitotenv.2022.154867
9. Sebastian M, Kanawade VP, Pierce JR (2021) Observation of sub-3nm particles and new particle formation at an urban location in India. Atmos Environ 256. https://doi.org/10.1016/j.atmosenv.2021.118460
10. Sharma M, Khare M, Mishra RK (2023) Air quality changes in Delhi due to open waste burning: an accidental fire in Bhalswa landfill. Int J Environ Sci Tech 21:655–664. https://doi.org/10.1007/s13762-023-04921-w
11. Sharma K, Goyal M, Mishra RK, Vijayakumar T, Kumar P, Rajagopal K (2025) Air quality status and estimated health exposure in major metropolitan Indian cities. Aerosol Sci Eng. https://doi.org/10.1007/s41810-024-00278-w
12. Sharma M, Khare M, Mishra RK (2025) Indoor particle number concentration measurements across different income group households in urban region. Urban Clim 102329. https://doi.org/10.1016/j.uclim.2025.102329
13. Yadav SK, Kompalli SK, Gurjar BR, Mishra RK (2021) Aerosol number concentrations and new particle formation events over a polluted megacity during the COVID-19 lockdown. Atmos Environ 259(March):118526. https://doi.org/10.1016/j.atmosenv.2021.118526

Impact of Solid Waste Dump Yards Fire on Air Quality of National Capital Delhi

Monika Sharma, Mukesh Khare, and Rajeev Kumar Mishra

Abstract Waste burning is one of the major issues in urban regions, which deteriorates the air quality of the area along with other complex sources of air pollution in the region. The study was conducted in 2022 when northern Delhi's solid waste dump yard underwent an accidental fire during the summer. The study region has three major landfills, and Bhalswa is one among them. Due to the burning of solid waste, the concentration of the particulate matter increased around 40 to 50% in the study region, and the concentration reached a maximum of about 338 μg/m^3 for PM_{10}. The burning emits particulates and other gaseous pollutants, which showed a moderate correlation ($R^2 = 0.5$ to 0.6) with the particulate matter. The generated fumes were dispersed to the surrounding regions during higher windspeed (>3 m/s), and less dispersion was seen during low and calm wind speed conditions (<3 m/s). Most Indian cities experience the problem of waste dumping, and it is necessary to process the solid waste from households to extract the energy byproducts to overcome it. The dumpyards generate aesthetic issues, leading to poor air quality in the surrounding regions. The study's outcomes help to understand the air quality changes due to open fires in municipal solid waste dump yards.

Keywords Open fires · Landfills · Particulate matter · Urban air pollution

M. Sharma (✉) · R. K. Mishra
Department of Environmental Engineering, Delhi Technological University, New Delhi, India

R. K. Mishra
e-mail: rajeevkumarmishra@dtu.ac.in

M. Khare
Department of Civil Engineering, Indian Institute of Technology, Delhi, New Delhi, India

S. M. Shiva Nagendra et al. (eds.), *Select Proceedings of the 8th Indian International Conference on Air Quality Management (IICAQM 2023)*, Lecture Notes in Civil Engineering 582, https://doi.org/10.1007/978-981-96-2359-4_9

1 Introduction

Delhi, the national capital of India, is a highly urbanized city that experiences continuous various development activities every day. The city has a population of about 26 million and the population rate is growing yearly [1–3]. The city's air quality remains poor most of the time in a year and its sources are complex. The air quality is influenced by various sources such as transportation, construction activities, industrial, household, road dust, diesel generators and open waste burning [4, 7, 16, 19]. Air quality maintenance is one of the key factors for sustainability, human health, and quality of life [10]. The concern for air quality is increasing daily, and the government is taking many initiatives to improve the air quality of Delhi [9]. The air quality is associated with health and causes various respiratory diseases and premature deaths [5, 6, 18]. The study region is one of the major cities in India, and globally, it is estimated that solid waste generation is around 2.01 billion tonnes per day, and it is estimated that by 2050, the value will reach 3.40 billion tonnes [8]. Delhi alone contributes to around 1 lakh tonnes of solid waste per day through regularised sources. Handling this solid waste is one of the tedious processes leading to dumping waste in dump yards without recycling or composting. The waste generated is not separated, and authorities find it difficult to process and segregate the huge quantity of waste. The waste generated under non-channeled sources accumulates in the corners of the street and is left unattended [11–13]. Solid waste collection efficiency ranges from 70 to 90% in urban regions, and in small towns, it will be around 70% maximum efficiency. Burning is considered one of the easiest and cheapest ways to handle solid waste from many unorganized sources, but it leads to serious environmental issues such as air pollution [5, 14]. Open burning releases a large quantity of fumes, which consist of particulate matter, gaseous pollutants, Dioxin, and furans. The contribution of open waste burning for poor air quality is 8% in India, and in China, it was around 22%. Studies show that waste burning can suddenly increase 40 to 80% aerosols. Waste burning is banned in India, but accidental fires such as landfills and forest fires are still uncontrollable.

The study analyses the air quality changes that occurred due to an accidental fire in the municipal solid waste landfill in the northern part of Delhi during the year 2022, and it lasted for around 12 days, which deteriorated the air quality in the nearby regions.

The study compared and analyzed the air quality changes before, during and after the fire which shows the impact of the solid waste burning on air quality. Along with particulate matter (PM_{10} and $PM_{2.5}$), gaseous parameter (SO_2) meteorological conditions such as wind speed and wind direction also studied which show the influence of meteorology conditions in the dispersion of pollutants which is not covered exclusively in the previous studies. The study's outcomes help to estimate the air quality changes in the surrounding regions, which will help to initiate necessary measures during such accidents through mitigative measures and policy formulations.

2 Description of the Study Area

The municipal landfill is located in northern Delhi (28.7395° N, 77.1573° E), commonly called the Bhalswa landfill. The city consists of around 1483 km^2 with a density of around 11 thousand persons per km^2. The landfill is surrounded by residential areas such as Smaypur, Bhalswa Village, and Jahangirpuri. The region experiences extreme summer and winter conditions, ranging from 5 ºC to 45 ºC throughout the year. The monitoring locations are considered around the landfill with an aerial distance of < 5 km. The monitoring locations considered for the study are Ashok Vihar, Jahangirpuri, Wazirpur, Sonai Vihar, Burari Crossing, and Rohini. The study regions come under the highly developed urban regions.

3 Materials and Methods

The air quality data was accessed from the ambient air quality monitoring stations in the study area, which are operational under different agencies such as CPCB, DPCC, and IMD. The data were collected for 19 days with three phases: Before Fire, During Fire, and After Fire (Fig. 1). The parameters considered for the study include particulate matter (PM_{10} and $PM_{2.5}$) and gaseous pollutant (SO_2) along with the local meteorological conditions such as wind speed and wind direction. The collected data was analyzed using Rstudio, and the statistical analysis was performed using SPSS software.

4 Results and Discussion

4.1 Box Whisker Plot Analysis

The Box Whisker plot analysed the particulate matter emitted during the fire. The analysis showed an average increase of around 60 to 70% of $PM_{2.5}$ concentration in the Ashok Vihar and Sonia Vihar during the fire phase. The high windspeed that occurred during that period caused high dispersion of the particles in the regions far away from the landfill. In regions like Rohini and Jahangirpuri, which received an increased concentration of around 50 to 60%, the lowest impact was observed at the Burari crossing (28%). During the after-fire phase, the concentration of $PM_{2.5}$ observed a reduction of around 40 to 50%. The before-fire and after-fire periods comparison showed an increased concentration of pollutants, nearly 20%, compared to the previous normal conditions. In terms of PM_{10}, the concentration during the fire events observed a maximum increase of 20% at almost all the locations except the Burari crossing and Jahangirpuri locations, and after fire, the reduction was found to be 25 to 45% of PM_{10} at the six different locations. The difference in

Fig. 1 Map displaying the monitoring stations

$PM_{2.5}$ and PM_{10} sizes reduction and increase was due to the smaller size variation's contribution to the larger particulate matter. The fire emits large amounts of gaseous pollutant (SO_2), which showed increased concentration around 50 to 100% at almost all the study locations. This showed that the open burning of solid waste releases particulates and gaseous pollutants in large quantities (Figs. 2 and 3).

The $PM_{2.5}$ concentration changed from 81.2 $\mu g/m^3$ to 137.3 $\mu g/m^3$ during the fire event, and after the fire event, it reached back to 83.4 $\mu g/m^3$ in the Ashok Vihar location, similary at the Burari crossing, before fire, $PM_{2.5}$ concentration was found to be around 105.7 $\mu g/m^3$ which reached to around 135.3 $\mu g/m^3$ during the fire and reached back to 72.1 $\mu g/m^3$. Similarly it was observed at almost all the locations. The difference in concentration of the pollutants was observed before, during, and after the fire events signifies the role of open waste burning in air quality. Waste-burning activities globally lead to around 24 to 29% of total emissions, and this emits harmful gaseous as well due to the burning of plastic and E-waste mixed with the municipal solid waste dump yards. The emitted pollutants are not only deteriorate air quality and health but also affect the surrounding region's visibility (Fig. 3).

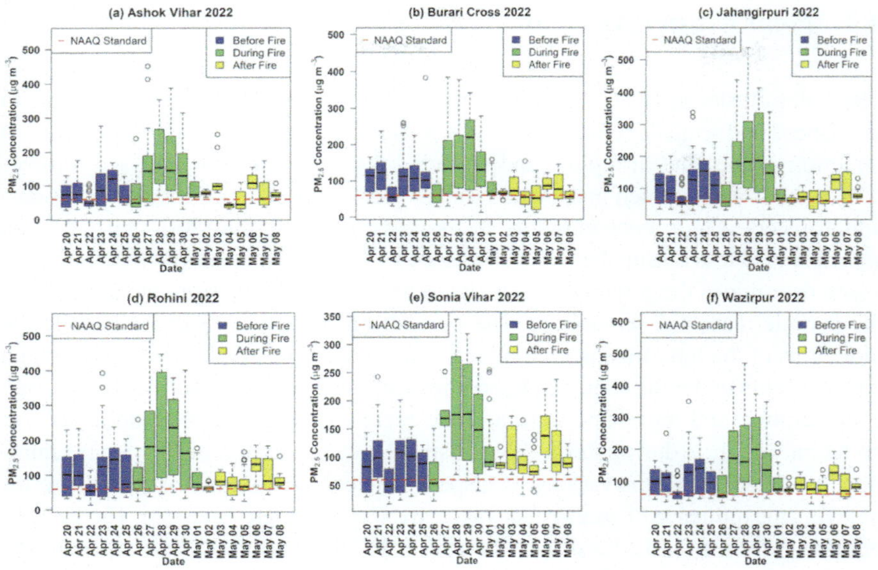

Fig. 2 $PM_{2.5}$ concentration in different locations during the study period

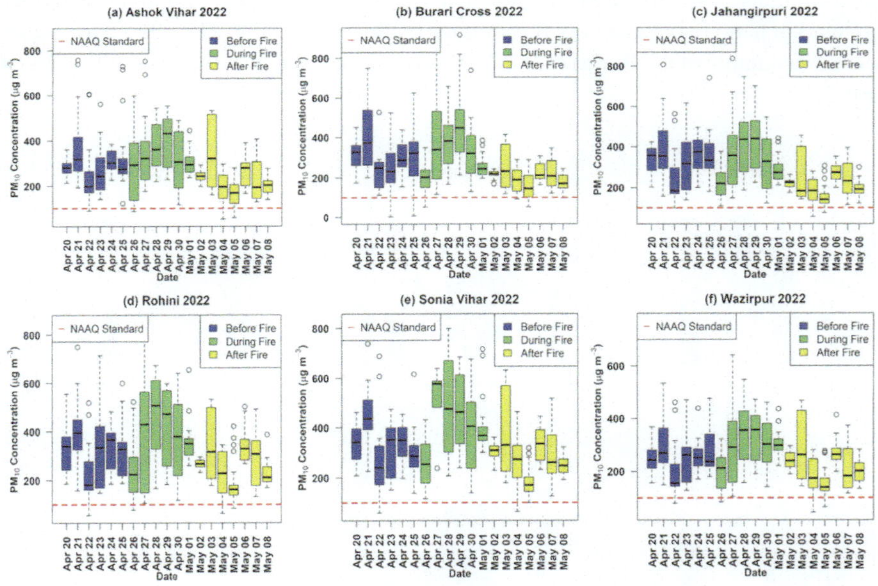

Fig. 3 PM_{10} concentration in different locations during the study period

4.2 Polar Cluster Analysis

The polar cluster analysis determines the dispersion of pollutants with respect to windspeed and wind direction. The Ashok Vihar location is situated in the southwest direction of the landfill region, and the wind direction was from the south direction of the landfill, which makes the Ashok Vihar location in downwind conditions, which received high concentration. In the same location, the PM_{10} concentration was observed less during the lower windspeed < 3 m/s, and it showed that the windspeed determines the dispersion of the particles, and the dispersion rate depends on the particle mass and windspeed. The monitoring stations are located in different directions of the landfill, and the regions in the downwind conditions received high concentration under the calm wind conditions, regions near the landfill received high concentration and less dispersion, and the regions away from the landfill receive high concentration during the high wind speed. The dispersion of $PM_{2.5}$ concentration fluctuated during the varying windspeed where the mass was lower, but in terms of PM_{10}, the influence of windspeed was high during the high windpeed. The gaseous pollutant was also influenced by wind speed. In 2016, the study region received a very high concentration of pollutants due to the stubble burning that occurred in the neighboring states during the event the meteorology played a major role in dispersing the pollutants from source to receptor. This showed that air quality and meteorology go hand in hand in determining the air quality of a region [17]. The wind parameters were unavailable in the Burari crossing area where the analysis is unavailable (Figs. 4 and 5).

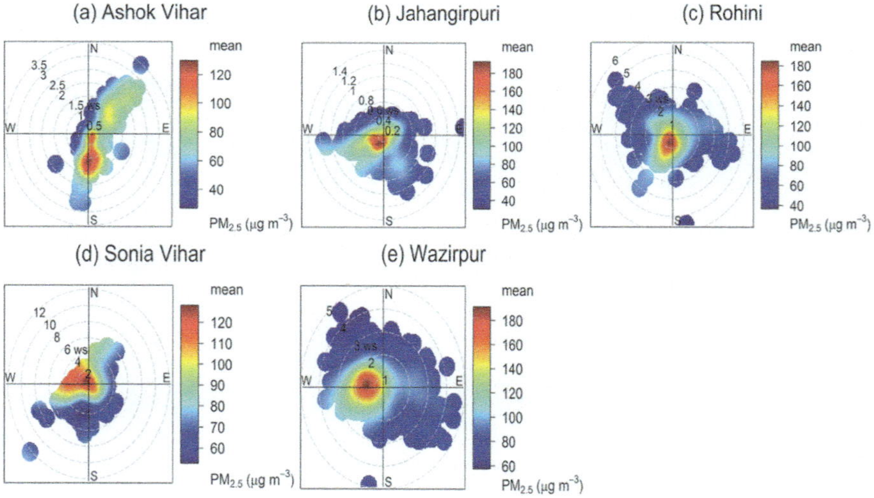

Fig. 4 Polar cluster analysis of $PM_{2.5}$ and windspeed, direction in different location

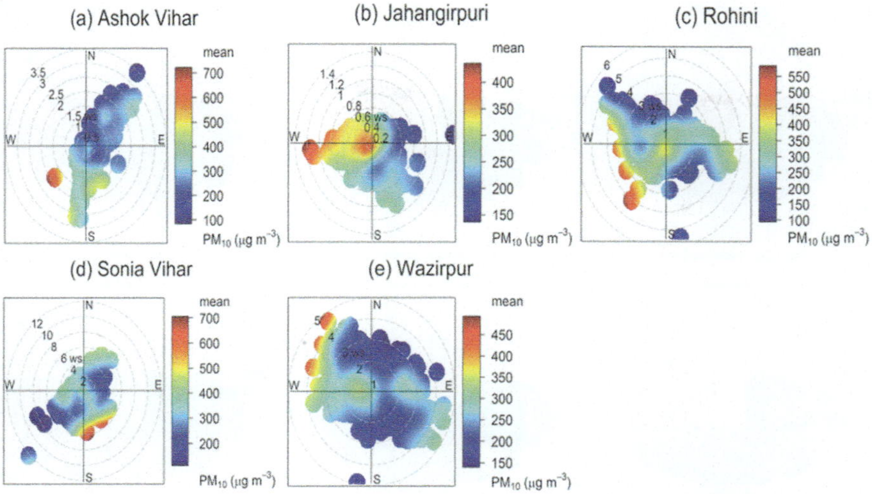

Fig. 5 Polar cluster analysis of PM_{10} and windspeed direction in different locations

4.3 *Percentage Contribution of $PM_{2.5}$ in PM_{10}*

The percentage contribution was analyzed before and after the fire event to know the contribution. The analysis showed that the nearby region, i.e., Janahgirpuri, showed a $PM_{2.5}$ contribution of around 54% in the PM_{10}, showing that most of the sources were from open fires in the region. Similarly, in almost all the locations during the fireworks event, the contribution of $PM_{2.5}$ concentration in PM_{10} ranges from 43 to 46%. The percentage contribution of open waste burning showed around half of the total concentration, showing that the air quality in urban regions is more sensitive to the intensity of sources, whether vehicular or waste burning or other sources. Various studies showed increased or decreased concentrations of air pollutants due to sudden stops or increases in anthropogenic activities such as lockdowns, fire events, dust storms, and odd–even vehicle policies [15]. The source reduction can contribute to bringing any region's air quality within the permissible range, but it requires participation from industries, policymakers, and the common public (Figs. 6 and 7).

4.4 *Correlation Analysis*

The waste burning emits a large number of SO_2 components due to the various mixtures of waste. Compared to PM_{10}, SO_2 showed more correlation with $PM_{2.5}$. The correlation values range from $R^2 = 0.4$ to 0.6, and the highest concentration was observed in Ashok Vihar, Jahangirpuri, and Rohini locations. The correlation of gaseous pollutants with particulate matter changes based on the composition of the waste. The incomplete combustion leads to carbonaceous emission. The SO_2 emitted in the atmosphere combines with water vapor in the atmosphere and converts into

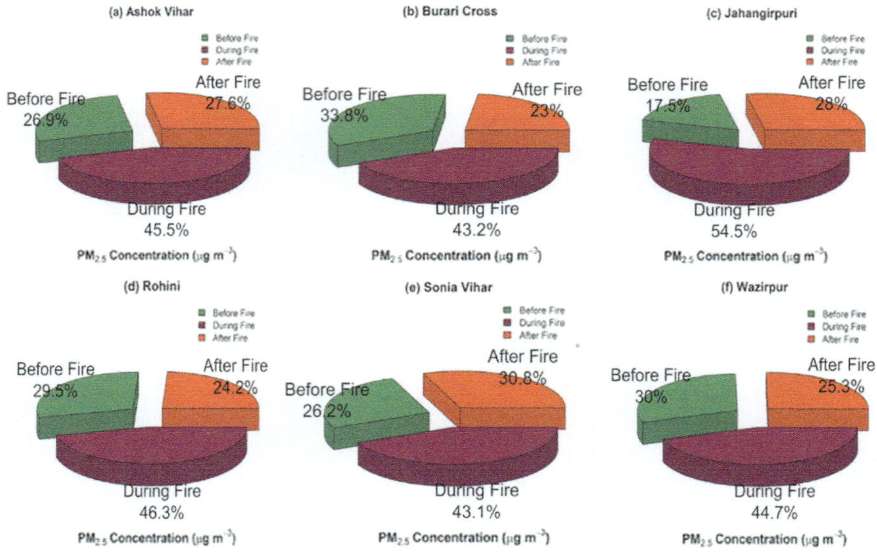

Fig. 6 Contribution of $PM_{2.5}$ concentration in study areas

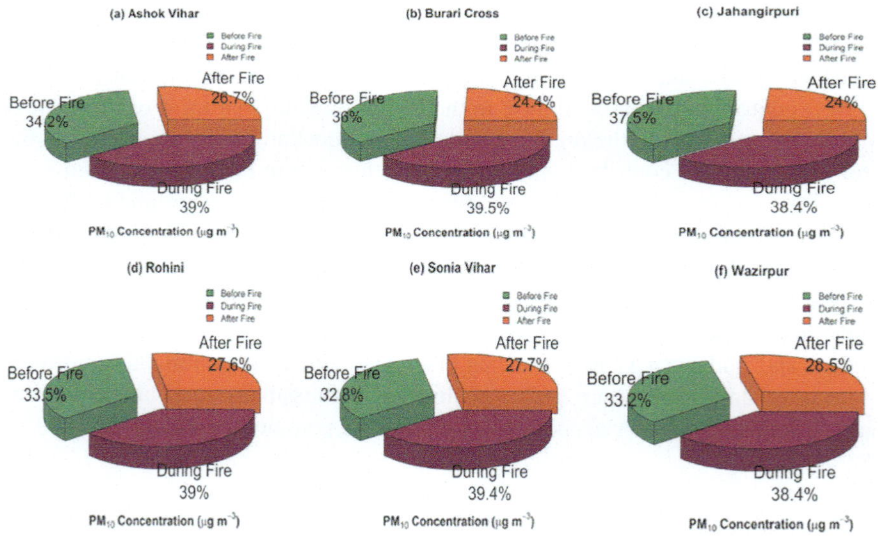

Fig. 7 Contribution of PM_{10} concentration in study areas

acid rain, affecting monuments, plants, animals, and human beings and causing skin damage. The studies revealed that the SO_2 emission caused corrosion of the famous Tajmahal. The acid rain causes corrosion and leads to skin irritation, skin cancer, and aesthetics problems [2]. The correlation of SO_2 with PM is insignificant except in the Jahangirpuri and Rohini locations, where the correlation value was $R^2 = 0.4$, and in other regions, the correlation was $R^2 = 0.2$ (Figs. 8 and 9).

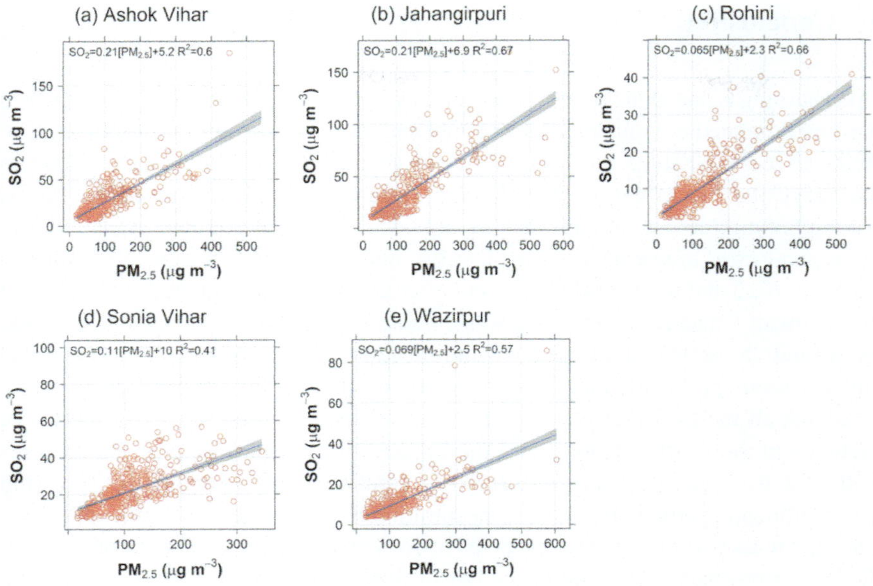

Fig. 8 Correlation of SO_2 and $PM_{2.5}$ in the study region

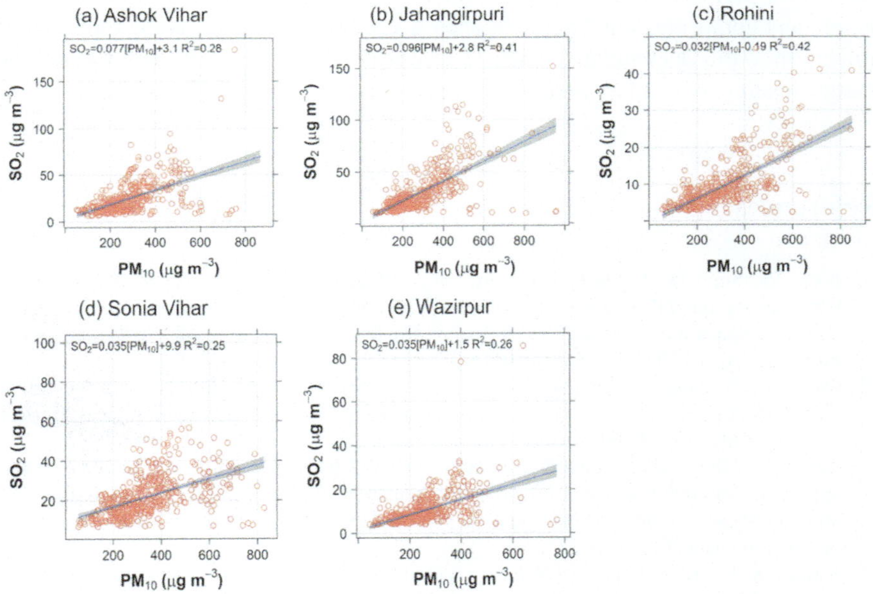

Fig. 9 Correlation of SO_2 and PM_{10} in the study region

5 Conclusion

Open burning is one of the major issues in many Indian cities. Even though open waste burning is banned in India, many accidental and illegal waste burning continues. The study aimed to analyze the air quality impact in the urban city of Delhi due to the accidental fire in the municipal solid waste landfill. Municipal solid waste is a complex mixture of sources such as organic, E-waste, hazardous waste, plastic waste, and even biomedical waste. The fire outbreak occurred from 26th April to 8th May, 2022 and the period before and after the fire was considered for analyzing the air quality changes due to open waste burning. The PM concentration increased by around 28 to 40% at all selected study locations. The open waste burning also emits gaseous pollutants and SO_2 concentration increased by around 50 to 100% in almost all the study regions. The lower wind speed caused the accumulation of particles in the nearby regions, and the higher wind speed dispersed the particles to the faraway places, observed during different windspeed conditions. The wind speed influence particulate matter concentration based on the mass of the particles. The high mass particles are not dispersed by lower wind speed, and on the other hand, the lower mass particles are influenced by the lower wind speed. The locations in the downwind conditions receive the maximum concentration of the pollutants. The SO_2 concentration correlated moderately with $PM_{2.5}$ concentration. Open waste burning for a short period can deteriorate air quality to a larger extent, and the common people and authorities must regulate open waste burning. Also, on the other hand, utilization of this accumulated waste is necessary to convert the waste to wealth or energy, leading to avoiding aesthetics issues as well.

References

1. Nagar PK, Singh D, Sharma M, Kumar A, Aneja VP, George MP, Agarwal N, Shukla SP (2017) Characterization of PM2.5 in Delhi: role and impact of secondary aerosol, burning of biomass, and municipal solid waste and crustal matter. Environ Sci Pollut Res 24(32), 25179–25189. https://doi.org/10.1007/s11356-017-0171-3
2. Okedere OB, Olalekan AP, Fakinle BS, Elehinafe FB, Odunlami OA, Sonibare JA (2019) Urban air pollution from the open burning of municipal solid waste. Environ Qual Manage 1–8. https://doi.org/10.1002/tqem.21633
3. Yadav A, Behera SN, Nagar PK, Sharma M (2020) Spatio-seasonal concentrations, source apportionment and assessment of associated human health risks of $PM_{2.5}$-bound polycyclic aromatic hydrocarbons in Delhi, India. Aerosol Air Qual Res 20(12):2805–2825. https://doi.org/10.4209/aaqr.2020.04.0182
4. Kumari K, Kumar S, Rajagopal V, Khare A, Kumar R (2019) Emission from open burning of municipal solid waste in India. Environ Technol (United Kingdom) 40(17):2201–2214. https://doi.org/10.1080/09593330.2017.1351489
5. Agarwal R, Shukla K, Kumar S, Aggarwal SG, Kawamura K (2020) Chemical composition of waste burning organic aerosols at landfill and urban sites in Delhi. Atmos Pollut Res 11(3):554–565. https://doi.org/10.1016/j.apr.2019.12.004
6. Cash JM, Langford B, Di Marco C, Mullinger NJ, Allan J, Reyes-Villegas E, Joshi R, Heal MR, Acton WJF, Hewitt CN, Misztal PK, Drysdale W, Mandal TK, Shivani, Gadi R, Gurjar BR,

Nemitz E (2021) Seasonal analysis of submicron aerosol in Old Delhi using high-resolution aerosol mass spectrometry: chemical characterisation, source apportionment and new marker identification. Atmos Chem Phys 21(13):10133–10158. https://doi.org/10.5194/acp-21-10133-2021

7. Goyal M, Sharma K, Mishra RK, Kumar TV, Rajagopal K (2025) Fuel usage policy implementation toward sustainability and household air pollution reduction in Indian subcontinent: prediction and analysis. In: Gautam S, Joshi SK, Ambade B (eds) Blue sky, blue water. Springer, Cham. https://doi.org/10.1007/978-3-031-82559-0_7
8. Krecl P, de Lima CH, Dal Bosco TC, Targino AC, Hashimoto EM, Oukawa GY (2021) Open waste burning causes fast and sharp changes in particulate concentrations in peripheral neighborhoods. Sci Total Environ 765:142736. https://doi.org/10.1016/j.scitotenv.2020.142736
9. Mishra RK, Rana R, Tomar S, Sidhant, Sharma M (2025) Air quality prediction using machine learning techniques. In: Gautam S, Joshi SK, Ambade B (eds) Blue sky, blue water. Springer, Cham. https://doi.org/10.1007/978-3-031-82559-0_15
10. Mohan V, Soni VK, Mishra RK (2024) Geographical variability of ultrafine particle concentrations in urban and background regions in India. Urban Clim 102066. https://doi.org/10.1016/j.uclim.2024.102066
11. Nagpure AS (2019) Assessment of quantity and composition of illegal dumped municipal solid waste (MSW) in Delhi. Resour, Conserv Recycl 141(July 2018):54–60. https://doi.org/10.1016/j.resconrec.2018.10.012
12. Peter AE, Nagendra SMS (2021) Dynamics of PM2.5 pollution in the vicinity of the old municipal solid waste dumpsite. Environ Monit Assess 193(5):1–16. https://doi.org/10.1007/s10661-021-09052-8
13. Ray MR, Roychoudhury S, Mukherjee G, Roy S, Lahiri T (2005) Respiratory and general health impairments of workers employed in a municipal solid waste disposal at an open landfill site in Delhi. Int J Hyg Environ Health 208(4):255–262. https://doi.org/10.1016/j.ijheh.2005.02.001
14. Ramaswami A, Baidwan NK, Nagpure AS (2016) Exploring social and infrastructural factors affecting open burning of municipal solid waste (MSW) in Indian cities: a comparative case study of three neighborhoods of Delhi. Waste Manage Res 34(11):1164–1172. https://doi.org/10.1177/0734242X16659924
15. Rajagopal K, Mohan V, Mishra RK (2024) Are Delhi residents exposed to lesser particle number concentration due to the firework ban in the city? Air Qual Atmos Health. https://doi.org/10.1007/s11869-024-01532-3
16. Rajagopal K, Ramachandran S, Mishra RK (2025) Traffc-induced nanoparticle emissions and associated respiratory risk analysis using measurements conducted in a roadside environment. Air Qual Atmos Health. https://doi.org/10.1007/s11869-024-01689-x
17. Rajagopal K, Ramachandran S, Mishra RK (2025) Influence of local meteorology and gaseous pollutant emissions on atmospheric nanoparticle concentrations in a pedestrian way in urban region. Atmos Pollut Res 102358. https://doi.org/10.1016/j.apr.2024.102358
18. Sharma K, Goyal M, Mishra RK, Vijayakumar T, Kumar P, Rajagopal K (2025) Air quality status and estimated health exposure in major metropolitan Indian cities. Aerosol Sci Eng. https://doi.org/10.1007/s41810-024-00278-w
19. Sharma M, Khare M, Mishra RK (2025) Indoor particle number concentration measurements across different income group households in urban region. Urban Clim 102329. https://doi.org/10.1016/j.uclim.2025.102329

Effect of Sulfur Content on Filterable and Condensable $PM_{2.5}$ Emissions from Diesel Generator

Sunil Kumar Gupta and Hsi-Hsien Yang

Abstract Particulate matter ($PM_{2.5}$) emission from diesel generators contains a filterable ($FPM_{2.5}$) and condensable (CPM) portion. In this study, five diesel fuel containing 10 ppm (GD10), 250 ppm (250 GD), 500 ppm (500 GD), 1000 ppm (1000 GD) and 2000 ppm (2000 GD) sulfur by weight were selected as generator fuel. U.S. EPA Method 201A/202 was followed to measure FPM and CPM emissions from the diesel generator and study the effect of sulfur content on $FPM_{2.5}$ and CPM emissions. $FPM_{2.5}$ concentrations for different sulfur-content diesel like GD10, GD250, GD500, GD1000, and GD2000 were 10.6, 12.8, 15.0, 15.1, and 17.4 mg/Nm^3, respectively. CPM concentrations were obtained to be 118.2, 145.0, 187.3, 281.4, and 334.6 mg/ Nm^3 for GD10, GD250, GD500, GD1000, and GD2000, respectively. CPM concentrations were also increased with the increase in sulfur content. The distribution of organic fraction in CPM accounted for 75.1%, 69.3%, 56.3%, 51.6%, and 51.4% for respective increasing sulfur concentrations.

Keywords Diesel generator · $FPM_{2.5}$ and CPM · Sulfur content · Emission

1 Introduction

Sulfur is found naturally in crude oil. Sulfur content of crude oil ranges from < 0.05 to > 10 wt% but typically falls in the range 1–4 wt%. As a result, when crude oil is processed in a refinery it turns into refined petroleum products like diesel, gasoline, etc. Diesel contains sulfur which is derived from the original crude oil source and can still be present after refining. After combustion in the engine, sulfur in the fuel

S. K. Gupta (✉) · H.-H. Yang
Department of Environmental Engineering and Management, Chaoyang University of Technology, Taichung, Taiwan
e-mail: snlenv@gmail.com

S. K. Gupta
Centre for Studies on Environment and Climate, Asian Development Research Institute (ADRI), Patna, Bihar, India

S. M. Shiva Nagendra et al. (eds.), *Select Proceedings of the 8th Indian International Conference on Air Quality Management (IICAQM 2023)*, Lecture Notes in Civil Engineering 582, https://doi.org/10.1007/978-981-96-2359-4_10

forms particulates which are a primary contributor to air pollutants. Sulfur content is responsible for increasing the particulate mass by oxidizing and then attaching to diesel particles resulting in increased particle size and mass [1]. Pollutants emitted from diesel engines mainly include carbon dioxide (CO_2), hydrocarbons (HC), oxide of nitrogen (NO_x), carbon monoxide (CO), particulate matter (PM), and others. Diesel engine exhaust PM, especially fine particles ($PM_{2.5}$) is getting more attention because of causing long-term damage to human health and the environment by regulatory and research agencies, like U.S. Environmental Protection Agency (EPA) and National Institute for Occupational Safety and Health [2]. The International Agency for Research on Cancer (IARC) has classified diesel exhaust emissions as probable carcinogens to humans (Group 1) based on sufficient positive evidence for lung cancer [3]. $PM_{2.5}$ (aerodynamic diameter $\leq 2.5\ \mu m$), typically contains carbonaceous materials (soot) and acid condensates [4, 5]. $PM_{2.5}$ emission from diesel engines consists of filterable particles ($FPM_{2.5}$), which are directly emitted in the solid/liquid phase and condensable particles (CPM) refer to materials in the vapor phase and condensed into solid/liquid phase during the dilution and cooling of flue gas in air [6, 4]. The physical and chemical properties of fuel have an important influence on exhaust emissions from diesel engines. Sulfur content is one important parameter among diesel fuel properties and is one key controlled factor of fuel standards. The knowledge behind this study comes from the use of high-sulfur fuel (2000–5000 ppm), which is used extensively in many developing countries such as Mongolia, Iran, Botswana, Ghana, Nigeria, Uganda, Rawanda, Iraq, Pakistan, and Afghanistan. Sulfur content limit is different because of different local diesel fuel standards. Therefore, the process of purifying diesel from sulfur requires new refineries and extensive possibilities for disposal of the extracted sulfur. In this study, five different sulfur-based diesel fuels were used to identify the effect of sulfur on filterable and condensable particulate matter ($FPM_{2.5}$ and CPM) emission from diesel generators.

2 Materials and Methods

2.1 Diesel Generator Specification

A diesel four-stroke engine (power generator) was used during this investigation to characterize the $FPM_{2.5}$ and CPM emission of the different sulfur content in diesel fuels. Table 1 shows the specifications of the diesel generator used in this study. The test engines included a Yanmar YDG-5500 generator which utilizes an L-100 air-cooled single-cylinder engine. The commercially available diesel (<10 ppm sulfur content) in Taiwan was used to mixed sulfur content in this study. The different sulfur content diesel fuel tests were performed at ambient temperature. Commercially available diesel fuel was tested as a baseline fuel.

Table 1 Specification of the diesel power generator used to generate condensable $PM_{2.5}$ samples

Power generator	Yanmar
Type	YDG, 5500
Manufacturing year	2001
Maximum output (kW)	5.5
Related output (kW)	5
Engine cooling	Vertical air-cooled
No. of cycle	4-cycle
Rated output (kW/rpm)	6.7/3600
Displacement (cm^3)	406
No. of cylinders	1
Combustion system	Direct injection
Compression ratio	20:01

2.2 Sulfur Content Diesel Preparation

Five different types of sulfur concentrations were used in the diesel fuel to evaluate the effect of sulfur content on the $FPM_{2.5}$ and CPM emission characteristics in this study. The diesel tested in this work was supplied by CPC Corporation, Taiwan (Taiwan Chinese Petroleum). The sulfur content in diesel fuel is not similar to the worldwide. The sulfur content limit of diesel in Taiwan is < 10 ppm. Inorganic sulfur (S_2) was purchased from Sigma Aldrich. After adding the inorganic sulfur to diesel samples, these were transferred to an ultrasonic bath (Branson Model 2510, Mexico) for a period of 15 to 30 min [7] and prepared the different fuel types. Five types of diesel fuels with different concentrations of sulfur by weight such as < 10 ppm, 250 ppm, 500 ppm, 1000 ppm, and 2000 ppm were used and designated as GD10, GD250, GD500, GD1000, and GD2000, respectively. These different types of sulfur-based diesel fuels can affect the emissions of pollutant compounds. The selected diesel was analyzed by ASTM method [2] and the characteristics are presented in Table 2.

2.3 Sampling Equipment and Methods

$FPM_{2.5}$ and CPM sampling methods were described more fully by Cano et al. [6] and Yang et al. [8] and are briefly summarized here. Most determinations of $FPM_{2.5}$ and CPM emissions have been carried out following the U.S. EPA Method 201A and Method 202, respectively. The APEX XC-5000 Automated Isokinetic Sampling Console system was used in this study. The sampling system included a nozzle, $PM_{2.5}$ cyclone, filter holder, pitot tube, sampling probe, spiral tube condenser, dry impingers, airtight vacuum pump, and computer control console (Figs. 1 and 2). During each sampling, the diesel generator was started and sampled after achieving

Table 2 Selected diesel characteristics used in diesel power generator

Item	Specification
Gross heating value (MJ/kg)	45.4
Net heating value (MJ/kg)	42.9
Specific gravity (at 20 °C)	0.87
Water and sediments (%)	Trace
Cetane N^0 (calculated) (Cst)	45–50 min
Viscosity (20 °C) (Cst)	4.1
Stoichiometric (fuel/air) ratio	0.077
Lubricity (mm)	332
Flashpoint (°C)	60
Pour point (°C)	−3
Cloud point (°C)	45
Aromatic (vol %)	22.7
Carbon residue (wt %)	0.09
Ash (%)	0.001
Carbon (wt %)	85.6
Hydrogen (wt %)	13.2
Sulfur (ppm)	10

a steady state. The leak was examined before each sampling. A total of six $FPM_{2.5}$ and CPM emission samples were collected simultaneously.

2.4 $FPM_{2.5}$ and CPM Analysis

The sampling cyclone was inserted into the exhaust pipe of diesel generator and preheated before sampling. The sampling rate was controlled to within ± 20% isokinetically. The cyclone head was heated to 10 °C above the temperature of the flue gas to prevent the condensation of vapors. Particulates of less than 2.5 μm diameter were sucked through the cyclone and $FPM_{2.5}$ samples were collected on a 47 mm quartz filter. Then, the sampled flue gas entered the CPM sampling trains. The CPM was collected in the water dropout impingers, connecting glassware, modified Greenburg Smith impinge, and CPM filter. To improve the collection efficiency of CPM, the CPM filter was placed between the second and third impingers. To reduce positive artefacts, all impinger contents were purged with nitrogen immediately after sampling to remove dissolved sulfur dioxide (SO_2) gases (flow rate at 14 L/min for 1 h). All sampling train components including the connecting glassware, condenser, impingers, and the front half of the CPM filter housing were rinsed twice each with water, acetone, and hexane. The CPM sample was extracted with water and hexane and named inorganic and organic fractions, respectively (U.S. EPA, Method 202).

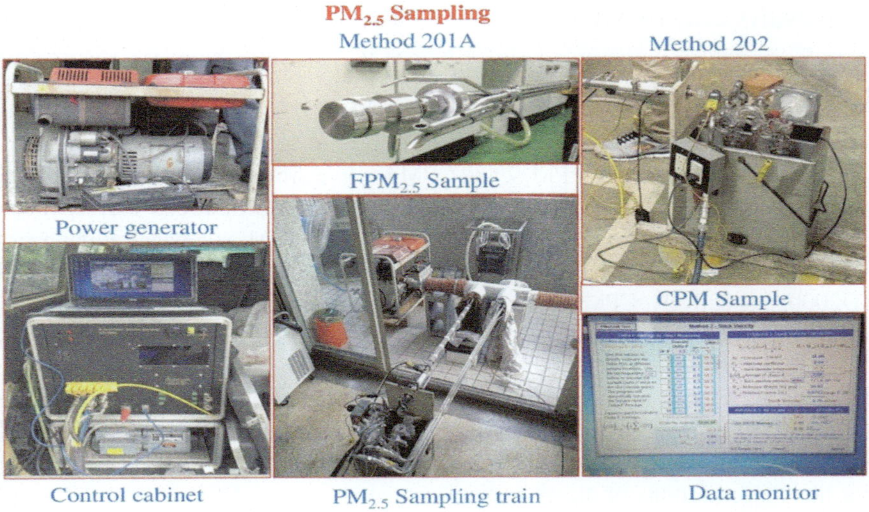

Fig. 1 Power generator and sampling system for $FPM_{2.5}$ (U.S. EPA Method 201A) and CPM (U.S. EPA Method 202) samples

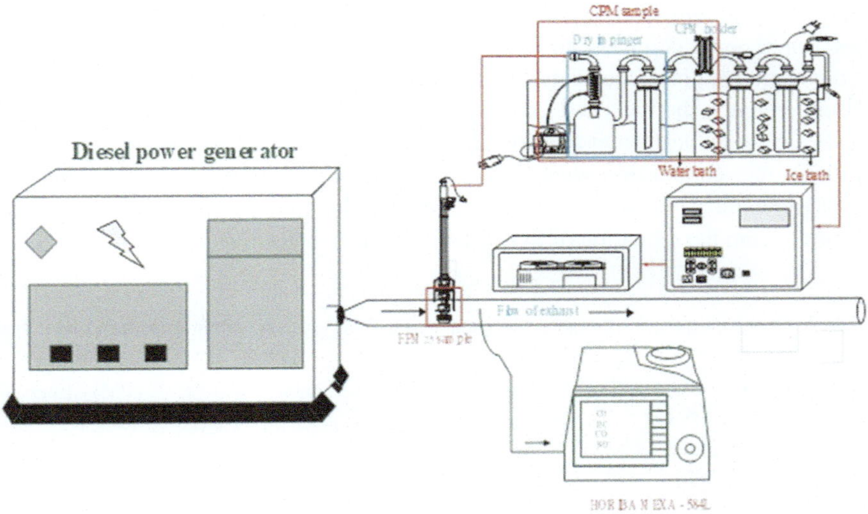

Fig. 2 Schematic diagram of the diesel generator and sampling system

Detailed descriptions of the sampling and analysis can be found in previous studies [5]. The R^2 values of all calibrations were above 0.99. A new calibration line was made when the bias was higher than ± 10%.

3 Results and Discussion

3.1 $FPM_{2.5}$ and CPM

The mass concentrations of $FPM_{2.5}$ and CPM for five different sulfur-content diesel emissions are shown in Table 3. $FPM_{2.5}$ emission concentrations corresponded to increases in the sulfur content in diesel fuel. The $FPM_{2.5}$ concentrations were 10.6, 12.8, 15.0, 15.1, and 17.4 mg/Nm^3 for GD10, GD250, GD500, GD1000, and GD2000, respectively. CPM concentrations were also increased with the increase of sulfur content in diesel fuel. They were obtained to be 118.2, 145.0, 187.3, 281.4, and 334.6 mg/Nm^3 for GD10, GD250, GD500, GD1000, and GD2000, respectively. Based on these, it is not surprising to see that the emissions of particles from the test diesel generator were directly proportional to the sulfur concentrations in diesel fuels. The results were comparable to those of a previous study by [2]. They reported that the concentrations of diesel particulate matter increased from 3.9 to 18.0 mg/m^3, 10.2 to 31.6 mg/m^3, and 15.6 to 36.8 mg/m^3 corresponding with sulfur content in diesel fuel of 500, 2100, and 3700 ppm, respectively.

The dominant contributors to CPM fractions in $TPM_{2.5}$ were 91.8%, 91.9%, 92.6%, 94.9%, and 95.1% for GD10, GD250, GD500, GD1000, and GD2000, respectively (Fig. 3). Yang et al. [8] reported higher CPM percentage (96.2%) emission for diesel-fired boiler compared to $FPM_{2.5}$ (3.8%) emissions. The formation of CPM in a combustion engine depends on the exhaust temperature, operating conditions, and sulfur content in the fuel used. Among them, fuel temperature and sulfur content in diesel fuel might be the major factors. In addition, higher fractions of CPM emissions in diesel fuel combustion were due to the lower ash content and higher combustion temperatures. CPM could be overestimated due to the positive sulfate artefact resulting from the oxidation of SO_2 present in the exhaust.

Table 3 Concentrations of filterable and condensable $PM_{2.5}$ versus sulfur content in diesel fuel samples

Sulfur content	$FPM_{2.5}$ (mg/Nm^3)	Condensable PM (mg/Nm^3)			$TPM_{2.5}$ (mg/Nm^3)
		Inorganic fraction	Organic fraction	Total CPM	
GD10	10.6	29.4	88.7	118.2	131.0
GD250	12.8	44.4	100.5	145.0	160.0
GD500	15.0	81.9	105.4	187.3	202.3
GD1000	15.1	136.2	145.2	281.4	296.4
GD2000	17.4	162.6	172.0	334.6	352.0

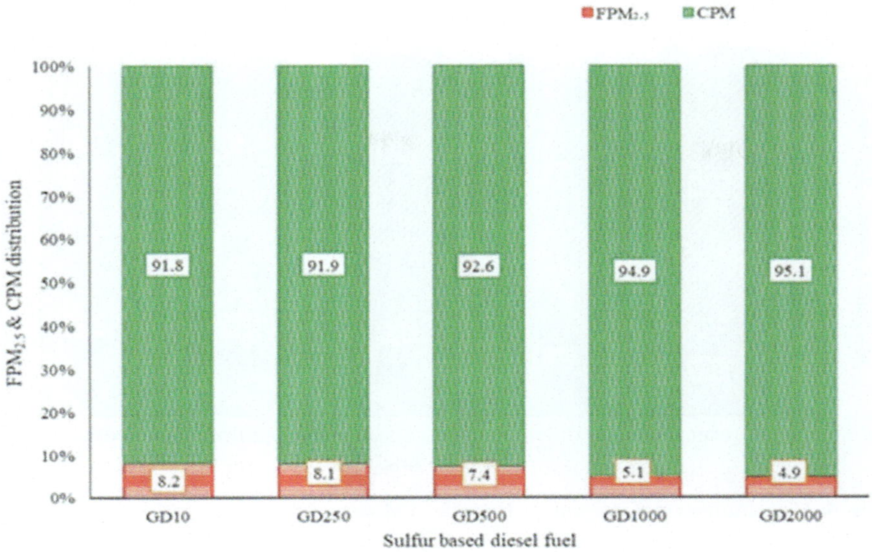

Fig. 3 Mass distribution of $FPM_{2.5}$ and CPM versus sulfur content in diesel fuel samples

3.2 *Inorganic and Organic Fractions in CPM*

The inorganic and organic fractions in CPM for sulfur-based diesel fuel are shown in Table 3 and Fig. 4. The organic fraction in CPM was obtained higher than the inorganic fraction for all types of sulfur-based diesel fuels. The organic concentrations were 88.7, 100.5, 105.4, 145.2 and 172.0 mg/Nm^3, however, inorganic concentrations were 29.4, 44.4, 81.9, 136.2, and 162.6 mg/Nm^3 for GD10, GD250, GD500, GD1000, and GD 2000 diesel fuels, respectively. The distribution of organic fraction in CPM accounted for 75.1%, 69.3%, 56.3%, 51.6% and 51.4% for GD10, GD250, GD500, GD1000, and GD 2000, respectively, mainly due to organic substances in diesel fuel. The organic fraction of CPM was gradually reduced by increasing the amount of sulfur in diesel fuel. Yang et al. [5] reported 62% and 72% organic CPM fractions for diesel-fired boilers and heavy oil boilers, respectively. Another study reported that the contribution of organic fraction of CPM for oil-fired boilers ranged from 20.4% to 80.5% [9]. The contribution of the organic CPM fraction was more pronounced for oil-fired sources. It is concluded that most of the diesel facilities emitted more organic compounds.

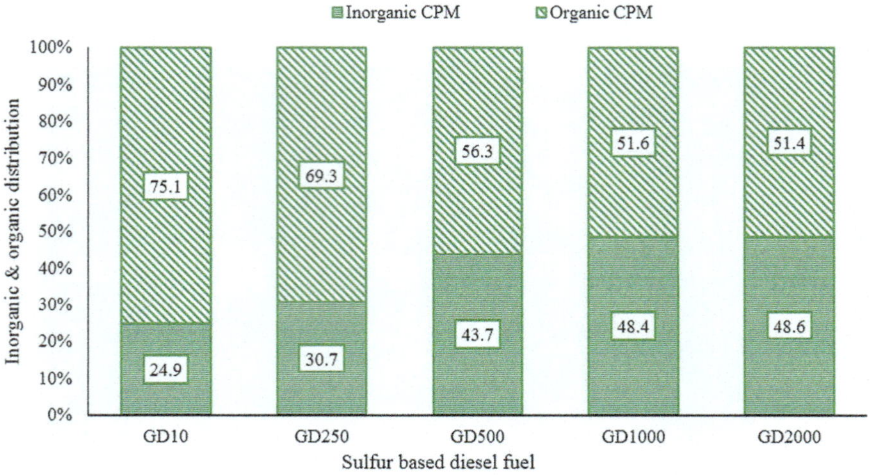

Fig. 4 Mass distribution of inorganic and organic CPM versus sulfur content in diesel fuel samples

4 Conclusions

With the increasing popularity of CPM, the attention of researchers has increased towards various aspects of CPM. The CPM concentrations were increased with the increase of sulfur content in diesel fuel. They were obtained to be 118.2, 145.0, 187.3, 281.4, and 334.6 mg/Nm^3 for GD10, GD250, GD500, GD1000, and GD2000, respectively. $FPM_{2.5}$ concentrations for different sulfur-content diesel like GD10, GD250, GD500, GD1000, and GD2000 were 10.6, 12.8, 15.0, 15.1, and 17.4 mg/Nm^3, respectively. The study shows that $FPM_{2.5}$ and CPM generated from diesel with high sulfur content are susceptible to being a major contributor to air pollution. The unsatisfactory state of fuel sulfur levels in many developing countries may be attributed to vehicle fuel standards that are several years behind the vehicle emissions standards of developed countries. Many countries with far higher fuel sulfur content have neither defined the resources nor policies necessary to match this global trend. These include Indonesia, Thailand, Nigeria and South Africa, all of which face significant air-quality challenges. A coordinated system approach of enforcing emission standards along with setting appropriate fuel standards will reduce pollution and improve air quality worldwide.

Acknowledgements The financial and technical support from Chaoyang University of Technology Taiwan is gratefully acknowledged.

References

1. Barry EG, McCabe LJ, Gerke DH, Perez JM (1985) Heavy-duty diesel engine/fuels combustion performance and emissions—a cooperative research program. SAE Technical Paper
2. Saiyasitpanich P, Lu M, Keener TC, Liang F, Khang S-J (2005) The effect of diesel fuel sulfur content on particulate matter emissions for a non-road diesel generator. J Air Waste Manag Assoc 55:993–998
3. IARC (2014) n.d. Diesel and gasoline engine exhausts and some nitroarenes/IARC working group on the evaluation of carcinogenic risks to humans
4. Feng Y, Li Y, Cui L (2018) Critical review of condensable particulate matter. Fuel 224:801–813
5. Yang H-H, Arafath SM, Lee K-T, Hsieh Y-S, Han Y-T (2018a) Chemical characteristics of filterable and condensable $PM_{2.5}$ emissions from industrial boilers with five different fuels. Fuel 232:415–422
6. Cano M, Vega F, Palomo D, Serrano J, Navarrete B (2019) Characterization of condensable particulate matter emissions in agricultural diesel engines using a dilution-based sampling train. Energy Fuels
7. Campos AT, Quintella CM, Meira M, Luna S (2018) Prediction of sulfur content in diesel/biodiesel blends using LED-induced fluorescence associated with multivariate calibration. J Braz Chem Soc 29:1367–1372
8. Yang H-H, Gupta SK, Dhital NB, Lee K-T, Hsieh Y-S, Huang S-C (2019) Establishment of indicatory metals for filterable and condensable $PM_{2.5}$ emitted from important stationary emission sources. Energy Fuels 33:10878–10887
9. Yang H-H, Arafath SM, Wang Y-F, Wu J-Y, Lee K-T, Hsieh Y-S (2018b) Comparison of coal-and oil-fired boilers through the investigation of filterable and condensable $PM_{2.5}$ sample analysis. Energy Fuels 32:2993–3002

Variation of Atmospheric Carbon Dioxide Over Maharashtra Using Remote Sensing Data

Digambar S. Londhe and Yashwant B. Katpatal

Abstract Carbon dioxide (CO_2) emissions are closely linked to climate change, primarily through the greenhouse effect. The Earth's atmosphere contains greenhouse gases, including carbon dioxide, that trap heat from the sun. Elevated CO_2 levels lead to an increase in global temperatures, which is a key aspect of climate change. This study aims to map the variation of atmospheric carbon dioxide from 2015 to 2021 and to explore the spatial and temporal variation of CO_2 over Maharashtra State using remote sensing data. Orbiting Carbon Observatory-2 (OCO-2) data has been used for the analysis. NASA's Orbiting Carbon Observatory-2 has been measuring carbon dioxide column-averaged dry air mole fraction, XCO_2, in the Earth's atmosphere. The OCO-2 data is analyzed and processed in the GIS environment. The results show the elevated atmospheric CO_2 concentration which was 397.383 ppm in 2015 and 418.183 ppm in 2021.

Keywords Atmospheric carbon dioxide · Orbiting Carbon Observatory-2 · Remote sensing · Geographic information system · Maharashtra

1 Introduction

The Earth's atmosphere is a complex fusion of various liquid, solid, and gaseous substances [1]. In the climate system, CO_2 concentration in the atmosphere exhibits significant influence as it is one of the major greenhouse gases that cause global warming. Spatial and temporal variations of CO_2 concentration due to natural and anthropogenic emissions have been observed in different parts of the world [2, 3]. Seasonal and temporal variations of CO_2 concentration have also been observed in different countries in the world [4–7]. XCO_2 is the column-averaged carbon dioxide in the atmosphere is represented in parts per million (ppm). The XCO_2 time series shows an evident seasonal cycle during winter and spring which is majorly driven by

---

D. S. Londhe (✉) · Y. B. Katpatal
Department of Civil Engineering, Visvesvaraya National Institute of Technology, Nagpur, India
e-mail: digambarlondhe@students.vnit.ac.in

S. M. Shiva Nagendra et al. (eds.), *Select Proceedings of the 8th Indian International Conference on Air Quality Management (IICAQM 2023)*, Lecture Notes in Civil Engineering 582, https://doi.org/10.1007/978-981-96-2359-4_11

emissions through anthropogenic activities and insignificant effects during summer due to biosphere activity for mid-latitude Northern Hemisphere locations [5]. OCO-2 retrievals revealed XCO_2 increases in the eastern United States that were up to 1 ppm above background values and coincided with significant fossil fuel emissions. Nearly twice as much XCO_2 has been increased in east China. The rise in XCO_2 linked to biomass burning in Africa moved north of the equator as the season changed [4].

Remote sensing and geographic information systems (GIS) are also used in the study and analysis of carbon dioxide (CO_2) concentrations in the atmosphere. These technologies offer valuable tools for monitoring, modeling, and understanding the spatial and temporal variations of CO_2 levels on Earth. Remote sensing satellites equipped with specialized sensors, such as spectrometers and radiometers, collect data on atmospheric composition, including CO_2 concentrations. Instruments like the Orbiting Carbon Observatory-2 (OCO-2) and the Tropospheric Monitoring Instrument (TROPOMI) on the Sentinel-5P satellite provide high-resolution and accurate measurements of CO_2 levels from space. These satellite-based remote sensing instruments capture data at different wavelengths, allowing for the retrieving CO_2 concentrations in the atmosphere. Models, such as the Community Earth System Model (CESM) [8] and the Weather Research and Forecasting with Chemistry (WRF-Chem) [9] model, simulate the transport and distribution of CO_2 in the atmosphere, considering various factors like emissions, meteorological conditions, and land surface interactions.

Remote sensing data, combined with ground-based measurements, is assimilated into atmospheric models. Data assimilation techniques help integrate satellite observations into models to create more accurate representations of CO_2 concentration patterns. The Orbiting Carbon Observatory (OCO) is a series of Earth-observing satellites developed and operated by NASA. The primary mission of the OCO series is to monitor and study carbon dioxide (CO_2) levels in Earth's atmosphere to better understand carbon cycling and its impact on climate change. The Orbiting Carbon Observatory-2 (OCO-2) was designed to measure atmospheric carbon dioxide (CO_2) with the precision, resolution, and coverage required to identify regional CO_2 sources and sinks at the global scale. OCO-2 is available for retrieving the column-averaged dry air mole fraction of CO_2 (XCO_2) in the atmosphere [10].

The main objective of this research study is to investigate and understand how the concentration of atmospheric carbon dioxide, represented by XCO_2, varies both in space and time specifically over the region of Maharashtra. To achieve this objective, the study utilizes data collected by the Orbiting Carbon Observatory-2 (OCO-2), a satellite equipped with advanced remote sensing capabilities, and leverages geographic information techniques for analysis and visualization.

2 Study Area and Data Used

Detailed information about the study area and data used in the analysis has been given in this section.

2.1 *Study Area*

In the context of the research, the study area encompasses the entire Maharashtra State in India. The study area encompasses the entirety of Maharashtra State, which is situated in Western India. Its geographical extends between the geographical coordinates of 15° 44′ to 22° 06′ North Latitude and 72° 56′ to 80° 54′ East Longitude. This geographical boundary covers a vast land area of approximately 3,07,713 km^2. The location map of the study area is shown in Fig. 1.

Maharashtra State exhibits a diverse range of climatic conditions. The mean annual temperature within this region fluctuates notably, ranging from 28.25 °C at its highest to 20.53 °C at its lowest. These temperature variations can have significant implications for atmospheric conditions, including the presence and distribution of greenhouse gases like carbon dioxide. The choice of Maharashtra State as the study area is driven by its extensive geographical coverage, its diverse climate conditions, and the significant variations in environmental factors across the region. These factors make it an ideal setting for studying the spatial and temporal variations of carbon dioxide (XCO_2) in the atmosphere.

Maharashtra, being one of India's industrial powerhouses, hosted a diverse array of industries encompassing manufacturing, chemicals, and power generation. Unfortunately, these industrial activities were associated with substantial CO_2 emissions. To address this, the state government implemented measures aimed at boosting energy efficiency and curbing emissions from industrial processes. The evolving landscape of CO_2 emissions in Maharashtra reflects the state's dynamic response to the imperatives of environmental sustainability and climate change mitigation.

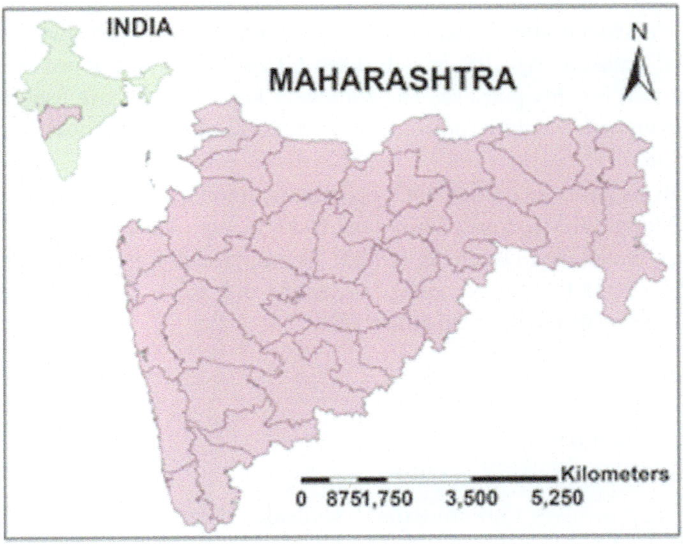

Fig. 1 Location map of the study area

2.2 Data Used—Orbiting Carbon Observatory-2

This is the Gridded Daily OCO-2 Carbon Dioxide assimilated dataset. The OCO-2 mission delivers the most accurate space-based XCO_2 measurements available thus far. Nonetheless, the data from the instrument exhibits significant interruptions in coverage owing to OCO-2's limited 10-km ground track and its inability to penetrate clouds and dense aerosol layers. This worldwide gridded dataset is generated using a data assimilation method often known as state estimation in the field of geophysics. Data assimilation combines simulations and observations, modifying the atmospheric constituents' state, such as CO_2, to align with observed values. In doing so, it effectively fills gaps in observations where they are missing, relying on prior observations and short transport simulations conducted by GEOS.

OCO-2 GEOS (Goddard Earth Observing System) Level 3 data is generated by incorporating OCO-2 L2 retrievals every 6h using the GEOS CoDAS system. This system is administered by NASA's Global Modeling and Assimilation Office (GMAO).

In this study, OCO-2 GEOS Level 3 daily, 0.5 $\times$ 0.625 assimilated CO_2 V10r (OCO2_GEOS_L3CO2_DAY) data has been used for the analysis. The data has been acquired from the Earthdata Web site. The spatial resolution of the data is 0.5 $\times$ 0.625 degrees, and data is available for daily and monthly temporal resolution.

3 Methodology

Certainly, analyzing geospatial data in a geographic information system (GIS) environment is a crucial step in gaining insights and deriving valuable information from spatial datasets. ArcGIS, widely used GIS software developed by ESRI, is a powerful tool for this purpose. The following key functions are used to explain how the downloaded data is processed in ArcGIS software for visualization and interpretation.

The analysis has been carried out by combining months during a year as January, February, March (JFM), April, May, June (AMJ), July, August, September (JAS), and October, November, December (OND). This approach was applied for the years from 2015 to 2021 to comprehensively examine the XCO_2 concentration patterns across all seasons throughout the entire year.

3.1 Data Import

The initial step involves importing the downloaded data into the ArcGIS software. The downloaded data is in the raster format which is imported in the ArcGIS software.

3.2 Data Exploration

Once the data is imported, it is important to explore its contents to understand its attributes and spatial relationships. This helps in identifying patterns and potential areas of interest for further analysis.

3.3 Data Pre-processing

Data preprocessing involves extraction of the data for the particular region and the conversion of the raster data into vector data, i.e., point shapefile data. For the extraction purpose, extract by mask tool, and for the conversion of raster to point shapefile, a conversion tool has been used in ArcGIS software.

3.4 Model Builder

ArcGIS includes a Model Builder tool that can create, manage, and document workflows. This feature enables the automation of repetitive tasks and the creation of reusable analysis workflows.

3.5 Mapping and Visualization

ArcGIS excels at creating maps and visualizing spatial data. It can be used to customize map layouts, symbology (e.g., colors, symbols, and labels), and cartographic elements to investigate the findings.

3.6 Spatial Statistics

ArcGIS offers statistical tools for analyzing spatial patterns and relationships. This tool helps in finding the output values of the XCO_2 values from the downloaded data.

4 Results and Discussion

The results section explains the spatial and temporal variation of the XCO_2 concentration over Maharashtra using remote sensing data. The focus on XCO_2 specifically means that the study is concerned with the concentration of CO_2 within a vertical column of the atmosphere, from the Earth's surface up to the top of the atmosphere. This measurement offers a comprehensive view of the CO_2 content in the region and is valuable for assessing the overall atmospheric CO_2 concentration.

4.1 Spatial Variation of XCO_2 Concentration

The monthly and spatial variation of XCO_2 concentration is presented in Fig. 2 (XCO_2 concentration is in PPM).

Elevated concentrations were consistently noted during the AMJ months in all the years that have been revealed in this analysis. Also, it has been observed that the eastern region in Maharashtra has higher XCO_2 concentration (412 ppm to 417 ppm) for all the months. One notable outcome of this analysis was the consistent presence of higher XCO_2 concentrations in the eastern part of Maharashtra.

The findings also indicate an alternating pattern of XCO_2 concentration in the western region of the state. Specifically, the XCO_2 concentration in the western region tends to be lower during AMJ and JAS months compared to the concentration observed during OND and JFM months. Spatial variation of XCO_2 concentration has also been observed that XCO_2 concentration in the eastern region during JFM and AMJ months is less as compared to the AMJ and JAS months.

The eastern and central regions in Maharashtra are showing low XCO_2 concentrations (397 ppm to 405 ppm) during all the months.

4.2 Temporal Variation of XCO_2 Concentration

The temporal variation of XCO_2 concentration is shown in Fig. 3 and Table 2.

Over the years under consideration, the amount of XCO_2 in the atmosphere is consistently growing. The XCO_2 concentration varies across different months during the year. The observations specifically mention four months: AMJ (April, May, June), JFM (January, February, March), JAS (July, August, September), and OND (October, November, December).

The key observation is that during the AMJ months of each year, the XCO_2 concentration consistently exceeds that of the other three periods (JFM, JAS, and OND). This suggests that there is a changing pattern in XCO_2 concentration, with the months of April, May, and June consistently experiencing higher XCO_2 levels compared to the other months.

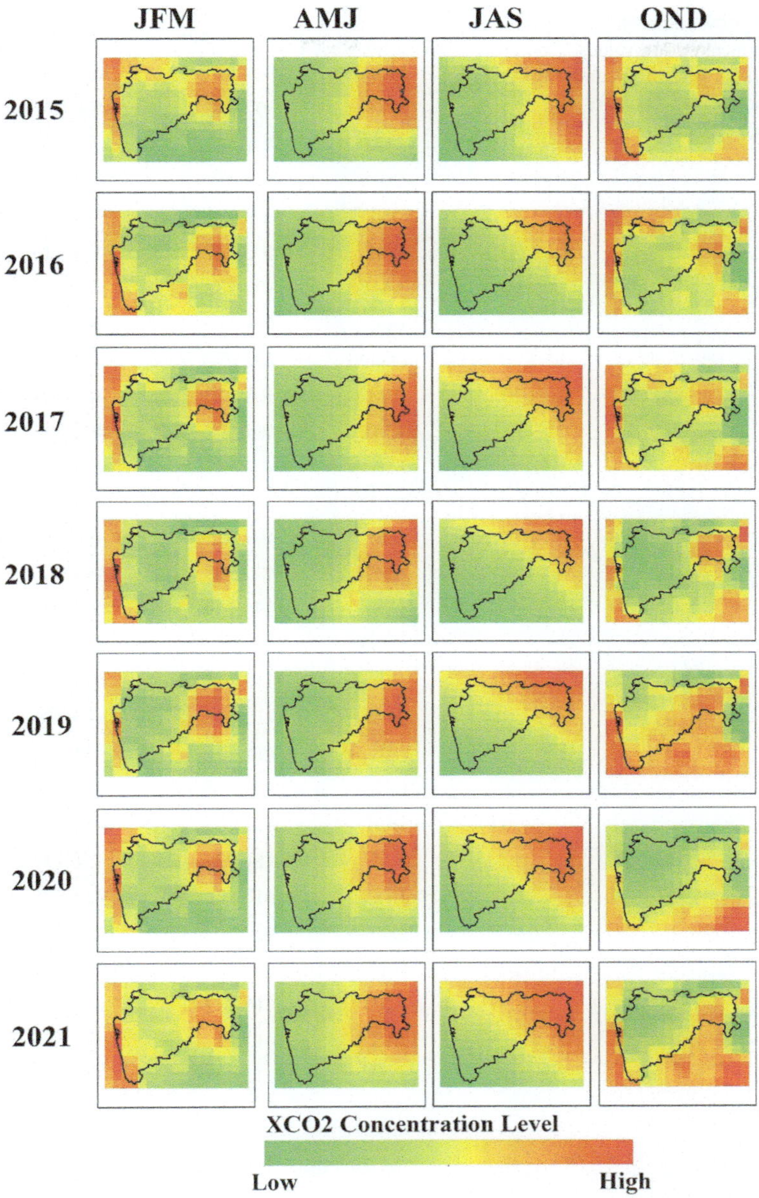

Fig. 2 Spatial variation of the XCO_2 concentration over Maharashtra (for low and high XCO_2 concentration values, refer Table 1)

Table 1 Low and high XCO_2 concentration values during JFM, AMJ, JAS, and OND months for the years 2015 to 2021

		XCO_2 concentration (PPM)	
Year	Month	Low	High
2015	JFM	397.383	400.034
	AMJ	401.46	402.677
	JAS	399.059	399.71
	OND	398.931	399.521
2016	JFM	402.727	403.326
	AMJ	405.288	406.5
	JAS	402.245	403.811
	OND	402.032	402.724
2017	JFM	405.472	406.086
	AMJ	407.291	408.558
	JAS	404.65	405.819
	OND	404.273	405.019
2018	JFM	407.333	408.01
	AMJ	409.705	410.736
	JAS	406.759	408.304
	OND	406.747	407.275
2019	JFM	410.008	410.541
	AMJ	412.446	413.326
	JAS	409.753	411.397
	OND	409.538	410.209
2020	JFM	412.174	412.763
	AMJ	414.924	416.147
	JAS	412.046	413.328
	OND	412.002	412.826
2021	JFM	414.732	415.603
	AMJ	416.831	418.183
	JAS	414.48	415.836
	OND	414.491	415.052

For instance, in 2015, during the JFM months (January, February, March), the XCO_2 concentration measured 397.9651 parts per million (PPM). However, by 2021, during the same JFM months, the XCO_2 concentration had increased to 415.0116 PPM. This shows the noticeable and significant rise in XCO_2 levels over a six-year period.

A similar rising pattern was seen in AMJ months also. The XCO_2 concentration during AMJ months of 2015 was 402.10347 PPM and that of AMJ month of 2021 was 417.3953 PPM.

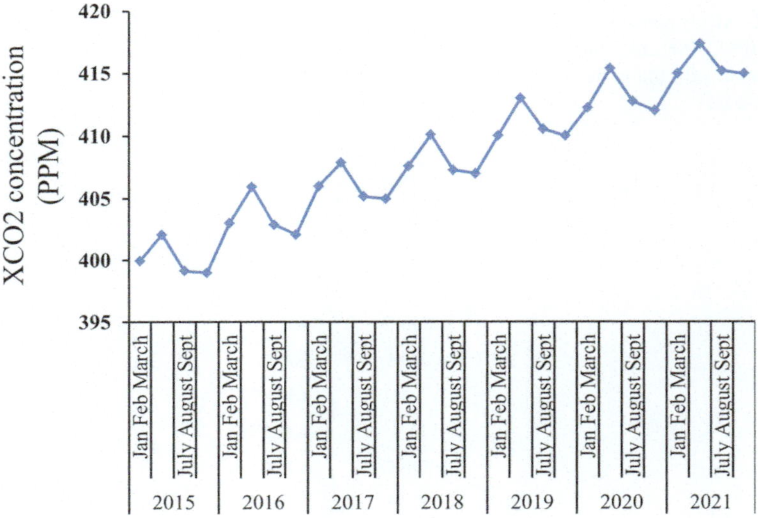

Fig. 3 Temporal variation of XCO_2 concentration (PPM)

Table 2 XCO_2 concentration during JFM, AMJ, JAS and OND months for the years 2015 to 2021

Year	Months	Average XCO_2 concentration (PPM)
2015	JFM	397.9651
	AMJ	402.1047
	JAS	399.1279
	OND	399
2016	JFM	403
	AMJ	405.9651
	JAS	402.8953
	OND	402.093
2017	JFM	406
	AMJ	407.8953
	JAS	405.1744
	OND	404.9884
2018	JFM	407.6047
	AMJ	410.1047
	JAS	407.2907
	OND	407
2019	JFM	410.0116
	AMJ	413
	JAS	410.5465
	OND	410
2020	JFM	412.2442
	AMJ	415.407
	JAS	412.7558
	OND	412.0116
2021	JFM	415.0116
	AMJ	417.3953
	JAS	415.2209
	OND	415

5 Conclusion

The XCO_2 concentration in the atmosphere has been analyzed by using remotely sensed OCO-2 Level 2 data. The analyses have been carried out for the spatial and temporal variation of XCO_2 concentration over Maharashtra, India. The following conclusions have been made from the analysis:

1. A persistent trend of elevated XCO_2 concentrations during the AMJ months across all considered years has been clearly noticed. This increase underscores the importance of understanding the regional dynamics of CO_2 emissions during specific times of the year.
2. The eastern region of Maharashtra has consistently exhibited higher XCO_2 concentrations ranging from 412 to 417 ppm throughout all months.
3. During the AMJ and JAS months, XCO_2 levels tend to be lower compared to the OND and JFM months. This cyclic variation highlights the complex interplay of factors influencing XCO_2 concentrations in different months.
4. The spatial analysis reveals that the XCO_2 concentration in the eastern and central regions of Maharashtra remains comparatively lower, ranging from 397 to 405 ppm across all months.
5. A prominent observation is the consistent rise in XCO_2 levels during April, May, and June (AMJ) each year, indicating a distinct variation. These rise consistently exhibit higher XCO_2 concentrations compared to the rest of the year (JFM, JAS, and OND).
6. In 2015, the XCO_2 concentration during JFM was 397.9651 PPM. However, by 2021, during the same JFM months, the XCO_2 concentration had notably increased to 415.0116 PPM. This substantial rise underscores the significance of monitoring and addressing the temporal variations in XCO_2 levels over time.

References

1. Hassan MS, Gomes RF, Bhuiyan MAH (2023) Seasonal distribution of AOT and its relationship with air pollutants in central Bangladesh using remote sensing and machine learning tools. Case Stud Chem Environ Eng 8:100399
2. Dhanu Radha SVV, Sabarathinam C, Al-Ayyadhi N, Al-Ajeel FK, Al-Qallaf H, Akber A (2022) Spatial and temporal variation of dissolved CO_2 in rainwater from an arid region with special focus on its association with DIC and pCO_2. Environ Earth Sci 81(4):113
3. Mitchell LE, Lin JC, Bowling DR, Pataki DE, Strong C, Schauer AJ et al (2018) Long-term urban carbon dioxide observations reveal spatial and temporal dynamics related to urban characteristics and growth. Proc Natl Acad Sci 115(12):2912–2917
4. Crisp D (2015) Measuring atmospheric carbon dioxide from space with the Orbiting Carbon Observatory-2 (OCO-2). In: Earth observing systems xx, vol 9607. SPIE, p 960702
5. Mermigkas M, Topaloglou C, Balis D, Koukouli ME, Hase F, Dubravica D et al (2021) Ftir measurements of greenhouse gases over Thessaloniki, Greece in the framework of Coccon and comparison with s5p/TROPOMI observations. Remote Sensing 13(17):3395
6. Wunch D, Wennberg PO, Osterman G, Fisher B, Naylor B, Roehl CM et al (2017) Comparisons of the orbiting carbon observatory-2 (OCO-2) XCO_2 measurements with TCCON. Atmosp Meas Tech 10(6):2209–2238
7. Zheng B, Chevallier F, Ciais P, Broquet G, Wang Y, Lian J, Zhao Y (2020) Observing carbon dioxide emissions over China's cities and industrial areas with the Orbiting Carbon Observatory-2. Atmos Chem Phys 20(14):8501–8510
8. Zhang H, Song H, Wang X, Wang Y, Min R, Qi M et al (2023) Effect of agricultural soil wind erosion on urban $PM_{2.5}$ concentrations simulated by WRF-Chem and WEPS: a case study in Kaifeng, China. Chemosphere 323:138250

9. Huang Y, Wei J, Jin J, Zhou Z, Gu Q (2022) CO fluxes in Western Europe during 2017–2020 winter seasons inverted by WRF-Chem/data assimilation research testbed with MOPITT observations. Remote Sens 14(5):1133
10. Eldering A, Wennberg PO, Crisp D, Schimel DS, Gunson MR, Chatterjee A et al (2017) The Orbiting Carbon Observatory-2 early science investigations of regional carbon dioxide fluxes. Science 358(6360):eaam5745

Initial Results of the Degradation of α-Pinene Utilising a Metakoalin-Based Geopolymer

Jamie Thomas Wilson, Xinyuan Ke, Daniel Maskell, and Richard James Ball

Abstract This study presents the initial results demonstrating the potential of the alkali-activated metakaolin geopolymer to degrade α-pinene in gas phase under UV activation. A metakaolin-based geopolymer sample pellet was synthesised with metakaolin and alkali activators (sodium hydroxide and sodium silicate), which contains naturally occurring photocatalysts TiO_2 (anatase). Photocatalytic degradation of gaseous α-pinene was studied in a novel flow cell and analysed using GC–MS. The α-pinene degraded to 55% of the total concentration, proving the photocatalytic reactivity of metakaolin-based geopolymer materials.

Keywords Indoor air quality · Geopolymer · Photocatalysis · Environmental remediation

1 Introduction

Geopolymers are an alternative to ordinary Portland cement which have up to 80% lower CO_2 emissions and therefore a lower environmental impact [1]. They have a structure of silica and alumina tetrahedrons with extra framework balancing cations around the aluminium ions [2]. Feedstocks include both industrial waste products such as fly ash and metal slags or from calcinated clays such as metakaolin [3].

J. T. Wilson · X. Ke · D. Maskell · R. J. Ball
Department of Architecture and Civil Engineering, University of Bath, Bath, United Kingdom

J. T. Wilson (✉) · R. J. Ball
Centre for Integrated Materials, Processes & Structures (IMPS), University of Bath, Bath, United Kingdom
e-mail: jtw57@bath.ac.uk

X. Ke · D. Maskell
Centre for Climate Adaptation & Environment Research, University of Bath, Bath, United Kingdom

S. M. Shiva Nagendra et al. (eds.), *Select Proceedings of the 8th Indian International Conference on Air Quality Management (IICAQM 2023)*, Lecture Notes in Civil Engineering 582, https://doi.org/10.1007/978-981-96-2359-4_12

Metakaolin was utilised in this study due to the more reactive platelet like structure and the inclusion photocatalytically active compounds such as titanium dioxide (TiO_2) and iron oxides (Fe_2O_3) [4].

The model air pollutant utilised was α-pinene due to the increasing prevalence in indoor spaces [5]. Pinene, like other terpenes, can be absorbed by internal surfaces which act as secondary sources but with increased heterogeneous reaction with ozone and other oxidising agents producing more hazardous volatile organic compounds (VOCs) [6]. The breakdown of VOCs is possible via photocatalytic reactions present on the geopolymer surface [7]. In this study, a novel experimental set-up enabled the effectiveness of a metakaolin geopolymer sample at abating α-pinene to be observed under controlled conditions.

2 Experimental Section

Metakaolin 1200S, sodium hydroxide pellets, and sodium silicate solution were purchased from Sigma-Aldrich. The sodium silicate solution contained 8.3% sodium hydroxide, 25.3% sodium silicate, and 66.4% water. The elemental composition of the metakaolin was determined via a 30-point EDX spectra and found to contain 41.93% alumina, 51.96% silica, 1.29% titanium dioxide, and 2.21% iron oxide amongst other trace elements.

The Al:Si:Na:H_2O molar ratio of the geopolymer was 1:2.1:1.02:9.94, respectively. The activator solution was produced by dissolving the appropriate amount of sodium hydroxide in distilled water and adding the solution to the sodium silicate solution. This was allowed to equilibrate over 24h. The activator solution was subsequently added to the metakaolin powder and mixed with a high-sheer overhead mixer until a homogenous paste was produced. The geopolymer paste was added to 15 mm × 15 mm × 5 mm moulds, sealed, and allowed to cure at 20 °C for 28d.

3 Characterisation

A Renshaw Quontor confocal Raman microscope was utilised to map the TiO_2 surface coverage of the geopolymer. The map was produced by component fitting with reference spectra from the RRUFF database (Anatase R060277, Rutile R050031). A 532 nm laser was utilised at 10% power, 65 μm slit opening and objective 50 × L. 71,765 points were taken to produce the TiO_2 surface coverage map.

The photocatalytic activity of the geopolymer sample was determined using an experimental set-up. This consisted of an air intake, flowmeter, a heated VOC chamber with a rubber regulator, a reaction chamber within a UV light box, sampling tenax tubes, and a filter. The reaction chamber contained temperature, pressure, and humidity sensors. A UV light of 253.7 nm wavelength, 9 Watt power was located

85 mm above the sample. The sensors recorded the data in real time and were connected to a Pico Technology USB TC-08 data logger. The tenax tubes were doped with 1 μL of 5% toluene in methanol solution to act as a surrogate standard and applied by passing nitrogen through the tube for 2 min at 100 mL/min. The α-pinene inlet concentration was 15 ug/m^3. The α-pinene concentration was calculated as a ratio to the toluene peak area.

The tenex tubes used for sampling were analysed via thermal desorption—gas chromatography mass spectrometry (TD-GCMS). A Markes TD100-xr automated thermal desorber (Markes International Ltd) was installed on a gas chromatograph (Agilent 8890 GC System, Agilent Technologies Inc.) coupled with a mass spectrometer equipped with an HP-5MS UI column (30 m, 0.25 mm diameter, 0.25 μm film). Thermal desorption followed these subsequent parameters: flow path temperature: 180 °C; tube desorption: 280 °C; trap temperature: 25 °C; helium carrier gas flow rate: 100 mL/min. The chromatographic temperature program was utilised as follows: 40 °C hold time of 2.25 min; ramp rate of 40 °C/min to 200 °C with a hold time of 2 min. The helium split slow rate was 10 mL/min. The mass spectra were in the m/z range of 30 to 300 amu. The ion sources were set at 245 °C. VOC identification was achieved utilising the NIST 14 Mass Spec Library.

4 Discussion

Photocatalysis occurs on the surface of the composite. The surface area of the photocatalytic geopolymer was 225 mm^2, which is 1/100th the surface area of the reaction chamber. Raman mapping was employed to show the average surface coverage of TiO_2 on a 201.5 μm × 600.6 μm section of the geopolymer with a resolution of 1.3 μm per pixel. The Raman mapping image was converted to binary, and a histogram of the images was used to identify the sections which were white that represented surface TiO_2 and black which represented geopolymer surface. The TiO_2 surface percentage was calculated to be 11.23% (Fig. 1).

The photoactivity of the geopolymer was determined over 3 regions. The larger peaks on the left of the gas chromatograms in Fig. 2 have been identified by mass

Fig. 1 Binary Raman mapping of geopolymer surface

spectroscopy as toluene and have been normalised. The smaller peaks on the right have been identified as α-pinene. The α-pinene peak area is compared to the toluene peak area, and the ratio between these is used to determine the pinene concentration. The doped air within the reaction chamber were sampled continuously for 30 min, and these sampling tubes were analysed. The spectra in Fig. 2 were used to create Fig. 3.

Region I shows the photodegradation of α-pinene when exposed to solely UV light with no geopolymer present. This shows that the UV light can breakdown 38% of the α-pinene before raising back to normal levels immediately after the UV is removed. Region II shows the interaction between the geopolymer and the α-pinene in the dark. The geopolymer absorbs 32% of the α-pinene before becoming saturated and returning to the original α-pinene levels. Region III shows the photocatalytic degradation of α-pinene by the geopolymer. The photocatalytic geopolymer consistently removed an average of 55% of the overall α-pinene level. This is a marked increase of 17% compared to the UV light solely.

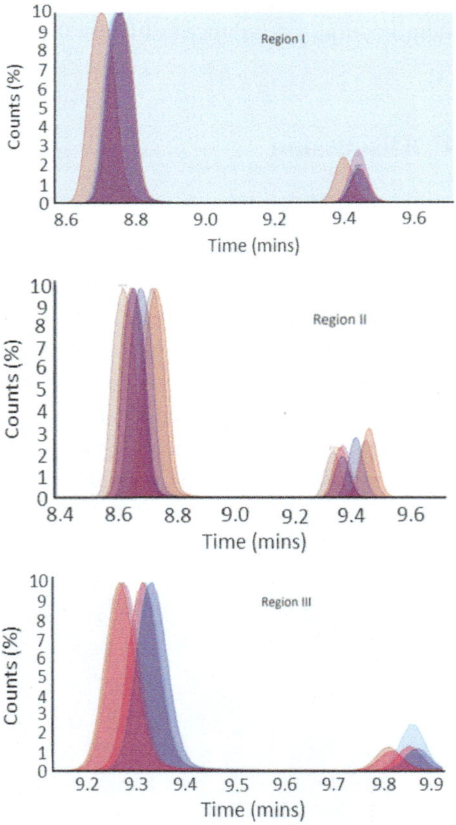

Fig. 2 Gas chromatogram of the regions tracking the photocatalytic activity

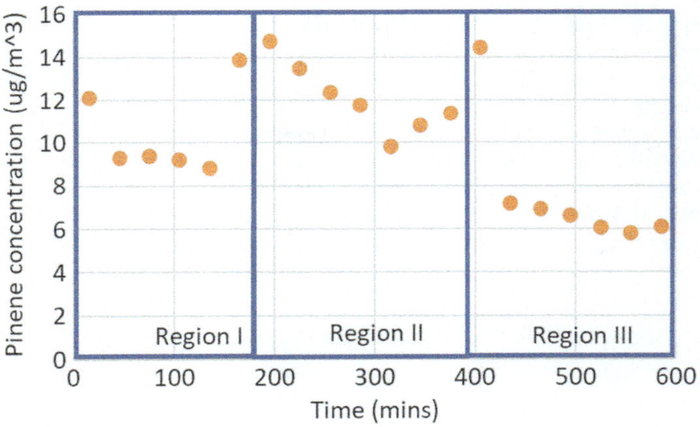

Fig. 3 Sequence of the photocatalytic activity determination

5 Conclusions

Degradation of α-pinene was observed on a metakaolin-based geopolymer in the presence of UV radiation of wavelength 253.7 nm. Photocatalytic activity of the geopolymer was confirmed by a decrease in α-pinene concentration of 17% compared to the initial concentration. Degradation of the α-pinene suggests that the geopolymer could degrade other similar VOCs enabling applications for improving indoor air quality. Results also validate the ability of the novel experimental set-up to provide reproducible data for the analysis of photocatalytic materials.

Acknowledgements The authors would like to acknowledge Ph.D. funding support through a University Research Studentships. Thanks are also extended to Will Bazeley for help setting up the experimental set-up and Philip Fletcher for Raman mapping and Kat Proctor for guidance on GCMS methods.

References

1. Duxson P, Provis JL, Lukey GC, Van Deventer JSJ (2007) The role of inorganic polymer technology in the development of green concrete [Article]. Cem Concr Res 37(12):1590–1597. https://doi.org/10.1016/j.cemconres.2007.08.018
2. Provis JL, Duxson P, Lukey GC, van Deventer JSJ (2005) Statistical thermodynamic model for Si/Al ordering in amorphous aluminosilicates [Article]. Chem Mater 17(11):2976–2986. https://doi.org/10.1021/cm050219i
3. Rozek P, Krol M, Mozgawa W (2019) Geopolymer-zeolite composites: a review [Review]. J Clean Prod 230:557–579. https://doi.org/10.1016/j.jclepro.2019.05.152
4. Sanalkumar KUA, Yang EH (2021). Self-cleaning performance of nano-TiO_2 modified metakaolin-based geopolymers [Article]. Cement Concr Compos 115:12, Article 103847. https://doi.org/10.1016/j.cemconcomp.2020.103847

5. Xu JL, Deng HF, Wang YQ, Li P, Zeng JQ, Pang HW, Xu X, Li X, Yang Y, Gligorovski S (2023) Heterogeneous chemistry of ozone with floor cleaning agent: implications of secondary VOCs in the indoor environment [Article]. Sci Total Environ 862:8, Article 160867. https://doi.org/10.1016/j.scitotenv.2022.160867
6. Thevenet F, Debono O, Rizk M, Caron F, Verriele M, Locoge N (2018) VOC uptakes on gypsum boards: sorption performances and impact on indoor air quality [Article]. Build Environ 137:138–146. https://doi.org/10.1016/j.buildenv.2018.04.011
7. Gao M, Mao MM, Shi JR, Liu YQ, Chen Q, Li J (2022) A review of the treatment techniques of VOC [Review; Early Access]. Appl Math Nonlinear Sci 12. https://doi.org/10.2478/amns.2021.2.00131

Measurement and Evaluation of Driving Patterns and Real-World Emissions from Heavy Tractor Vehicles Under Varied Load Conditions

Saket Ranjan, S. P. S. Prabhakaran, and S. M. Shiva Nagendra

Abstract While constituting a minority of the vehicle population, heavy tractor vehicles have been recognized as significant contributors to air pollution. Despite the existing apprehension, there is a scarcity of published data related to emissions from heavy-duty vehicles on Indian roads, and obtaining real-world measurements of emission factors for regulated and unregulated pollutants is even more challenging. Precise estimation of vehicle activity is of paramount importance for accurately estimating emissions. This paper demonstrates a method for quantifying the emission patterns of NOx, unburned HC, CO, and CO_2 coming from diesel tractor vehicles that comply with the Indian emissions standards to evaluate the real-world emissions, and heavy tractor vehicle was tested on different loading conditions using portable emission measurement system. The on-road emission factors (EFs) operating under empty load on trailer and without trailer were compared with the EFs under full load capacity. The analysis revealed that the driving patterns under different loading conditions resulted in increased emissions. Specifically, emissions were approximately 12% higher for CO, 19% for HC, 38% for NOx, and 34% for CO_2 when transitioning from empty trailer to a full load capacity. The findings encompass vehicle attributes, acceleration and speed activity profiles derived from real-world data, and potential factors influencing the emissions factor.

Keywords Heavy tractor vehicle · RDE · Gaseous emission · PEMS · Diesel emission

---

S. Ranjan · S. M. Shiva Nagendra (✉)
Department of Civil Engineering, Indian Institute of Technology Madras, Chennai, India
e-mail: snagendra@iitm.ac.in

S. Ranjan
e-mail: ce20d095@smail.iitm.ac.in

S. P. S. Prabhakaran
Department of Energy and Environment, National Institute of Technology, Tiruchirappalli, India
e-mail: sathiyanitt1491@gmail.com

S. M. Shiva Nagendra et al. (eds.), *Select Proceedings of the 8th Indian International Conference on Air Quality Management (IICAQM 2023)*, Lecture Notes in Civil Engineering 582, https://doi.org/10.1007/978-981-96-2359-4_13

1 Introduction

Air pollution has an adverse effect on human health, natural and built environment, economy, and quality of life. Throughout history, road vehicles have consistently stood as one of the primary sources of pollutants impacting urban air quality and have been the leading catalyst for establishing progressively stringent emissions standards [8]. Enhancing urban air quality has served as the primary motivation behind numerous policies, initiatives, and actions in every nation. Diesel engines drive the majority of heavy-duty vehicles (HDVs) in India, comprising just 9% of the vehicle fleet. However, diesel-based trucking significantly contributes to concerns regarding air pollution, greenhouse gas emissions, freight costs, trade balance, and energy security [18]. Tractors are an indispensable component of India's agricultural and rural landscape, playing a multifaceted role in enhancing productivity, income, and overall well-being in rural communities. Addressing exhaust emissions from these vehicles is essential for a sustainable and healthy future for India, necessitating investments in cleaner fuels, advanced vehicle technologies, and robust emission control measures. However, their combustion engines are a major source of air pollutants such as particulate matter (PM), nitrogen oxides (NOx), hydrocarbons (HC), and carbon monoxide (CO) [11]. The high levels of pollution generated by these emissions contribute to the worsening air quality in many Indian cities, leading to health problems like respiratory diseases and reduced quality of life [2]. Furthermore, the greenhouse gases emitted by heavy tractor vehicles, particularly carbon dioxide (CO_2), exacerbate climate change, making it crucial for the transportation sector to adopt cleaner and more efficient technologies.

In numerous studies, the utilization of portable emission measurement systems (PEMS) for on-road vehicle emission monitoring has been embraced because of its dependable nature, precision, and capacity to assess emissions across diverse ambient and operational scenarios [21]. The speed and acceleration of heavy-duty vehicles are influenced by several critical factors. Engine power plays a significant role, with higher horsepower and torque allowing for faster acceleration and higher top speeds [16]. The weight and load of the vehicle are crucial factors, as heavier loads can reduce acceleration and top speed due to increased inertia [4, 20, 25]. Aerodynamics also come into play, with sleek designs reducing wind resistance and increasing speed [7]. Tire quality and grip affect both acceleration and speed, as tires with better traction can provide improved performance [19]. Additionally, road conditions, such as grade and surface quality, have a notable impact on heavy-duty vehicle performance, particularly during acceleration on inclines. Lastly, maintenance and mechanical condition are essential, as well-maintained engines, transmissions, and drivetrains can optimize a heavy-duty vehicle's speed and acceleration capabilities [5].

Obtaining accurate emission factor estimates is crucial for the successful implementation of measures to reduce emissions and strategies for managing air quality. This encompasses tasks like formulating guidelines for regional transportation planning and creating emission inventories. Therefore, this study showcases an approach for generating emission estimates for tractor vehicles operating under various loading conditions on Indian roads.

2 Experimental Methodology

2.1 Site Description

This study was conducted on a congested urban thoroughfare in IIT Madras, India, measuring 8.25 km in length and 16 m in width, with two lanes dedicated to traffic. The climate in this region is classified as 'Tropical,' marked by lengthy and intense summers from March to June, followed by the rainy season from mid-October to early November, and experiencing milder winters in January. Figure 1 illustrates the study area's geographic boundaries, including the designated routes.

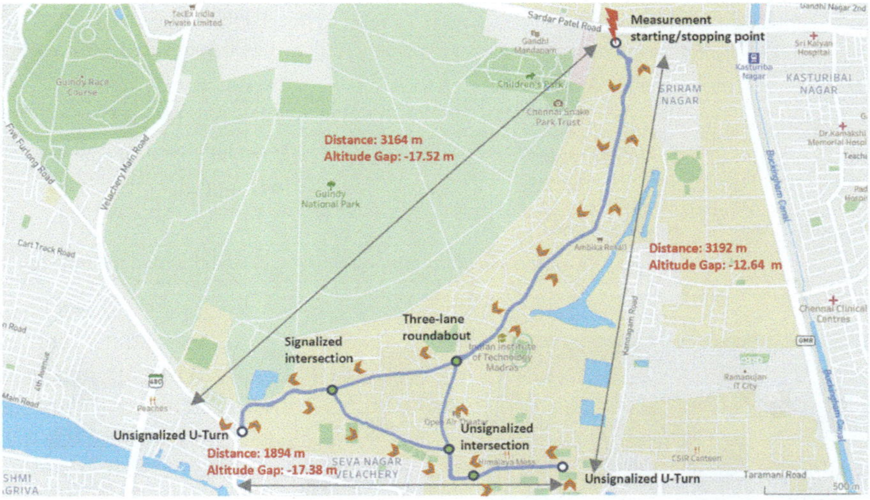

Fig. 1 Section of test route in the IIT Madras perimeter, Chennai. *Source* Google map, accessed on 27 September 2023

3 Materials and Methods

The speed and time data during on-road driving were gathered through GPS technology, utilizing the handheld Garmin e-Trex 20 GPS device. During each test run, the vehicle's speed and emissions were recorded on a per-second basis (Table 1). Concurrently, on-board emission data in real-world conditions were acquired through the AVL Ditest 1000 gas analyzer as shown in Fig. 2.

Table 1 Specification table of the gas analyzer [1]

Measured quantities	Measuring range	Resolution	Accuracy
CO	0–15% vol	0.01% vol	< 0.6% vol: ± 0.03% vol
			0.6% vol: ± 5% v. M
CO_2	0–20% vol	0.01% vol	< 10% vol: ± 0.5% vol
			≥ 10% vol: ± 5% v. M
HC	0–30,000 ppm vol	1 ppm vol	< 200 ppm vol: ± 10 ppm vol
			< 200 ppm vol: ± 5% v. M
			≥ 10,000 ppm vol: ± 10% v. M
O_2	0–25% vol	0.01% vol	< 2% vol: ± 0.1% vol
			≥ 2% vol: ± 5% v. M
NOx	0–5000 ppm vol	1 ppm vol	< 500 ppm vol: ± 50 ppm vol
			≥ 500 ppm vol: ± 10% v. M
Response time	$t_{95} \leq 10$ s		
	t_{95}: 95% of the gas concentration must be detected within the time t_{95}		

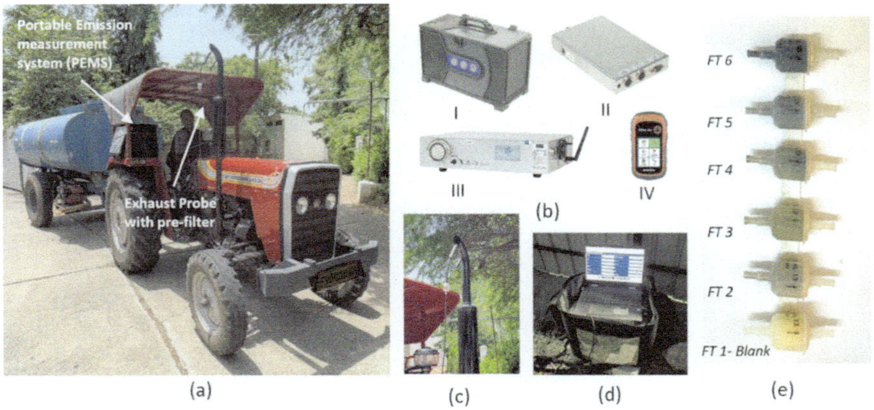

Fig. 2 Test setup for collecting on-board emissions data from tractor at different loading conditions. **b** Measurement assembly (I) AVL Ditest 1000 (II) RPM Sensor (III) Gas analyzer (IV) GPS (Garmin E-trex 20×). **c** Exhaust probe. **d** Operating system diagnosis. **e** Exhaust pre-filter used during measurements

Table 2 Classification of vehicle operating modes

Operating mode	Definitions
Idle	$v < 2.5$ kmph, $-0.1 \leq \lvert a \rvert \leq 0.1$ m/s^2
Deceleration	$a < -0.1$ m/s^2
Cruising	$v \geq 2.5$ kmph, $-0.1 \leq \lvert a \rvert \leq 0.1$ m/s^2
Acceleration	$a > 0.1$ m/s^2

The gas analyzer makes use of the non-dispersive infrared (NDIR) technique to record emissions of CO, CO_2, and HC on a per-second basis. It also utilizes an electrochemical cell for the measurement of NOx and O_2. Prior to the study, the equipment was calibrated, and accuracy checks were performed before each test to ensure accuracy.

The vehicles were loaded with varying cargo loads, specifically full weight on the trailer (i.e., 4000 kgs), empty trailer, and without trailer, as shown in Fig. 2. Subsequently, the PEMS test cycle was executed, and the corresponding data were gathered to conduct emission tests on actual roads with different load conditions, followed by the analysis and discussion of the test results.

The instantaneous emission rate was calculated based on exhaust concentration, velocity, and exhaust flow in the subsequent manner:

$$\mathrm{ER}_{(P)} = P \times M_{\mathrm{Exhaust}} \times \rho_{\mathrm{fraction}} \tag{1}$$

$$\mathrm{EF\ (g/km)} = \sum_{0}^{t} \mathrm{ER}(p) \times \frac{1000}{v} \tag{2}$$

where $[P]$ is the instantaneous concentration of pollutant P (% volume or ppm), M_{Exhaust} is the exhaust volumetric flow rate in l/s, and ρ_{fraction} is the density of pollutant in g/l. The emission factor was determined by dividing the total emissions by the distance traveled (2).

Table 2 summarizes the classification of vehicle operating mode was based on the established definitions as outlined in previous research on on-board emission studies by Saleh et al. [22] and Huo et al. [13].

4 Results and Discussion

4.1 Effect of Load and Driving Characteristics

Table 3 presents the driving cycle of the heavy tractor vehicle tested under various load conditions. The mean and median of speeds during without trailer conditions is higher as compared to the loaded and unloaded. Also, significantly higher time is spent in speed below 10 kmph during load conditions. A study by Bishop et al. [3] found that the emissions of pollutants, such as CO_2 and NOx, are closely related to driving patterns, load conditions, and driving speeds. The average acceleration was observed to be at its peak (0.47 ± 0.015 m-s^{-2}) when the vehicle was operating without any load, marking a 56% increase compared to full load conditions. Additionally, the average time in transit decreased by 41%, while idle duration increased by 2.4 times compared to no load conditions.

4.2 Real-World Driving Profiles on Cycle-Average Emissions Rates

The driving patterns and resulting emissions on Indian roads exhibit dynamic and inconsistent behavior due to the absence of lane discipline, interruptions caused by pedestrians, and random roadside parking [6].

When considering the presence of loading conditions, with on-road measured emission rates (ERs) in g/s for CO, HC, NOx, and CO_2, compared to the absence of load conditions, a deviation of 52% for CO, 19% for HC, 43% for NOx, and 17% for CO_2 was observed. However, when conducting the test run with a full load, the vehicle idles 2.7 times more frequently in comparison with when it is unloaded. Likewise, vehicle cruising increases by a factor of 1.9 under loading conditions. The outcomes of the comparative analysis between vehicle operating parameters and emissions are illustrated in Fig. 3. The speed distribution of tractors without a load showed greater dispersion, affecting the distribution of samples across speed categories.

4.3 Comparison of Real-World Measured EFs When Operated Under Different Load Conditions

The gaseous pollutant emission rate, as depicted in Fig. 4, exhibits variations throughout the entire journey. These fluctuations can be linked to factors such as engine load, aerodynamics, rolling resistance, road gradient, and vehicle acceleration, as discussed in previous studies [12, 24]. It was observed that the driving profiles during each loading conditions were emission intensive for CO, HC, NOx, and CO_2, around 12, 19, 38, and 34%, higher from full weight to empty trailer, respectively. While NOx was 64% higher during full weight operations as compared to without trailer, which might be because it has a higher functional relationship with the higher speed and accelerations.

Emission factors (EFs) were estimated to have the highest for HC + NOx emission for full load than empty load on trailer and without trailer operations, i.e., 3.22, 1.98, and 1.44 g/km, whereas for CO emissions were observed to 3.73, 3.15, and 1.91 g/

Table 3 Driving characteristics on different load conditions

Vehicle loading condition	Weight (kg)	Idle time (min)	Duration[b] (s)	Avg speed (km/h)	Max speed (km/h)	Distance (km)	Avg. accel. (ms^{-2})	Max accel. (ms^{-2})	Avg. decel. (ms^{-2})	Max decel. (ms^{-2})
Trailer with max load	1915 + 1720 + 4000[d]	927	2916|2237	9.08 ± 0.13	20.05	8.25	0.30 ± 0.012	1.9	− 0.29 ± 0.01	− 1.7
Unloaded trailer	1915 + 1720[c]	884	2857|1913	9.84 ± 0.13	24.08	8.25	0.43 ± 0.015	2.4	− 0.44 ± 0.013	− 2.1
Without trailer	1915[a]	385	2416|1429	12.46 ± 0.12	25.63	8.25	0.47 ± 0.015	2.3	− 0.43 ± 0.017	− 2.7

[a] Gross weight of vehicle
[b] Total time\| time in movement
[c] Weight of empty trailer
[d] Operating under full load on trailer

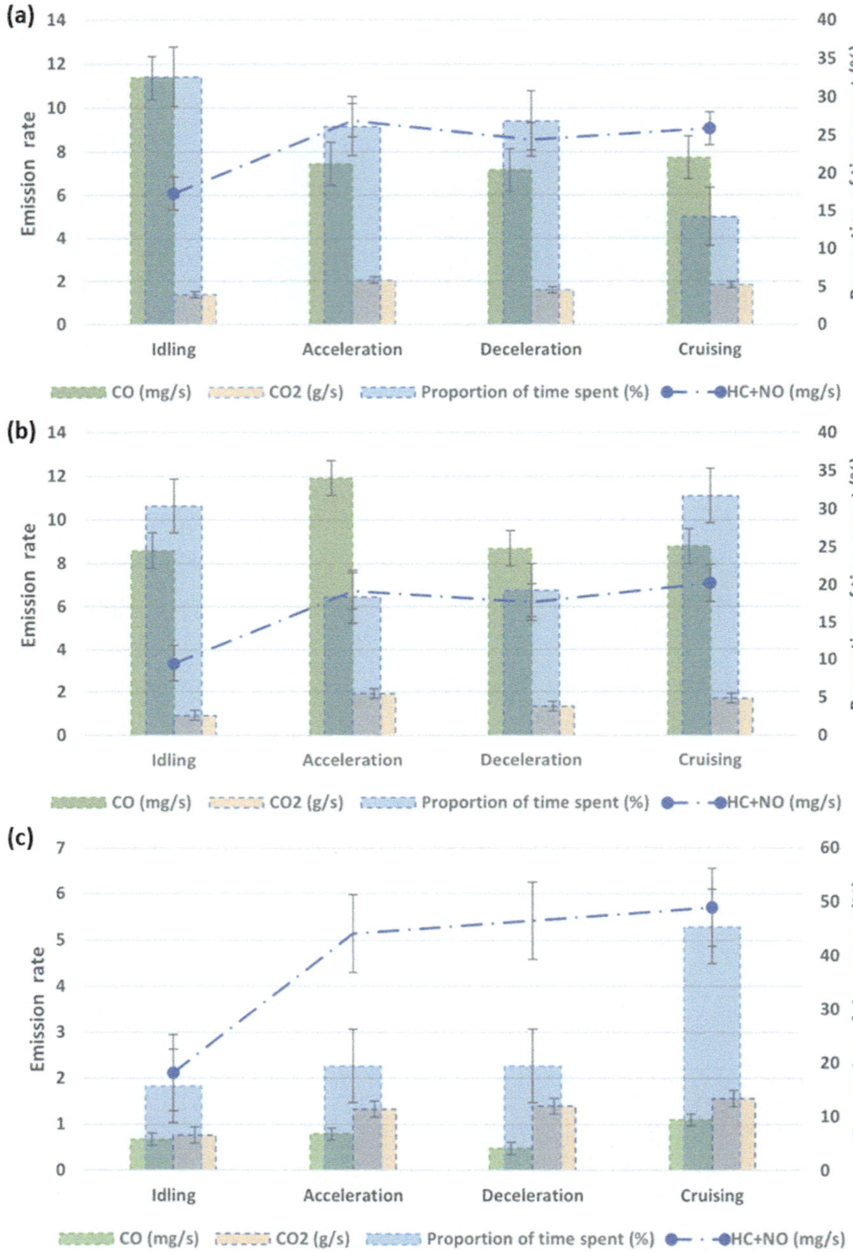

Fig. 3 Share of proportion of time spent during each vehicle driving operation and consequent emission rates (g/s) **a** without trailer, **b** unloaded trailer, **c** trailer with max load

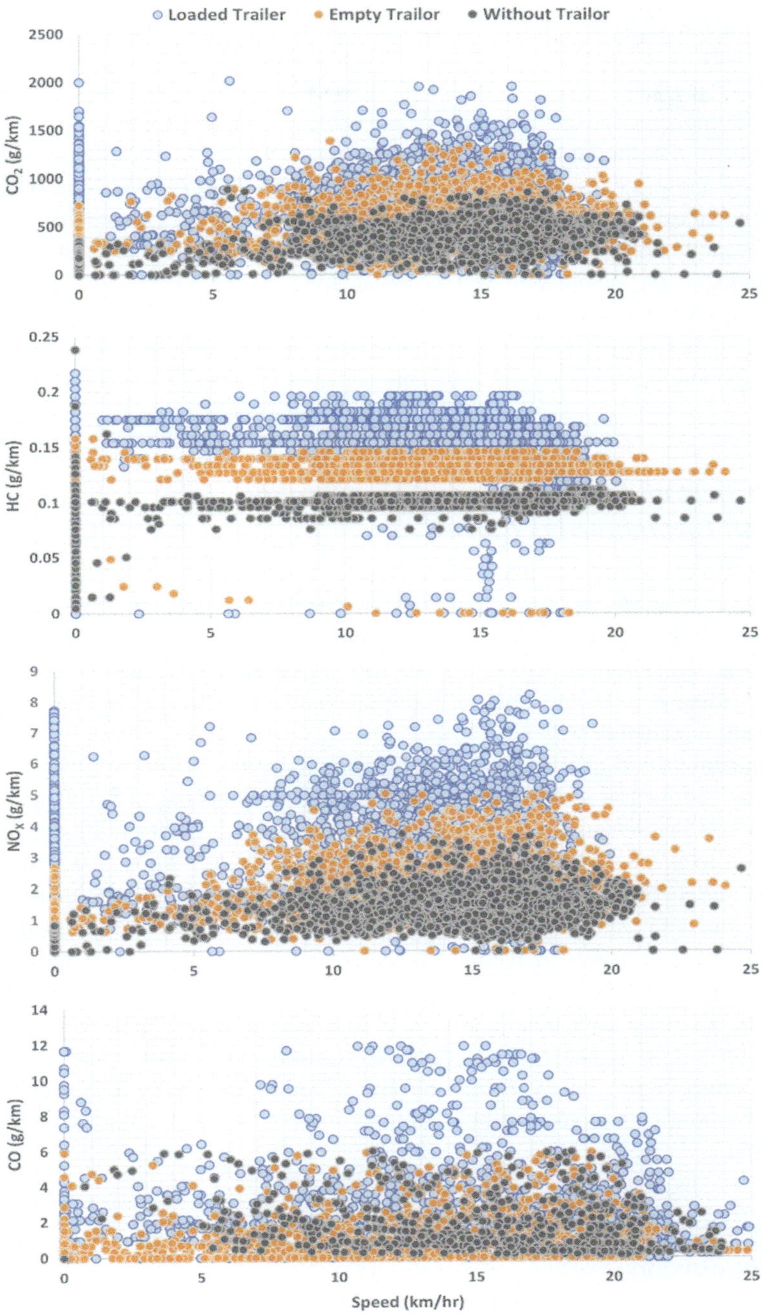

Fig. 4 Contrasting emissions factors observed in real-world settings while operating under varying load conditions

km; similarly, CO_2 emissions had the highest EF, i.e., 675.7, 498.6, 393.1 g/km, respectively.

Among the same make for diesel HDV, older models exhibit higher emission factors, a characteristic that can be attributed to the engine's condition and its combustion efficiency. A similar trend was observed earlier [9, 23, 25]. The contrast between this study and prior research lies in the increased levels of hydrocarbon (HC) and nitrogen oxides (NOx) emissions. The variance in findings is expected due to differences in methodology, the equipment used for real-time vehicle exhaust emissions measurement, the quantity of test drives conducted, the variety and quantity of passenger cars, the study location, built environment, traffic density, flow conditions (ranging from free and smooth flow to unstable and congested), meteorological factors, and numerous other variables that vary from one study to another.

4.4 Effect of Acceleration and Deceleration on Instantaneous Emissions

The effect of acceleration on speed and combined emission rate of HC and NOx is shown in Fig. 5. It can be seen that significant emissions are likely to result from rapid accelerations exceeding 0.5 m/s^2. Encouraging smoother traffic flow and providing eco-driving training represent potential strategies to mitigate high acceleration incidents [15].

In the absence of load and trailer weight, the reduction of 14–39%, respectively, was observed for HC + NOx emission rates (g/s) as compared to full weight operations while the deviation of average vehicle speed running with full load capacity increased to 9.25% when the trailer was empty and 26% when it was operated without trailer. This could be attributed to frequent alterations in operational circumstances, including sharp accelerations, decelerations, and stop-and-go occurrences [14], which result in increased engine load, instances of engine misfire, and at times, fuel injection cut-off. These phenomena have been documented in previous studies [10, 17]. Since no after treatment control technologies have been installed in the tractor, the tailpipe exhaust is being operated at sufficiently high temperatures to efficiently reduce engine-out hydrocarbons (HC) and nitrogen oxides (NOx) during low-speed conditions, as opposed to the conditions observed during medium to high-speed operation.

5 Conclusion

Although the emission patterns and the impact of varying load capacities show promising outcomes. They reveal that during specific instances of rough terrains and congestion, accurately evaluating air quality in rural areas can be a complex

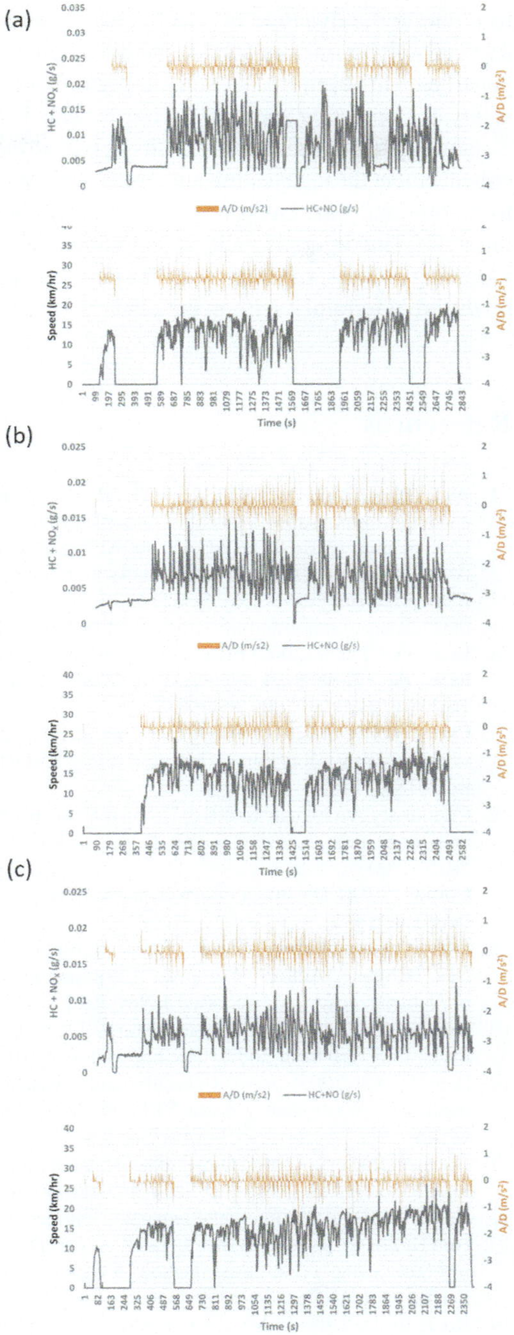

Fig. 5 Temporal variation of the instantaneous HC and NOx emission rates and vehicle velocity for the driving profiles of the tractor vehicle **a** without trailer, **b** unloaded trailer, **c** trailer with max load

task due to the dynamic off-road activities and road characteristics of rural roads. Typically, these tractors operate on unrestricted-access roads to avoid administrative penalties. However, due to vehicle performance constraints and prioritizing driver safety, the operating speed of such overloaded vehicles remains relatively low.

Consequently, the model results serve as indicators and have the potential for enhancement through testing numerous vehicles on the roads and off-road conditions to capture real-time emissions. Additionally, these outcomes may be further refined through (i) incorporating a large, measured dataset and enhancing input data for the modeling process. (ii) Improving the emission inventory by conducting experiments on different terrain, environment, and off-road conditions.

References

1. AVL (2014) AVL DITEST GAS 1000. Retrieved from https://www.avlditest.com/index.php/en/supportoptionsdownloads.html?file=files/content/downloads/handbecher/handbuecher-en/User-manual-AVL-DITEST-GAS1000AT7651E.pdf
2. Abhyankar N et al (2022) Freight trucks in india are primed for electrification
3. Bishop JD, Hargreaves DM, Ball DR (2016) The effects of driving conditions and payload on the fuel consumption and emissions of a diesel-powered vehicle. Sci Total Environ 550:504–513
4. Brodrick CJ et al (2002) Effects of engine speed and accessory load on idling emissions from heavy-duty diesel truck engines. J Air Waste Manag Assoc 52(9):1026–1031. https://doi.org/10.1080/10473289.2002.10470838
5. Chong HS et al (2018) Analysis of real driving gaseous emissions from light-duty diesel vehicles. Transp Res Part D Transp Environ 65(October):485–499. https://doi.org/10.1016/j.trd.2018.09.015
6. Choudhary A, Gokhale S (2016) Urban real-world driving traffic emissions during interruption and congestion. Transp Res Part D Transp Environ 43:59–70. https://doi.org/10.1016/j.trd.2015.12.006
7. Corp C (2018) The effect of aerodynamic drag changes on the exhaust emissions of a typical large car, p 412
8. DG Climate Action (2018) Road transport: reducing CO_2 emissions from vehicles. Climate Action—European Commission. https://ec.europa.eu/clima/policies/transport/vehicles/heavy_en. Accessed 1 Nov 2018
9. Dou Y, Li Y, Ling J (2021) Research on real-driving emissions of heavy tractor under different loads. E3S Web Conf 268:01041. https://doi.org/10.1051/e3sconf/202126801041
10. Frey HC et al (2003) On-road measurement of vehicle tailpipe emissions using a portable instrument. J Air Waste Manag Assoc 53(8):992–1002. https://doi.org/10.1080/10473289.2003.10466245
11. Harrison RM et al (2018) Diesel exhaust nanoparticles and their behaviour in the atmosphere. Proc R Soc A Math Phys Eng Sci 474:20180492. https://doi.org/10.1098/rspa.2018.0492
12. Heywood JB (1988) Internal combustion engines fundamentals. McGraw-Hill, Inc.
13. Huo H, Zhang Q, He K, Wang Q, Yao Z, Streets DG (2009) High-resolution vehicular emission inventory using a link-based method: A case study of light-duty vehicles in Beijing. Environ Sci Technol. 43(7):2394–2399. https://doi.org/10.1021/es802757a
14. Ko J, Jin D, Jang W, Myung C-L et al (2017) Comparative investigation of NOx emission characteristics from a Euro 6-compliant diesel passenger car over the NEDC and WLTC at various ambient temperatures. ApEn 187:652–662
15. Ko S et al (2020) NOx emissions from Euro 5 and Euro 6 heavy-duty diesel vehicles under real driving conditions. Energies 13(1):1–13. https://doi.org/10.3390/en13010218

16. Mahesh S, Ramadurai G, Nagendra SMS (2019) On-board measurement of emissions from freight trucks in urban arterials: effect of operating conditions, emission standards, and truck size. Atmos Environ 212(May):75–82. https://doi.org/10.1016/j.atmosenv.2019.05.030
17. Marsden G, Bell M, Reynolds S (2001) Towards a real-time microscopic emissions model
18. Ministry of Road Transport and Highways (MoRTH) (2019) Annual Report 2019–20. Government of India. Available at: https://www.morth.gov.in/sites/default/files/AnnualReport2019-20.pdf [Accessed 27 July 2022]
19. Mordi N et al (2023) Estimation of wheel slip in 2WD mode for an agricultural tractor during plowing operation using an artificial neural network. Sci Rep 1–16. https://doi.org/10.1038/s41598-023-32994-7
20. Rasmussen RE (1990) Effect of fuel properties on mutagenic activity in extracts of heavy-duty diesel exhaust particulate. J Air Waste Manag Assoc 40(10):1391–1396. https://doi.org/10.1080/10473289.1990.10466791
21. Ropkins K et al (2009) Real-world vehicle exhaust emissions monitoring: review and critical discussion. Crit Rev Environ Sci Technol. https://doi.org/10.1080/10643380701413377
22. Saleh W, Kumar R, Kirby H, Kumar P (2009) Real world driving cycle for motorcycles in Edinburgh. Transportation Research Part D: Transport and Environment. 14(5):326–333. https://doi.org/10.1016/j.trd.2009.03.003
23. Selleri T et al (2022) Measuring emissions from a demonstrator heavy-duty diesel vehicle under real-world conditions—moving forward to Euro VII. Catalysts 12(2):1–27. https://doi.org/10.3390/catal12020184
24. Vasconcellos EA (2001) Urban transport, environment and equity. the case for developing countries. Earthscan, London
25. Wang X et al (2021) Effects of vehicle load on emissions of heavy-duty diesel trucks: a study based on real-world data. Int J Environ Res Public Health 18(8):3877. https://doi.org/10.3390/ijerph18083877

Road Traffic-Borne $PM_{2.5}$ and Noise Indices in Urban Roadside Microenvironments

Argha Kamal Guha and Sharad Gokhale

Abstract Plethora of research has been labeled urban road traffic air and noise pollution are serious environmental burdens to the urban population and traffic corridor users. This research discusses the significant findings in terms of both ambient fine particulate matter ($PM_{2.5}$) and noise indices to explore the combined exposure near an urban traffic corridor in developing Guwahati city. The year-long (Jan, 2019–Dec, 2019) traffic-borne $PM_{2.5}$ and noise sampling were conducted near different microenvironments along the road traffic corridor on both weekdays and weekends. The results indicate that in winter, roadside $PM_{2.5}$ concentrations on weekdays and weekends were 131 ± 39 μg/m^3 and 110 ± 2 μg/m^3, respectively. Inside a roadside building on weekdays and weekends, the observed $PM_{2.5}$ concentrations were 39 ± 3 μg/m^3 and 43 ± 14 μg/m^3, respectively. The obtained equivalent noise (L_{eq}) was 62–80 dB (A), and the noise pollution level (L_{NP}) was ~ 80 dB (A), which exceeded the WHO safety level of 70 dB (A) in all roadside and indoor environments. The traffic noise index (TNI) obtained for the outdoor roadside and inside the indoor building were > 70 dB (A) and > 50 dB (A), respectively, causing noise annoyance to the receptors. The $PM_{2.5}$-noise index for the roadside and inside the building was observed to be 0.19 ± 0.05 and 0.25 ± 0.06, respectively. This elucidates the degradation of environmental quality in terms of air and noise pollution which may cause deleterious psychological and physiological health issues. These results could be used for developing road traffic air and noise abatement strategies and urban policy.

Keywords $PM_{2.5}$ · Equivalent noise · Noise pollution level · Traffic noise index · City noise-air index · Vehicular pollution · Microenvironment · Urban population

---

A. K. Guha (✉) · S. Gokhale
Department of Civil Engineering, Indian Institute of Technology Guwahati, Guwahati, India
e-mail: arghakguha@iitg.ac.in

S. M. Shiva Nagendra et al. (eds.), *Select Proceedings of the 8th Indian International Conference on Air Quality Management (IICAQM 2023)*, Lecture Notes in Civil Engineering 582, https://doi.org/10.1007/978-981-96-2359-4_14

1 Introduction

Urban road traffic-borne air pollution and noise have been a major environmental threats [4, 10, 17]. It is paramount to study these urban pollutants to delineate human health effects (psychological and physiological) [10]. Research related to the burden of disease in six European countries revealed that 80% of the disease is attributed to $PM_{2.5}$, noise, lead, and benzene, among which transportation is one of the key contributors [18]. Among different air pollutants, $PM_{2.5}$ is a respirable fraction of air particulate that can penetrate the human respiratory system and is associated with various respiratory, cardiovascular, and psychological health problems [15]. Due to various urban developments, $PM_{2.5}$ pollution has been abruptly increasing worldwide, and road traffic activities contribute a significant portion of $PM_{2.5}$ in urban air. Similarly, traffic activities emanate a significant portion of environmental noise which is also associated with several auditory and non-auditory ailments (annoyance, stress, fatigue, cardiovascular, neuronal disorders, etc.) [3, 15, 18, 20]. About 40% of the E.U. population lives in the area where the day–evening–night equivalent noise (L_{den}) exceeds the WHO permissible limit of 55 dB (A) [18]. In Hong Kong, Fung and Lee [7] observed that the 30-min average traffic noise (L_{10}) was observed at 55–75 dB (A), and 10-min average PM_{10} was 0.24–0.68 ppm. This study concluded that in the study area, traffic noise and PM_{10} linearly correlated with the logarithm of their corresponding distance from the road ($\log R$). The result highlighted that '$\log R$' could be used as a common parameter for estimating the combined effect of road traffic on a residential unit's noise and air pollution. In London, the daytime noise level (L_d) was 55.6 dB (A), which was linked with the risk of hospital admission [8]. In Indian cities, equivalent noise level (L_{eq}) often exceeded the 70 dB (A) permissible limit of WHO [2, 5, 16]. Prasad et al. [16] observed that traffic noise contributes to annoyance, sleep disturbance, and other physiological problems, including blood pressure for motorcycle riders. Traditionally, equivalent noise (L_{eq}), day–evening–night noise level (L_{DEN}), day–evening time equivalent noise (L_{DE}), day–night noise level (L_{DN}), and daytime and nighttime equivalent noise (L_D and L_N) have been used to describe noise pollution of a particular area [14, 15]. Minimal study used statistical noise (L_{10}, L_{50}, L_{90}), noise pollution level (L_{NP}) denotes the overall noise level, including the noise fluctuation throughout the study period, and traffic noise index (TNI), indicating annoying noise [5, 11]. Lim et al. [13] observed the $PM_{2.5}$ concentration of $47.6 \pm 16.5\ \mu g/m^3$ inside an office building located near the roadside, and at the roadside area, the same was $37.7 \pm 17.2\ \mu g/m^3$. This study concluded that indoor $PM_{2.5}$ mainly depended on the outdoor particle dispersion into the building and, to some extent, on the indoor sources, environmental tobacco smoke (ETS). Outdoor $PM_{2.5}$ contributes 54–63% of indoor $PM_{2.5}$ concentration when windows are closed and > 92% when windows are open [9]. The indoor sources contribute approximately 4% when windows are open and 37–46% when windows are closed [9]. The outdoor $PM_{2.5}$ concentration and $PM_{2.5}$ deposition rate are the two significant factors influencing indoor $PM_{2.5}$ concentration [9]. Silva and Mendes [19] reported that the city noise-air index indicates a combined air-noise quality of an area varied between 0 and 0.6 in kerb side environments of primary and secondary roads (depends on traffic volume) of the city which represented very poor to acceptable air quality in terms of both air and noise pollution status. Else commonly, the city area had a noise-air index between 0.6 and 1, which denoted good to very good air quality in terms of air and noise pollution. In Kolkata, Kundu Chowdhury et al. [12] determined that the noise-air index was ranging 0.17–0.5 in consideration of $PM_{2.5}$, NO_2, and noise, indicated very poor to relatively fair noise-air quality near the urban roadside area. However, this research evidence is very poor till date in Global and Indian context.

Therefore, the documented literature related to $PM_{2.5}$ and noise highlighted that a wide array of research is needed to explore these pollutants in urban road environments. To contribute to this undue need, the present research has aimed to investigate $PM_{2.5}$ and noise pollution in various road environments by formulating different pollution indices. This research has been conducted in Guwahati city of north-east India, the busiest developing urban city, which accommodates different urban activities, including transportation, industries, and office, business, education, and health units.

2 Methodology

2.1 Study Area

Guwahati city (26.1445°N, 91.7362°E) is the heart of the north-east (N.E.) India, situated in Brahmaputra River Valley (BRV), by the hills on the three sides and river Brahmaputra on one side, has humid climatic conditions and hilly fogs most of the year. This study was conducted inside and outside of an office building with brick at Bhangagarh location (26.1709° N, 91.7673° E) along the busiest road stretch of Guwahati city, i.e., Guwahati-Shilong road (G.S. road) where the more than 10,000 vehicle operation takes place per hour throughout the day was observed. Many people share this road stretch for the different needs of their daily life, including study, office, shopping, business and living, etc. In this location, several heights of different types of commercial, residential, and institutional buildings are situated in both directions of the roadway irregularly. Due to the heterogeneity of this urban area and heavy traffic operations, air and noise pollution act as detrimental stressors for hampering human beings surrounding that urban environment. So, it is assumed that choosing this location may satisfy these study objectives (Fig. 1).

2.2 $PM_{2.5}$ and Noise Sampling

$PM_{2.5}$ and noise data were collected on weekdays and weekends (10 A.M.–8 P.M.) from January-December, 2019. Bruel and Kjaer Class-1 two sound level meters

Fig. 1 Study area

(SLM) (Model: SVANTEK 977) were used for indoor and outdoor noise measurement of office and street side. On the street side, SLM is placed at a 1.2 m distance from the kerb side and 5 m from the front façade of the office building and inside the indoor, and SLM is kept at a 1.2 m distance from the indoor walls and furniture. SLM Microphone was placed 1.5 m above the ground in both study areas. Two fine particulate air samplers monitored $PM_{2.5}$ data in indoor (Model: Omni FT, Sr No. 2434) and outdoor (Model: Envirotech APM 550 EL) with the inlet height 1.7 m above the ground for indoor sampling and approximately 16.7 m above the ground (at rooftop) for outdoor sampling. Airflow was drawn by the samplers with the flow rates of 5 l/min and 16.667 l/min, respectively, through a particle size separator, then through a 24-h desiccated and pre-weighted PTFE filter medium with 47 mm diameter and 2 μm pore size. The impaction principle separates 10 μm and 2.5 μm particles. Then again, the exposed and re-desiccated filter paper was re-weighted in a microbalance (SARTORIOUS, model no. CD-225D). Then, the weight difference was divided by the average initial and final airflow volume drawn by the pump associated with the ambient air sampler to calculate the concentration of $PM_{2.5}$ in $\mu g/m^3$.

3 Data Analysis

3.1 $PM_{2.5}$ Data Analysis

For indoor and outdoor, 10-h and different hourly average concentration of $PM_{2.5}$ was obtained by simply dividing the weight difference between final and initial filter paper by the total drawn air flow volume.

$$C = \frac{W_2 - W_1}{\frac{V_1 + V_2}{2}} \tag{1}$$

where C is the concentration of $PM_{2.5}$, W_1 is the initial weight of the desiccated filter paper before sampling, and W_2 is the final weight of the desiccated filter paper after sampling.

3.2 Noise Data Analysis

After extracting raw noise data from SLM, the logarithmic average of sound pressure level on an hourly basis was carried out with the following equation:

$$L_{\text{eq}} = 10 \cdot \log \sum_{i=1}^{n} f_i \left(10^{L_i/10} \right) \tag{2}$$

where f_i is the time fraction, and L_i is the noise level at each time fraction.

To quantify noise annoyance-related parameters in ambient and indoor environment, traffic noise index (TNI) and noise pollution level (L_{NP}) were evaluated by the following equations:

$$\text{TNI} = 4 \cdot (L_{10} - L_{90}) + L_{90} - 30 \tag{3}$$

$$L_{\text{NP}} = L_{\text{eq}} + L_{10} - L_{90} \tag{4}$$

3.3 $PM_{2.5}$-Noise Index

The $PM_{2.5}$-noise index has been calculated by city air-noise index developed by Silva and Mendes [19] which indicates the pollution status of the study area.

$$\text{City air-noise index} = 0.5 \cdot \text{city air} + 0.5 \cdot \text{city noise} \tag{5}$$

where city air index is the weighted linear combination of normalized concentration of air pollutant which is subjected to the product of dummy variables [19],

$$\text{City air} = \sum_i w_i c_i \cdot \prod_i v_i \tag{6}$$

where w_i is the weightage of each air pollutant, and c_i is the normalized concentration, expressed as

$$c_i = \cos^2 \alpha \tag{7}$$

$$\alpha = \frac{(x - x_a)}{(x_b - x_a)} \cdot {}^{\pi}/_{2} \tag{8}$$

where v_i, is the dummy variable, if $c_i \leq L_i$, then $v_i = 1$, and if $c_i > L_i$, then, $v_i = 1$.

Here, L_i is the prescribed concentration limit of air pollutant by the regulatory body. In this case, x_a and x_b are the control points.

Now, city noise is the normalized concentration of noise level, which can be expressed as

$$\text{city noise} = w \cdot c \cdot v \tag{9}$$

where w is the weightage of the noise level in that study area, c is the normalized noise level, and v is the dummy variable. In this case, v equals zero when $c \leq L$, and v equals one if $c > L$. Here, L is the prescribed noise level value by the regulatory body.

The numerical value of the city noise-air index varies between 0 and 1. For the very poor or unacceptable quality of environment, this index becomes 0, and in the case of acceptable or healthy environmental conditions, this index becomes 1 [19].

4 Results and Discussion

4.1 $PM_{2.5}$ Pollution

4.1.1 Roadside $PM_{2.5}$ Pollution

Figure 2 reveals the fact that in the winter (Jan–Feb) season, the $PM_{2.5}$ level increases and violates the NAAQS prescribed standard for the daily average concentration of 60 $\mu g/m^3$ due to random vehicular activities throughout the day and biomass burning in the early morning (e.g., garbage incineration and night time firework for heat, etc.).

For pre-monsoon season (March–May) at the weekend, due to the increasing number of vehicular activities near field Big Bazar shopping complex and other

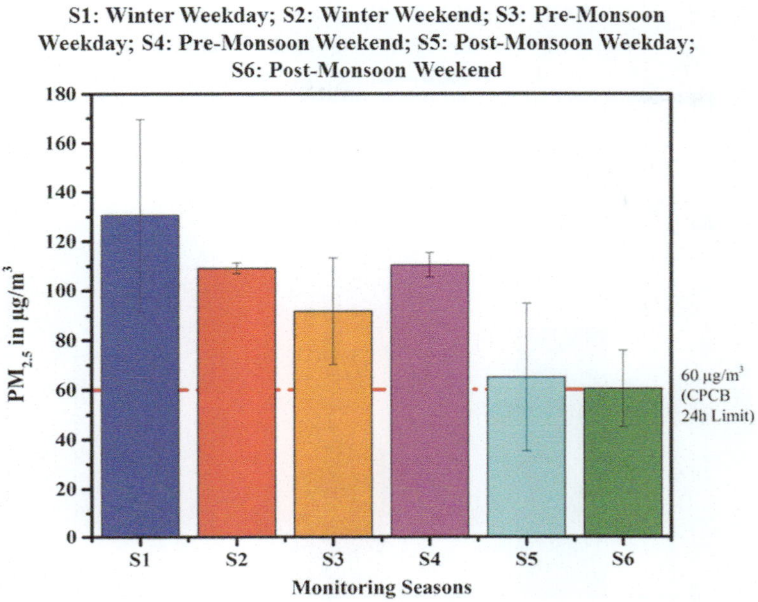

Fig. 2 Seasonal variation of 10-h average concentration of $PM_{2.5}$ distribution in roadside

shopping areas, exhaust emission from some street-side food stalls may be the precursors of this higher $PM_{2.5}$ concentration. There is some similar observation drawn by Tiwari et al. [21] in their study of $PM_{2.5}$ pollution in Guwahati city.

4.1.2 $PM_{2.5}$ Pollution of Office Indoor

Figure 3 shows that indoor $PM_{2.5}$ was higher on a weekday with the window partly open. It can be concluded that outdoor vehicular activities contribute significantly to indoor $PM_{2.5}$ within the study durations.

But, on the weekend, the indoor $PM_{2.5}$ level was elevated in window-closed condition due to room cleaning activities. Datta et al. [6] found a similar trend of $PM_{2.5}$ in the non-residential building in India.

4.2 Noise Pollution

Figures 4 and 5 show that on weekdays L_{eq} was 77.35 ± 0.92 dB (A) on the street side and 68.12 ± 1.10 dB (A) at the office. On the weekend, L_{eq} was 76.90 ± 2.22 dB (A) on the street side and 64.08 ± 2.84 dB (A) at the office.

The street-side L_{eq} exceeded the Central Pollution Control Board (CPCB) and WHO prescribed limits of 65 dB (A) and 70 dB (A), while for office, it exceeded

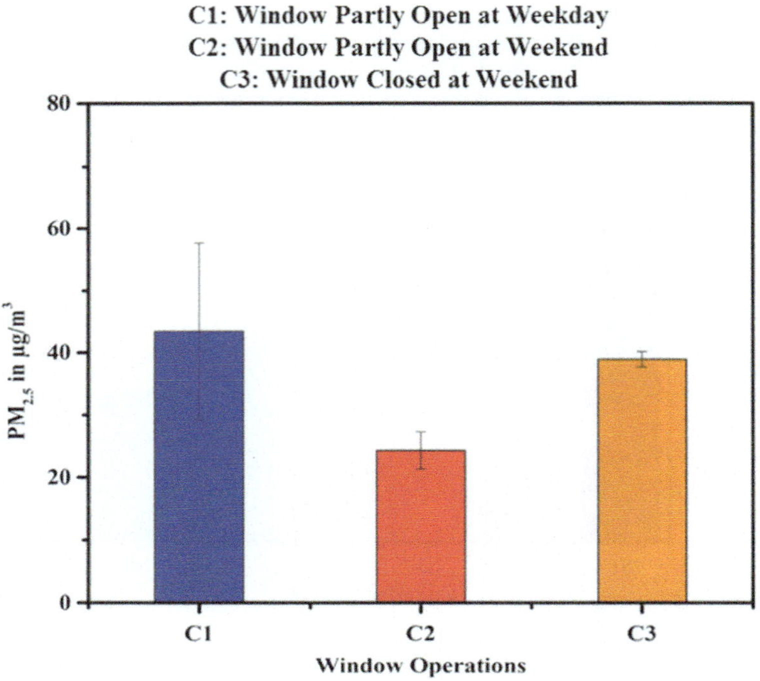

Fig. 3 Variation of $PM_{2.5}$ concentration in weekday and weekend with different window operations in the pre-monsoon season (March to May)

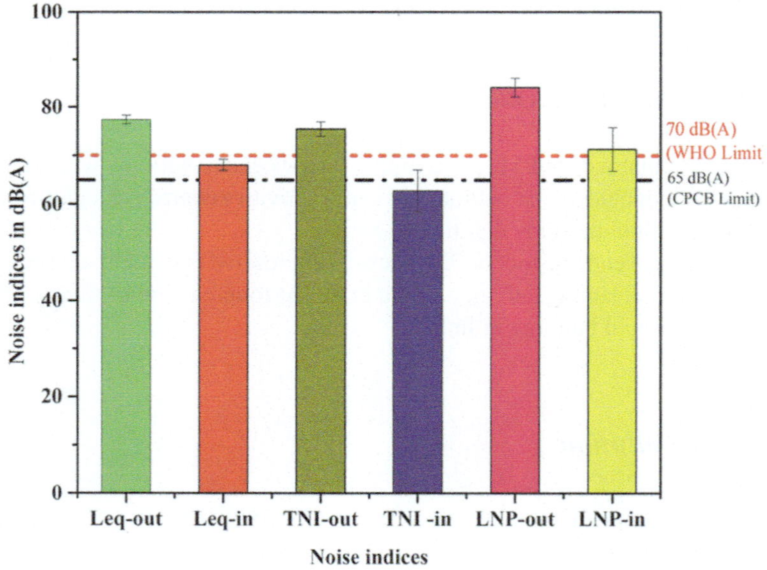

Fig. 4 Different noise indices for street roadside and office indoor of weekdays

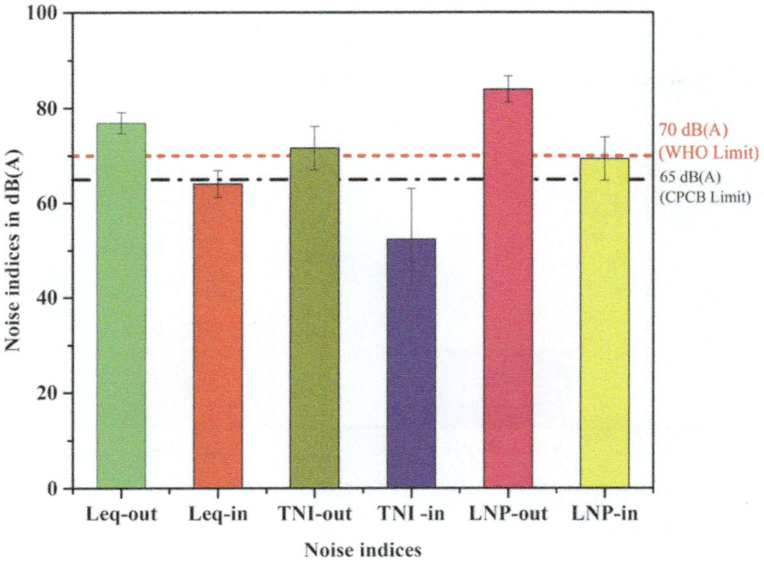

Fig. 5 Different noise indices for roadside and office indoor of weekend

the CPCB allowable noise limit. On weekday, L_{NP} and TNI for street side and office indoor were 84.15 ± 1.98 dB (A), 71.33 ± 4.44 dB (A), 75.41 ± 1.46 dB (A), and 62.77 ± 4.31 dB (A), respectively. Similarly, on the weekend, the same was 71.63 ± 4.57 dB (A), 52.36 ± 10.69 dB (A), 84 ± 2.70 dB (A), and 69.39 ± 4.56 dB (A), respectively. This finding indicates severe deterioration of acoustic quality in both the study areas.

Traffic activities, honking events, loudspeaker operations, and other anthropogenic activities are the key sources of noise pollution in that area. Similar findings were reported in Indian contexts by Balasubramanian and Jagannath [1] and Kundu Chowdhury et al. [12] for the Indian street-side area.

4.3 $PM_{2.5}$-Noise Index

Figure 6 shows that $PM_{2.5}$-noise index of office and street side was 0.19 ± 0.05 and 0.25 ± 0.06, respectively. It indicates the deterioration of environmental quality in respect of air-noise quality. Silva and Mendes [19] and Kundu Chowdhury et al. [12] reported similar findings for the roadside area. This index attributes the fact that environmental air and noise quality has been tremendously deteriorated in that study area.

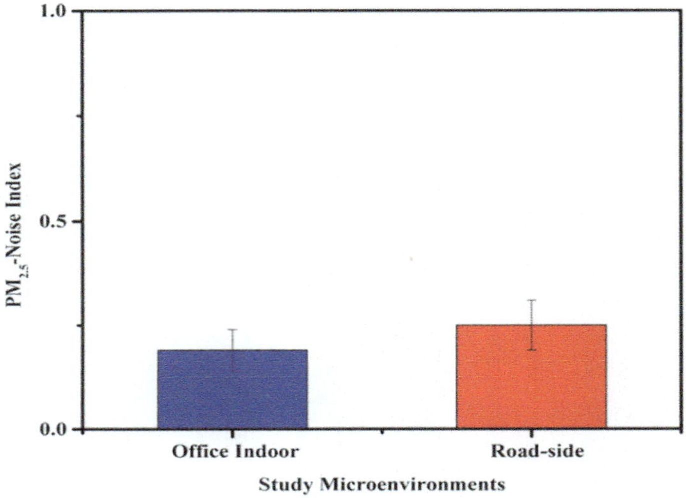

Fig. 6 $PM_{2.5}$-noise index

5 Conclusion

The growing urbanization-induced traffic-borne $PM_{2.5}$ and noise study was done on urban street side and inside an office building throughout the weekdays and weekends to assess the environmental air and noise quality in terms of $PM_{2.5}$ and noise. The results indicated that street-side and indoor office $PM_{2.5}$ levels exceeded the NAAQS limit for 24-h, but noise pollution exceeded the WHO and CPCB prescribed limit in the street-side area and indoor. Besides traffic, biomass burning, ETS, cooking, and other anthropogenic activities contribute to the $PM_{2.5}$ pollution in that area. While traffic operations, loudspeakers, and honking contribute to noise problems in those areas. TNI and L_{NP} were very high, both outdoor and indoor, indicating the annoying acoustic quality in the study area. $PM_{2.5}$-noise index based on $PM_{2.5}$ and noise were close to 0 for both the study areas, although for indoors, it was relatively higher. However, this index indicates severe degradation of atmospheric environmental pollution based on $PM_{2.5}$ and noise. This pollution could be detrimental to human health (e.g., psychological problems like annoyance, stress, mental fatigue, and physiological ailments like cardiovascular, respiratory, cerebral problems, etc.) that used to share that traffic corridor daily. There is a sincere need for continuous monitoring of other air pollutants, including $PM_{2.5}$, with noise and meteorological parameters for developing integrated environmental quality and its' mitigation strategies. The observations of this research could help in urban planning and policy-making for developing and to be developed cities. Implanting tree canopies along the road, vertical gardening on building façade, constructing noise barrier, and developing clean air zone (CAZ) could improve the scenario.

Acknowledgements The authors acknowledge the Air and Noise Pollution Research Lab IIT Guwahati for technical and instrumentation support, and Employee Provident Fund Office Guwahati, for providing space for the study.

Funding This work was funded by Ministry of Earth Science, Government of India.

References

1. Balasubramanian V, Jagannath M (2014) Detecting motorcycle rider local physical fatigue and discomfort using surface electromyography and seat interface pressure. Transp Res F Traffic Psychol Behav 22:150–158. https://doi.org/10.1016/j.trf.2013.12.010
2. Banerjee D, Das PP, Fouzdar A (2014) Urban residential road traffic noise and hypertension: a cross-sectional study of adult population. J Urban Health 91(6):1144–1157. https://doi.org/10.1007/s11524-014-9916-1
3. Banks J, Fink D (2022) Noise as a public health hazard. J Bone Joint Surg 75(5):6. https://doi.org/10.1097/01.HJ.0000831140.87139.eb
4. Bloemsma LD et al (2019) Green space, air pollution, traffic noise and cardiometabolic health in adolescents: the PIAMA birth cohort. Environ Int 131:104991. https://doi.org/10.1016/J.ENVINT.2019.104991
5. Compendex S et al (2016) Evaluation and forecasting of highway traffic noise in the City of Nagpur , India
6. Datta A et al (2017) Indoor air quality of non-residential urban buildings in Delhi, India. Int J Sustain Built Environ 6(2):412–420. https://doi.org/10.1016/j.ijsbe.2017.07.005
7. Fung YW, Lee WL (2011) Identifying a common parameter for assessing the impact of traffic-induced noise and air pollutions on residential premises in Hong Kong. Habitat Int 35(2):231–237. https://doi.org/10.1016/j.habitatint.2010.09.007
8. Halonen JI, Hansell AL, Gulliver J, Morley D, Blangiardo M, Fecht D, Toledano MB, Beevers SD, Anderson HR, Kelly FJ, Tonne C (2015) Road traffic noise is associated with increased cardiovascular morbidity and mortality and all-cause mortality in London. Eur Heart J 36(39):2653–2661. https://doi.org/10.1093/eurheartj/ehv216
9. Ji W, Zhao B (2015) Contribution of outdoor-originating particles, indoor-emitted particles and indoor secondary organic aerosol (SOA) to residential indoor $PM_{2.5}$ concentration: a model-based estimation. Build Environ 90:196–205. https://doi.org/10.1016/J.BUILDENV.2015.04.006
10. Khan J et al (2018) Science of the total environment road traffic air and noise pollution exposure assessment—a review of tools and techniques. Sci Total Environ 634:661–676. https://doi.org/10.1016/j.scitotenv.2018.03.374
11. King EA, Murphy E, Mcnabola A (2009) Reducing pedestrian exposure to environmental pollutants: a combined noise exposure and air quality analysis approach. Transp Res Part D 14(5):309–316. https://doi.org/10.1016/j.trd.2009.03.005
12. Kundu Chowdhury A, Debsarkar A, Chakrabarty S (2016) Assessment of combined noise-air quality and combined exposure to noise-air pollution at the curbside open-air microenvironment of Kolkata city, India. Pollution 2(1):35–47. https://doi.org/10.7508/pj.2016.01.004
13. Lim JM et al (2011) The analysis of $PM_{2.5}$ and associated elements and their indoor/outdoor pollution status in an urban area. Indoor Air 21(2):145–155. https://doi.org/10.1111/j.1600-0668.2010.00691.x
14. Moshammer H et al (2019) Acute effects of air pollution and noise from road traffic in a panel of young healthy adults. Int J Environ Res Public Health 16(5):788. https://doi.org/10.3390/ijerph16050788

15. Pitchika A et al (2017) Long-term associations of modeled and self-reported measures of exposure to air pollution and noise at residence on prevalent hypertension and blood pressure. Sci Total Environ 593–594:337–346. https://doi.org/10.1016/J.SCITOTENV.2017.03.156
16. Prasad C et al (2022) Association between traffic noise-induced psychophysiological, and socio-demographic factors of motorcycle riders. Appl Acoust 196:108898. https://doi.org/10.1016/j.apacoust.2022.108898
17. Ross Z et al (2011) Noise, air pollutants and traffic: continuous measurement and correlation at a high-traffic location in New York City. Environ Res 111(8):1054–1063. https://doi.org/10.1016/J.ENVRES.2011.09.004
18. Shamsipour M et al (2022) Burden of diseases attributed to traffic noise in the metropolis of Tehran in 2017. Environ Pollut 301:119042. https://doi.org/10.1016/J.ENVPOL.2022.119042
19. Silva LT, Mendes JFG (2012) City noise-air: an environmental quality index for cities. Sustain Cities Soc 4:1–11. https://doi.org/10.1016/j.scs.2012.03.001
20. Thiesse L et al (2020) Transportation noise impairs cardiovascular function without altering sleep: the importance of autonomic arousals. Environ Res 182:109086. https://doi.org/10.1016/j.envres.2019.109086
21. Tiwari S et al (2016) Assessment of $PM_{2.5}$ and PM_{10} over Guwahati in Brahmaputra River Valley: temporal evolution, source apportionment and meteorological dependence

Air Quality Health Impacts and Policy Interventions

Seasonal and Temporal Distribution of Urban Fine Particulates and Associated Health Risk Analysis: A Long-Term Measurement (2018–2020) Analysis in Megacity Delhi

Vignesh Mohan, Rajeev Kumar Mishra, and Vijay Kumar Soni

Abstract Understanding the dynamics of urban air pollution demands a multifaceted approach due to its complex nature. The present research studied various aspects of air pollution from 2018 to 2020 in the megacity of Delhi and estimated particulate matter-based negative impacts on human health and its contribution toward mortality. This study aims to identify the seasonal fluctuations of PM_{10}, $PM_{2.5}$, and PM_1 and gaseous pollutants for a long term in the megacity of Delhi and to evaluate the health impact assessment using AirQ+ software. Seasonal and temporal profiles of various pollutants are explored, and estimation of long and short-term effects such as stroke, lung cancer, cardiovascular diseases, and other major diseases is analyzed using AirQ+ software for Delhi city using population and annual mean concentration of pollutants. The maximum and minimum annual mean concentration for $PM_{2.5}$ and PM_{10} was recorded in 2018 (145.75 and 228.82) and 2020 (74.896 and 119.97), respectively. The seasonal pattern in 2018 and 2019 was winter > Post-Monsoon > Summer > Monsoon. In 2020 the seasonal pattern was Post-Monsoon > Winter > Summer > Monsoon due to the effect of restriction and seasonal factors. The annual mean concentration health impact followed All-Cause > IHD > Stroke > LC > COPD > ALRI for long-term exposure and for short-term exposure, and it was HA-RD > All causes > HA-CVD. The year 2018 records higher mortality, and the long-term exposure of all causes reached maximum mortality of 1,44,449 individuals. Air quality improvement helps improve human health in polluted cities and can be achieved by implementing strict policies to help combat air quality and fight against climate change.

Electronic Supplementary Material The online version of this chapter (https://doi.org/10.1007/978-981-96-2359-4_15) contains supplementary material, which is available to authorized users.

V. Mohan · R. K. Mishra (✉)
Department of Environmental Engineering, Delhi Technological University, New Delhi, India
e-mail: rajeevkumarmishra@dtu.ac.in

V. K. Soni
India Meteorological Department, New Delhi, India

S. M. Shiva Nagendra et al. (eds.), *Select Proceedings of the 8th Indian International Conference on Air Quality Management (IICAQM 2023)*, Lecture Notes in Civil Engineering 582, https://doi.org/10.1007/978-981-96-2359-4_15

Keywords AirQ+ model · Health impacts · Health risk analysis · $PM_{2.5}$ · PM_{10} · PM_1

1 Introduction

Air quality is a significant concern for human health [3]. The addition of pollutants in ambient air due to various activities deteriorates air quality, leading to severe health hazards. Urbanization and industrial development play a major role in air pollution, and the concentration of pollutants exceeds the prescribed limits in developing countries [28]. Therefore, measuring and analyzing the various pollutants globally is necessary, especially in urban regions. Measurements are done through different government bodies and regulatory agencies per their standards and requirements. Numerous continuous ambient air quality monitoring stations are established for measuring pollutants along with meteorological parameters [17]. Concerns about potential sensitivity to air pollution have increased in recent years, and as a result, these challenges hastened the need for technological improvement in many countries [3]. As per 2011 census, 31% population lives in urban areas. In India, 410 million people were living in urban areas in 2014, it is estimated that about 814 million people will be living in urban areas by 2050 [5]. An urban agglomeration is occurring rapidly in several Indian cities, which caused a few cities to rank on the top list of cities with higher particulate matter concentration, per the World Health Organization (WHO) [16].

Delhi, the national capital of India, has been listed as one of the most polluted cities in the world. The pollutant concentrations exceeded the permissible limits prescribed by the WHO and the Indian National Ambient Air Quality Standards (NAAQS) in the National Capital Territory of Delhi (NCT) [1]. The complex mixture of pollutants such as particulate matter (PM), nitrogen dioxide (NO_2), sulfur dioxide (SO_2), ozone (O_3), black carbon (BC), carbon monoxide (CO), heavy metals (HM), or black smoke (BS) makes it tedious to mitigate and control [21]. Particulate matter (PM) comprises suspended solid and liquid particles, whereas aerosols include a collection of suspended particles and gaseous matter.

Air pollution yields numerous cases of premature morbidity and mortality globally. A cocktail of emissions from ambient air pollution and indoor air pollution seems to be the blame for the majority of premature mortality compared to other environmental pollution [27]. Southeast Asia and Western Pacific regions contribute to around 7 million deaths caused by rising PM concentration worldwide, as per WHO. These seven million deaths, ischemic heart disease (IHD) contributes to 36%, stroke 33%, chronic obstructive pulmonary disease (COPD) 17%, acute lower respiratory infection (ALRI) 8%, and 6% of lung cancer (LC) [12, 16]. According to several studies, a 2 $\times$ increase in particle exposure can result in a 30 to 50% rise in lung cancer [10]. In recent decades, several studies provided the importance of studying human exposure to air pollution and premature risk [10, 19]. World Health Organization (WHO) stated that approximately 24% of world mortality or 13.7 million deaths annually is linked to environmental causes [30].

The effects of changes in air quality and its associated health and economic impacts can be assessed using various tools. Still, air quality health impact assessment (AirQ+) is one of the best tools for obtaining good and effective results. AirQ+ is a more detailed tool for evaluating air pollution health impact than other computer-based tools. It was developed by WHO to quantify the health impact assessment due to air pollution, and it is a very user-friendly and open-access tool. It can be used in assessing the morbidity and mortality effects of a variety of pollutants, including PM_{10}, $PM_{2.5}$, black carbon (BC), NO_2, and O_3 [20].

This research study utilizes the long-term measurement of PM_{10}, $PM_{2.5,}$ and PM_1 from 2018 to 2020 in Delhi. The objectives of the study are as follows: (1) To analyze the particulate matter load in Delhi city for PM_{10}, $PM_{2.5,}$ and PM_1 and gaseous pollutants. (2) Seasonal variation of particulate matter. (3) Health impact assessment using AirQ + software. The importance of the study is that, it provides information regarding the potential health hazard of using PM_1 concentration, which is rarely reported for the study area for long-term measurements. The study's outcomes will be helpful in decision-making, policy formulation, and initiating control measures for reducing air pollution.

2 Methodology

2.1 Study Area

The National Capital Territory of Delhi, India, has comprised of approximately more than 17 million population and is located (28°35′N; 77°12′E) at 300 m above mean sea level (Fig. 1). Thar Desert encompasses Delhi in the west, plains in the south and east, and the mighty Indo-Gangetic plains in the north. Delhi has semi-arid climate conditions with extreme summer (max 45–48 °C), and minimum temperature of around 1–4 °C receives medium rainfall of around 600–800 nm. The major seasons are winter (December–March), pre-monsoon or summer (April–June), monsoon (July–September), and post-monsoon (October–November) [29]. Delhi is classified as an urban region with urban agglomeration occurring at a higher rate. The monitoring location is located in the center of the city and represents the proper urban region surrounded by various road networks. The region experiences frequent pollution events and extreme climatic conditions. The major sources of pollutants in the study area include transportation, road dust, biomass burning, and construction activities (Fig. 1).

Fig. 1 Geographical depiction of study location (marked with a symbol, red marker) situated at India Meteorological Department, Lodhi Road, New Delhi, India

2.2 Data Collection and Analysis

The particulate matter (PM_{10}, $PM_{2.5}$, and PM_1) was measured from January 1, 2018, to December 31, 2020, using an suspended particulate monitor (MP101M) with CPM option on daily bases in India Meteorological Department, Lodhi Road, Delhi. The gaseous parameter such as NO, NO_2, NO_x, CO, and O_3 have also been analyzed. The demographic information, such as total population and other parameters required for AirQ+ software, is taken from the Indian Census data and various reports from the World Health Organization (WHO) for health impact assessment of air pollution. The data interpretation was done using RStudio version 2022.07.2 and MATLAB version 2016a. The outliers in the data are removed and used for analysis (Fig. 2).

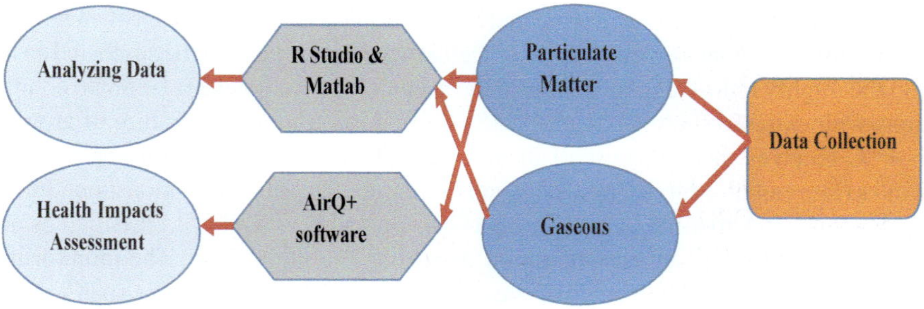

Fig. 2 Performance of steps in this study

2.3 Instrumentation

The suspended particulate monitor (MP101M) coupled with CPM, is the instrument used for the measurement. It works based on the principle of beta attenuation. In the beta attenuation technique, the samples measured through the fiber tape are exposed to a radioactive source and then based on the quantity of radiation absorbed by the sample used to estimate concentration. Continuous particle measurement (CPM) works on the principle of light scattering. The two technologies make the real-time monitoring of particles with simultaneous direct detection of PM_{10}, $PM_{2.5}$, and PM_1. The flow rate of the instrument is 1 m^3/h. The measurement ranges are up to 10,000 $\mu g\ m^{-3}$ (customizable) with a detection limit of 0.5 $\mu g\ m^{-3}$. The measurement cycles vary from 0.25 h to 24 h, and measurement time ranges from 10 to 300 s. The operating temperature of the instrument varies from 10 to 40 °C with automatic calibration. The sampling inlets are based on the US EPA recommendations.

2.4 Baseline Mortality and Population Data

The population data of Delhi, including data for different age groups people (all ages, $\leq$ 5 years of age, $\geq$ 25, and $\geq$ 30) were obtained from the literature [1], and Indian census data for the years 2018 to 2020. The model values such as relative risk (RR) and baseline incident (BI) were taken from WHO data files and other similar studies. The model values used are provided in supplementary material (STable 1) [11, 25].

2.5 Health Risk Assessment of Air Pollution

In the present study, AirQ + software tool was used to quantify the health effects of PM_{10} and $PM_{2.5}$ on the risk population using the annual mean concentration of particulate matter (PM_{10} and $PM_{2.5}$), total risk population of the city, baseline incidence (per 10^5 people), a cutoff value for consideration 10 $\mu g\ m^{-3}$ as recommended by the WHO, and relative risk (RR). It is used to quantify the long-term mortality caused due to COPD, IHD, LC, and stroke, respectively. The attributable proportion (AP) is the portion of the health effect attributed to exposure to air pollution in at-risk populations. The model calculations are based on the equations used in previous studies done by Refs. [1, 11, 13, 14, 23–25].

3 Results and Discussion

3.1 Seasonal Contribution of Particulate Matter and Gaseous Pollutants in NCT Delhi

The annual average concentration of PM_{10}, $PM_{2.5}$, and PM_1 was highest in 2018, followed by 2019 and 2020. The concentration in all the years exceeded the national standards (National Ambient Air Quality Standards $PM_{2.5}$ 40 $\mu g\ m^{-3}$ and PM_{10} 60 $\mu g\ m^{-3}$ as well as WHO standards, i.e., 5 and 15 $\mu g\ m^{-3}$). The concentration of PM_{10} was 228.82, 176.35, and 119.97 $\mu g\ m^{-3}$. The concentration of PM_{10} in 2018 was higher due to maximum emissions in the city, and the subsequent reduction in 2019 of around 22% was achieved through the initiation of various mitigation measures such as the Commission for Air Quality Management in NCR and Adjoining Areas (CAQM). In the year 2020, a 47% decline in concentration was observed on comparing 2018 due to the effect of restrictions imposed for various anthropogenic emissions due to the COVID-19 pandemic [22]. The concentration in 2020 was 31% lesser than the 2019 concentration. Similarly, for $PM_{2.5}$ in 2019, the reduction was 31% and 48% in 2020 compared to 2018, and in 2020, a 25% reduction was observed compared to 2019. Compared to the year 2018, PM_1 concentration was seen a reduction of about 32% in 2019 and 38% in the year 2020 in the study area. The annual average concentration of the particles is shown in Table 1. The inhabitants of Delhi have seen a growth of about 0.8–1.1% of the total population in the years 2019 and 2020.

The study period is classified into different seasons, such as winter (December–March), pre-monsoon or summer (April–June), monsoon (July–September), and post-monsoon (October– November) for all three consecutive years. The season-wise distribution of different pollution (PM_{10}, $PM_{2.5}$, and PM_1) and their contribution is analyzed. The study shows that the seasonal mean concentration of PM_{10} was higher during the winter season compared to all the other seasons in the year 2018 and 2019 with an average concentration of 258.48 ± 88.80 $\mu g\ m^{-3}$ in 2018, 228.98 ± 115.61 $\mu g\ m^{-3}$ in 2019 and 192.26 ± 58.49 $\mu g\ m^{-3}$ in 2020. The order of concentration in 2018 and 2019 was similar, i.e., Winter > Post monsoon > Summer and Monsoon. In 2020, the order of concentration was Post monsoon > Winter > Summer > and monsoon (Fig. 3). The change in the concentration pattern was due

Table 1 Total inhabitation of Delhi along with the annual concentration for $PM_{2.5}$ and PM_{10}

Year	Inhabitation	PM_{10}	$PM_{2.5}$	PM_1
2018	1,94,83,678	228.82	145.74	52.22
2019	1,97,06,881	176.35	100.08	35.09
2020	1,98,68,163	119.97	74.896	32.37

Inhabitation data was collected from the literature review for 2018, and the remaining years 2019 and 2020 were collected from the census for India. The unit of PM_{10}, $PM_{2.5}$, and PM_1 is $\mu g\ m^{-3}$

to the restrictions imposed in March 2020, which falls under the winter season. The monsoon season receives the lowest concentration in all three consecutive years. The precipitation helps in the removal of pollutants through a deposition. The concentration in monsoon seasons was seen in the range of 183.16 ± 135.8 μg m^{-3}, 114.29 ± 92.32 μg m^{-3}, and 45.82 ± 44.12 μg m^{-3} in 2018, 84.64 ± 52.30 μg m^{-3}, 41.98 ± 24.76 μg m^{-3}, and 14.28 ± 12.84 μg m^{-3} in 2019 and 66.88 ± 36.56 μg m^{-3}, 141.84 ± 72.84 μg m^{-3}, and 42.11 ± 36.75 μg m^{-3} in 2020 for PM_{10}, $PM_{2.5}$, and PM_1 (Fig. 4). The summer season records the second lowest concentration throughout the study period. Seasonal factors such as temperature, precipitation, and humidity influence the concentration of the pollutants. The major sources of the region include vehicular emissions, biomass burning, and construction sites. The concentration of $PM_{2.5}$ in different seasons of the year 2018 to 2020 and their seasonal contribution to the annual average concentration is shown in Fig. 3. Similarly, for PM_{10} and PM_1, the same is provided in the supplementary material (SFigure 1 and 2). The seasonal variation of mean and standard deviation from 2018 to 2020 for particulates PM_{10}, $PM_{2.5}$, and PM_1 and gaseous pollutants NO, NO_2, NO_x, CO, and O_3 are provided in STable 3.

Apart from the particulate matter, the gaseous pollutant concentration is also influenced by different emissions and seasonal factors. The gaseous pollutants play a significant role in forming secondary pollutants, which acts as precursor component. Ground-level ozone concentration shows maximum concentration during summer every year compared to other seasons due to the role of solar intensity in the formation of ozone from the VOCs. The concentration of O_3 was 35.70 ± 10.09 μg m^{-3}, 37.26 ± 9.11 μg m^{-3}, and 40.52 ± 10.55 μg m^{-3} in the consecutive years 2018, 2019, and 2020 (STable 3 and Fig. 5e). The CO concentration was seen to be high during the

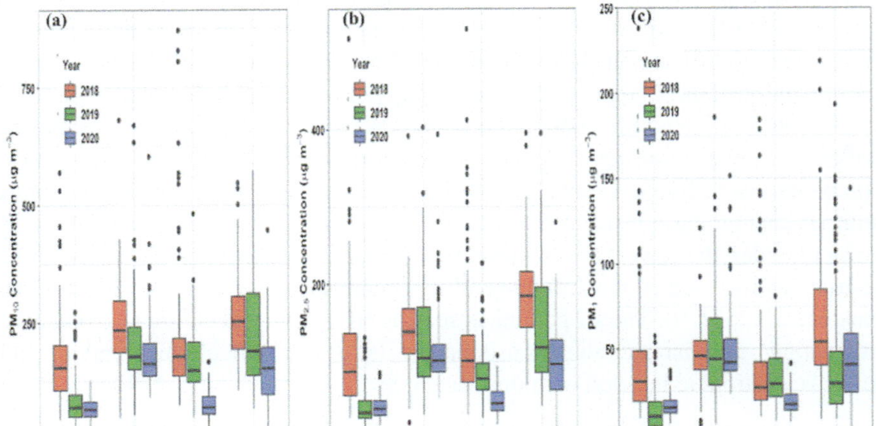

Fig. 3 Box plot representing the seasonal concentration (monsoon, post-monsoon, summer, winter) of particulate matter PM_{10}, $PM_{2.5}$, and PM_1 in 2018, 2019, and 2020

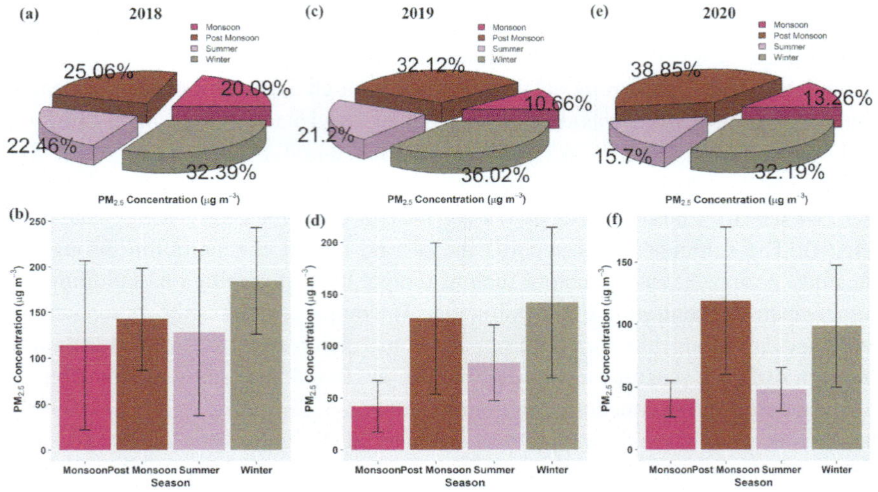

Fig. 4 Pie chart representing the seasonal percentage of particulate matter $PM_{2.5}$ in 2018, 2019, and 2020. Bar plot depicting the seasonal concentration contribution of particulate matter $PM_{2.5}$ in 2018, 2019, and 2020

summer, followed by winter (Fig. 5d). The higher concentration in summer is driven by the two-wheeler emissions. In urban regions, two-wheelers contribute more to CO emissions [26]. In winter, the concentration of CO is driven by biomass and crop residue burning in the neighboring states of Delhi (Fig. 5d). The monsoon season records the lowest concentration of CO in all three years. The recorded annual average concentration of rainfall in the study period was about 546.9, 678.5, and 687.3 mm from 2018 to 2020. The precipitation leads to be wipeout of pollutants from the atmosphere [17], so this season records the minimum concentration of all pollutants. During the summer season, the concentration of NO, NO_2, and NO_x is seen less due to the utilization of these compounds in secondary pollutants formation such as ground-level ozone. These compounds act as a precursors for the formation of secondary pollutant. Similarly, during winter and post-monsoon season, the NO, NO_2, and NO_x compounds are observed more due to the presence of these primary compounds without undergoing transformation (STable 3 and Fig. 5a–c). On average, the year 2020 recorded the lowest concentration of gaseous pollutants due to the effect of restrictions imposed on the different anthropogenic sources due to the spread of the pandemic [2, 6, 7, 9]. The air pollutants such as NO_X, CO, and O_3 are also potential of creating health hazards such as asthma, dizziness, and difficulty in breathing and lead to catastrophic injuries, including death.

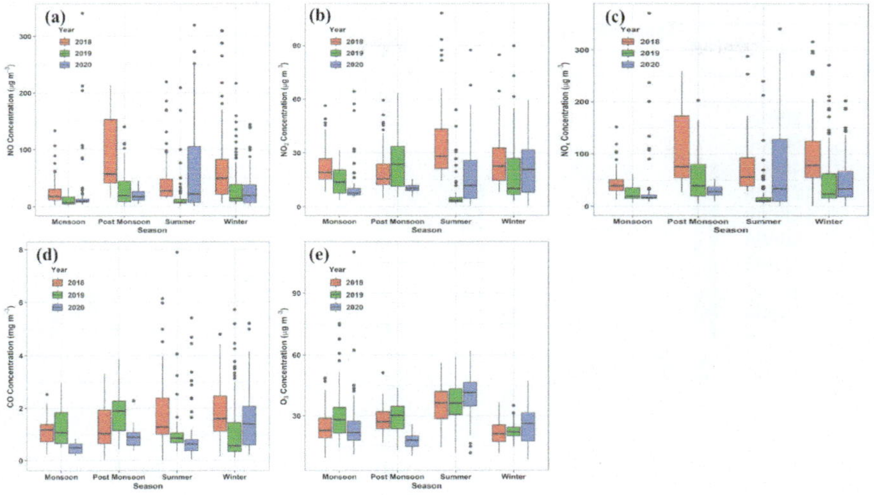

Fig. 5 Box plot representing the seasonal concentration (monsoon, post-monsoon, summer, and winter) of NO, NO_2, NO_x, CO, and O_3 in 2018, 2019, and 2020

3.2 *Health Impacts*

3.2.1 Long and Short-Term Effect of $PM_{2.5}$

The AirQ+ uses the annual average concentration to analyze long- and short-term effects. The excess number of cases, i.e., the increase in patients due to the long-term impact of $PM_{2.5}$ for all causes, ALRI, COPD, LC, IHD, and Stroke, along with lower and upper confidence intervals for 2018, 2019, and 2020, as well as the short-term effect of $PM_{2.5}$ for all causes, HA-RD, and HA-CVD with lower and upper intervals for the years 2018, 2019, and 2020 are depicted in Figs. 6 and 7.

All Causes—The long-term effects of air pollution due to all the causes are mentioned as all causes. Due to higher concentrations in 2018, the long-term effect peaked at a value of 144,449 cases with a 95% confidence level (Fig. 6 and STable 2). The lower and upper confidence limits are 113,702 and 162,891. Similarly, in 2019 and 2020, the values were 126,239, and 97,398 with a 95% confidence level; the lower and upper limit corresponds to 95,668–146,642 and 70,512–117,520. The short-term exposure to $PM_{2.5}$ also followed the same pattern: the concentration was more in 2018 and less in 2019 and 2020. But when compared to all causes, it is the highest recorded in long-term effects, but in the short term, the concentration of HA-RD, i.e., hospital admission due to respiratory diseases, is more than all causes (Fig. 7 and STable 2).

ALRI—Acute lower respiratory infection is the least impact seen due to the long-term effects of air pollution, with a concentration of about 5211, 4993, and 4528 mortalities recorded in 2018, 2019, and 2020.

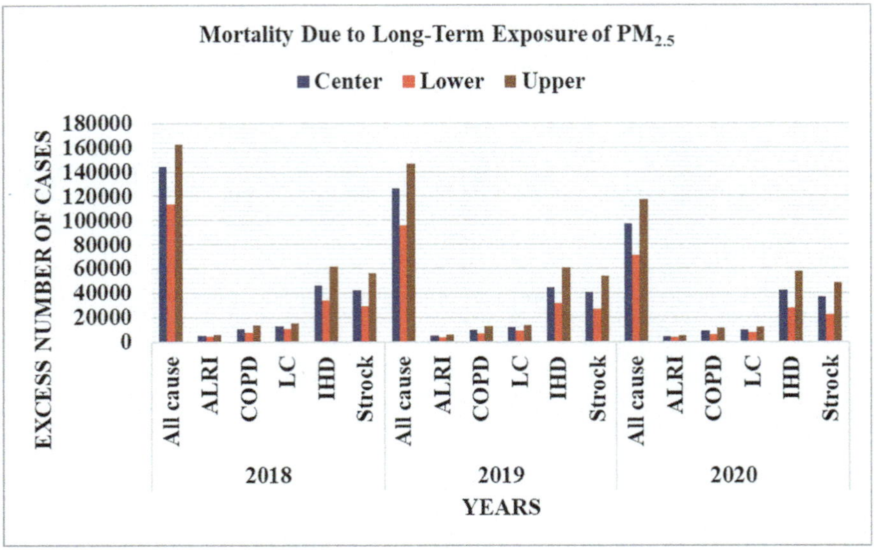

Fig. 6 Excess number of cases due to the long-term effect of $PM_{2.5}$ of total mortality, ALRI, COPD morbidity (hospital admissions, LC, IHD, and Stroke in 2018, 2019, and 2020 in Delhi). *Note*: *ALRI* Acute lower respiratory infection, *COPD* Chronic obstructive pulmonary disease, *LC* Lung cancer, and *IHD* Ischemic heart disease

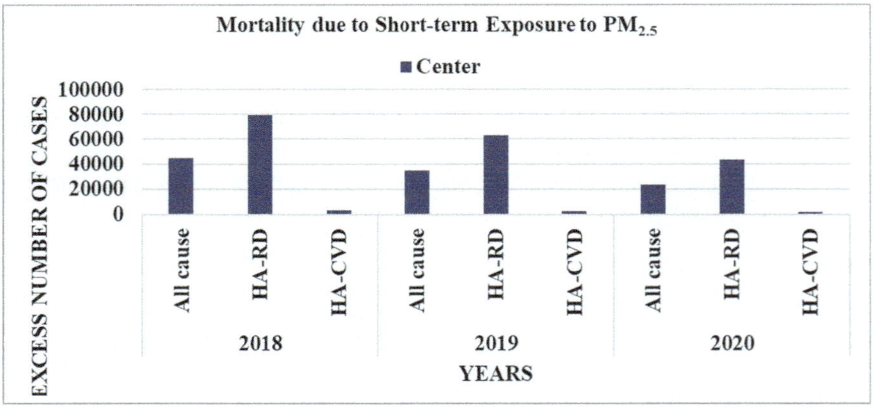

Fig. 7 Excess number of cases due to the long-term effect of $PM_{2.5}$ on total mortality, HA-RD, and HA-CVD in 2018, 2019, and 2020 in Delhi. *Note*: *HA-RD* Hospital admission for Respiratory Diseases, *HA-CVD* Hospitalization for cardiovascular disease

COPD—Chronic obstructive pulmonary disease is the second recorded lowest concentration due to long-term exposure to $PM_{2.5}$. The mortality was high in 2018 and followed by 2019 and 2020, respectively, with cases accounting for mortality of

10,346, 9678, and 8566 individuals. The contribution of AALRI and COPD is less for all causes of health impacts due to long-term air pollution exposure.

LC—Lung cancer mortality caused by long-term exposure to $PM_{2.5}$ above 30 years of age shows that the cause of the highest values was observed in 2018, followed by 2019 and 2020. The total mortality above 30 years of age was 13,007, 11,834, and 9,995 in 2018, 2019, and 2020, respectively.

IHD—Ischemic heart disease is the second-highest health impact due to long-term exposure to $PM_{2.5}$, and the concentration trend was similar to other effects. IHD is a commonly known health impact due to exposure to air pollution. In 2018, the mortality was 46,055. In 2019, it was 44,584, and 41,838 in 2020 (Fig. 6 and STable 2).

Stroke—is the most common health impact seen in people living in highly polluted regions and contributes to maximum mortality cases for long-term exposure to air quality. Annually stroke contributes 36,000 to 42,000 mortality in Delhi from 2018 to 2020.

HA-RD and HA-CVD—Hospital admission for respiratory diseases and hospitalization for cardiovascular disease are the two significant symptoms due to short-term exposure to air pollution. The pattern of cases was similar to the long-term exposure, but compared to all causes of mortality, hospital admission due to respiratory diseases are more for the short-term exposure. The mortality range was from 79,785 to 63,301. Hospital admission due to cardiovascular diseases ranges from 3000 to 1500 from 2018 to 2020 (Fig. 7 and STable 2).

3.3 *Health Implications*

Globally, every year, the mortality due to air pollution is increasing gradually, and around 2,27,000 cases of mortality were recorded due to exposure to outdoor air pollution. About 30.2% of deaths are caused by IHD, COPD, stroke, lung cancer, and mortality, and all are caused due to a 10 $\mu g\ m^{-3}$ increase in $PM_{2.5}$ concentration [12]. In 2015, 4.2 million mortality were recorded due to $PM_{2.5}$ exposure in southern and eastern parts of Asia, and it is around 59% contribution to global mortality due to air pollution. Around 59% contribution of 4.2 million global mortality was recorded due to ambient $PM_{2.5}$ in 2015 from South and East Asia. South Asia alone contributes to around 1.36 million mortalities [8]. All growing urban cities are located in south Asia and the study's health impact estimation helps estimate the increase in mortality rates in growing cities like Delhi. This particular study estimates mortality of around 1.45×10^5 and 4.3×10^4 in 2018 for the different types of mortality causes, including both long and short-term effects. Similarly, in 2020 the mortality rate decreased to – 32% and – 47%, respectively (STable 2). The decrease in annual mean concentration from 2018 to 2020 was the primary reason. This shows the role of anthropogenic sources and the importance of implementing stringent policies toward pollution combination. The study found the order of mortality due to different cases for all three years, i.e., All-Cause > IHD > Stroke > LC > COPD > ALRI, which is similar to other global studies which estimate the impacts of long-term exposure to air pollution [1, 12]. Similar other studies in the Indian subcontinent also emphasized that Delhi records higher mortality compared to all other cities in India, which is proved again through this study [18, 25].

Besides all-natural mortality, the study exhibits the second-highest mortality observed from IHD annually. It contributes to around one-quarter of death compared to all other means. In the year 2014–2019 in India, the IHD contributes to around 17.12% of total mortality, and it can rise to 25% in total mortality when there is a rise in 10 $\mu g\ m^{-3}$ concentration of pollution [12]. Most premature mortality deaths were also recorded in the Indian states due to poor air quality [15]. Recent studies revealed that air pollution also increases stroke cases [4, 16]. Only very few studies estimated the ALRI mortality and highlighted the mortality growth through ALRI, especially in children below 5 years [1].

Lung cancer is another leading mortality cause in polluted cities, which produces cytokines and oxidative stress in the lungs. Around 17.12% of lung cancer (LC) mortality was caused by long-term exposure to $PM_{2.5}$ [12]. Compared to other Indian megacities, NCT Delhi observed higher lung cancer (LG) mortality cases.

The short-term effect of $PM_{2.5}$ shows that hospital admission due to respiratory diseases is high compared to cardiovascular diseases, similar to previous studies [17]. This is due to the more penetration of $PM_{2.5}$ in the alveolar region, which triggers alveolar inflammation and precipitates acute respiratory and cardiovascular illness.

4 Conclusions and Recommendations

The current study narrates the seasonal variations of PM_{10}, $PM_{2.5}$, PM_1, NO, NO_2, NO_x, O_3, and CO in Delhi city from 2018 to 2020, along with its health impact analysis using AirQ+ software developed by WHO. The long and short-term health effects and mortality rates are calculated using the estimation of the excessive number of cases recorded. It helps to evaluate the impact of air pollution on human health in polluted cities. The main findings of this research study are as follows:

1. The highest annual mean concentration for $PM_{2.5}$ and PM_{10} was recorded in 2018 (145.75 $\mu g\ m^{-3}$ and 228.82 $\mu g\ m^{-3}$, respectively), and the lowest was found in 2020 (74.89 $\mu g\ m^{-3}$ and 119.97 $\mu g\ m^{-3}$, respectively) due to a reduction of emission sources. This shows that better air quality can also be obtained in polluted cities by properly implementing policies.
2. The seasonal order of concentration for pollutants in 2018 and 2019 was Winter > Post-Monsoon > Summer > Monsoon. In 2020, due to monsoon changes, the order was Post-Monsoon > Winter > Summer > Monsoon due to the effect of the lockdown in NCT Delhi and changes in the monsoon pattern.

3. Ozone concentration shows higher concentration during summer every year, and the concentration of ground-level ozone not only depends on the direct emissions but also due to secondary formation. To reduce ozone concentration, the reduction of precursor molecules is also necessary.
4. The health impact analysis for long-term exposure, the order of mortality was All-Cause > IHD > Stroke > LC > COPD > ALRI, and for short-term exposure, it was HA-RD > All causes > HA-CVD. The mortality rate is also directly related to the mean concentration of the pollutants.
5. A significant reduction in air pollution would result in substantial health benefits that can be achieved by properly implementing policies and strategies.

Further, the study scope can be expanded to all the other highly motorized cities to analyze the potential health impacts and help achieve the standard of living, contributing to sustainable development goals and helping in fighting climate change.

Acknowledgements The authors are thankful to the India Meteorological Department (IMD) for providing the data facilities and to members of the Advanced Air and Acoustics Research Laboratory, Department of Environmental Engineering, Delhi Technological University (DTU), for their support to conduct this study.

References

1. Afghan FR, Patidar SK (2020) Health impacts assessment due to $PM_{2.5}$, PM_{10} and NO_2 exposure in National Capital Territory (NCT) Delhi. Pollution 6(1):115–126. https://doi.org/10.22059/POLL.2019.286454.662
2. Agarwal A, Kaushik A, Kumar S, Mishra RK (2020) Comparative study on air quality status in Indian and Chinese cities before and during the COVID-19 lockdown period. Air Qual Atmos Health 13(10):1167–1178. https://doi.org/10.1007/s11869-020-00881-z
3. Aliyu YA, Botai JO (2018) An exposure appraisal of outdoor air pollution on the respiratory well-being of a developing city population. J Epidemiol Glob Health 8(1–2):91–100. https://doi.org/10.2991/j.jegh.2018.04.002
4. Ansari M, Ehrampoush MH (2019) Meteorological correlates and AirQ+ health risk assessment of ambient fine particulate matter in Tehran, Iran. Environ Res 170:141–150. https://doi.org/10.1016/j.envres.2018.11.046
5. Bhagat RB (2014) World migration report : urban migration trends, challenges and opportunities in India. International Organization for Migration, pp 1–35
6. Chelani AB, Gautam S (2022) The influence of meteorological variables and lockdowns on COVID-19 cases in urban agglomerations of Indian cities. Stoch Env Res Risk Assess. https://doi.org/10.1007/s00477-021-02160-4
7. Chelani A, Gautam S (2021) Lockdown during COVID-19 pandemic: A case study from Indian cities shows insignificant effects on persistent property of urban air quality. Geosci Front 13(6):101284. https://doi.org/10.1016/j.gsf.2021.101284
8. Cohen AJ, Brauer M, Burnett R, Anderson HR, Frostad J, Estep K, Balakrishnan K, Brunekreef B, Dandona L, Dandona R, Feigin V, Freedman G, Hubbell B, Jobling A, Kan H, Knibbs L, Liu Y, Martin R, Morawska L et al (2017) Estimates and 25-year trends of the global burden of disease attributable to ambient air pollution: an analysis of data from the Global Burden of Diseases Study 2015. Lancet 389(10082):1907–1918. https://doi.org/10.1016/S0140-6736(17)30505-6

9. Gautam S (2020) COVID-19: air pollution remains low as people stay at home. Air Qual Atmos Health 13(7):853–857. https://doi.org/10.1007/s11869-020-00842-6
10. Goel A, Saxena P, Sonwani S, Rathi S, Srivastava A, Bharti AK, Jain S, Singh S, Shukla A, Srivastava A (2021) Health benefits due to reduction in respirable particulates during covid-19 lockdown in India. Aerosol Air Qual Res 21(5):1–16. https://doi.org/10.4209/aaqr.200460
11. Gurjar BR, Jain A, Sharma A, Agarwal A, Gupta P, Nagpure AS, Lelieveld J (2010) Human health risks in megacities due to air pollution. Atmos Environ 44(36):4606–4613. https://doi.org/10.1016/j.atmosenv.2010.08.011
12. Hajizadeh Y, Jafari N, Mohammadi A, Momtaz SM, Fanaei F, Abdolahnejad A (2020) Concentrations and mortality due to short- and long-term exposure to $PM_{2.5}$ in a megacity of Iran (2014–2019). Environ Sci Pollut Res 27(30):38004–38014. https://doi.org/10.1007/s11356-020-09695-z
13. Hussein T, Sogacheva L, Petäjä T (2018) Accumulation and coarse modes particle concentrations during dew formation and precipitation. Aerosol Air Qual Res 18(12):2929–2938. https://doi.org/10.4209/aaqr.2017.10.0362
14. Kumar A, Mishra RK (2018) Human health risk assessment of major air pollutants at transport corridors of Delhi, India. J Transp Health 10(June):132–143. https://doi.org/10.1016/j.jth.2018.05.013
15. Maheshwarkar P, Sunder Raman R (2021) Population exposure across central India to $PM_{2.5}$ derived using remotely sensed products in a three-stage statistical model. Sci Rep 11(1):1–13. https://doi.org/10.1038/s41598-020-79229-7
16. Manojkumar N, Srimuruganandam B (2021) Health benefits of achieving fine particulate matter standards in India—a nationwide assessment. Sci Total Environ. https://doi.org/10.1016/j.scitotenv.2020.142999
17. Manojkumar N, Srimuruganandam B (2021) Health effects of particulate matter in major Indian cities. Int J Environ Health Res 31(3):258–270. https://doi.org/10.1080/09603123.2019.1651257
18. Manojkumar N, Srimuruganandam B (2022) Spatio-temporal health benefits attributable to $PM_{2.5}$ reduction in an Indian city. Int J Environ Health Res 33(6):552–562. https://doi.org/10.1080/09603123.2022.2040959
19. Mehmood T, Tianle Z, Ahmad I, Li X (2019) Integration of AirQ+ and particulate matter mass concentration to calculate health and ecological constraints in Islamabad, Pakistan. In: Proceedings of 2019 16th International Bhurban Conference on Applied Sciences and Technology, IBCAST 2019, pp 248–254. https://doi.org/10.1109/IBCAST.2019.8667203
20. Mirzaei A, Tahriri H, Khorsandi B (2021) Comparison between AirQ+ and BenMAP-CE in estimating the health benefits of $PM_{2.5}$ reduction. Air Qual Atmos Health 14(6):807–815. https://doi.org/10.1007/s11869-021-00980-5
21. Mishra RK (2017) Air pollution health risk based on AirQ+ software tool. Int J Appl Res Technol 2(3):190–199. https://doi.org/10.24163/ijart/2017/2(3):190-199
22. Mohan V, Kumar R (2022) A picture of Delhi's regional air quality during diminished anthropogenic activities in the COVID-19 era Indian Council of Medical Research. Arab J Geosci 2:1–14. https://doi.org/10.1007/s12517-022-10567-8
23. Nabizadeh R, Yousefi M, Azimi F (2018) Study of particle number size distributions at Azadi terminal in Tehran, comparing high-traffic and no traffic area. MethodsX 5(November):1549–1555. https://doi.org/10.1016/j.mex.2018.11.013
24. Nabizadeh R, Yousefian F, Moghadam VK, Hadei M (2019) Characteristics of cohort studies of long-term exposure to $PM_{2.5}$: a systematic review. Environ Sci Pollut Res 26(30):30755–30771. https://doi.org/10.1007/s11356-019-06382-6
25. Nagpure AS, Gurjar BR, Martel JC (2014) Human health risks in national capital territory of Delhi due to air pollution. Atmos Pollut Res 5(3):371–380. https://doi.org/10.5094/APR.2014.043
26. Nagpure AS, Sharma K, Gurjar BR (2013) Traffic induced emission estimates and trends (2000–2005) in megacity Delhi. Urban Climate 4:61–73. https://doi.org/10.1016/j.uclim.2013.04.005

27. Ravindra K, Singh T, Mor S (2022) Preventable mortality attributable to exposure to air pollution at the rural district of Punjab, India. Environ Sci Pollut Res 29(21):32271–32278. https://doi.org/10.1007/s11356-022-19668-z
28. Sasmita S, Kumar DB, Priyadharshini B (2022) Assessment of sources and health impacts of PM_{10} in an urban environment over eastern coastal plain of India. Environ Chall 7(January):100457. https://doi.org/10.1016/j.envc.2022.100457
29. Tiwari S, Srivastava AK, Bisht DS, Parmita P, Srivastava MK, Attri SD (2013) Diurnal and seasonal variations of black carbon and $PM_{2.5}$ over New Delhi, India: influence of meteorology. Atmos Res 125–126:50–62. https://doi.org/10.1016/j.atmosres.2013.01.011
30. WHO (2022) World Health Statistics. World Health, pp 1–177

Elemental Composition, Sources and Health Impact of PM_{10} Over Nainital, India

Sakshi Gupta, Priyanka Srivastava, Preeti Tiwari, Manish Naja, Nikki Choudhary, Manoj Kumar, and Sudhir Kumar Sharma

Abstract Coarse mode particulate matter (PM_{10}) samples were collected at Nainital (29.35°N, 79.45°E, 1958 m amsl) from January to December, 2021 to study their elemental composition, sources and health impact. The annual average concentration of PM_{10} was estimated as $64 \pm 6\ \mu g\ m^{-3}$ during the sampling period at Nainital. The concentrations of 36 different elements were analyzed using WD-XRF, with higher levels of Ca ($1.789 \pm 0.351\ \mu g\ m^{-3}$), Al ($1.498 \pm 0.272\ \mu g\ m^{-3}$), Fe ($1.479 \pm 0.207\ \mu g\ m^{-3}$), S ($1.303 \pm 0.142\ \mu g\ m^{-3}$), and K ($1.175 \pm 0.152\ \mu g\ m^{-3}$) followed by other trace elements. PCA identified five primary sources of these elements present in PM_{10}: natural crustal and soil particulates, vehicle-generated emissions, fossil fuel combustion, biomass burning, and industrial emissions (Mo-rich). Furthermore, health risk assessment (HRA) was identified as a non-carcinogenic risk linked to manganese (Mn) exposure in both adults and children, carcinogenic risks associated with chromium (Cr) exposure in adults and non-carcinogenic risks in children were also identified.

Keywords PM_{10} · Elements · Nainital · Principal component analysis · Health risk assessment

S. Gupta (✉) · P. Tiwari · N. Choudhary · M. Kumar · S. K. Sharma
Environmental Science and Biomedical Metrology Division, CSIR-National Physical Laboratory, New Delhi, India
e-mail: sakshigupta21096@gmail.com

Academy of Scientific and Innovative Research (AcSIR), Ghaziabad, India

P. Srivastava · M. Naja
Atmospheric Division, Aryabhatta Research Institute of Observational Sciences (ARIES), Nainital, Uttrakhand, India

S. M. Shiva Nagendra et al. (eds.), *Select Proceedings of the 8th Indian International Conference on Air Quality Management (IICAQM 2023)*, Lecture Notes in Civil Engineering 582, https://doi.org/10.1007/978-981-96-2359-4_16

1 Introduction

The ecology of the Himalayas is facing imminent and substantial threats from a diverse array of pollutants [4]. The Himalayan region is witnessing rapid urbanization and development, fueled by increasing anthropogenic activities. This impacts the local land use pattern and raises concerns about the atmospheric environment. Primary issues involve the increased usage of vehicles driven by tourism, as well as the use of biomass and fuelwood for cooking and heating, especially in the elevated regions of the Himalayas [11, 12]. Consequently, the study of Himalayan aerosols has gained significant importance in recent decades, as it grapples with the intricate relationship between human activities and the fragile mountain environment, with profound implications for regional and global ecosystems [2, 3, 23]. The substantial elevation of the Himalayan range naturally functions as a distinct demarcation for the central Himalayas. Thus, conducting analysis at a high-altitude region like Nainital can offer insights into particulate matter (PM) spanning the entire Indian sub-continent [23]. Given the climatic significance of trace and heavy elements within particulate matter (PM) in Indian Himalayan Region (IHR), current study was undertaken.

This paper presents the extractions of the elemental constituents of PM_{10}, obtained from the Wavelength Dispersive X-ray Fluorescence (WD-XRF) method from January to December, 2021. To gain deeper insights into the sources and cancer-related risks associated with the measured heavy elements found in PM_{10} within the central Himalayan region of India, we utilized Principal Component Analysis (PCA) and Health Risk Assessment (HRA). Such analysis is of paramount importance to assess the pollution levels of various elements in the atmosphere, pinpoint their primary sources, and develop robust control measures to mitigate the adverse impacts of air pollution on both environmental quality and public health, particularly in many developing nations.

2 Methodology

2.1 Aerosol Sampling and Sampling Site

PM_{10} samples were collected at the Aryabhata Research Institute of Observational Science (ARIES) is located in Manora Hills of Nainital, India (29.35°N, 79.45°E, 1958 m amsl) from January to December 2021 (Fig. 1). PM_{10} samples were not collected during monsoon months (July, August, and September) due to rain. The detailed sampling procedure and instrumentation can be found in prior publication [2].

Fig. 1 Sampling site: ARIES, Nainital (*Source* Google Earth)

2.2 Elemental Composition

Total of 40 PM_{10} samples (January–December, 2021) were collected and analyzed for their elemental concentrations. The PM_{10} samples were subjected to quantitative elemental analysis using a Wavelength Dispersive X-ray Fluorescence Spectrometer (WD-XRF), specifically the ZSX Primus model from Rigaku in Japan. Study of the elements within the PM_{10} samples followed standard reference conditions, including a vacuum, a constant temperature of 36 °C, and a tube rating of 2.4 kW. Quantitative results were obtained using the parameter method approach integrated into the software package of the WD-XRF instrument [10]. A total 36 elements (Ca, Al, Fe, S, K, Mg, Pd, B, Mo, Zn, Ag, Nb, Ga, Cl, Ti, Zr, Na, Cr, Cu, Mn, Cs, P, Y, Sb, Ni, Sn, F, Sr, Br, Pm, U, Pb, Th, Br, Rb, and V) were quantified during the analysis.

2.3 Enrichment Factor (EF) and PCA

To discern the likely source of elemental species in PM_{10}, the EF for PM_{10} was calculated, following the method outlined in Taylor and McLennan's [28] work from 1985, where they used Eq. (1) for this purpose.

$$EF = \frac{El_{sample}/X_{sample}}{El_{crust}/X_{crust}}. \quad (1)$$

where El_{sample} and X_{sample} represent the elemental's mass concentration (El) and reference element's mass concentration (X) within the sample. Aluminum (Al) was selected as the reference element due to its stability and abundant presence in the Earth's crust. El_{crust} and X_{crust} refer to the elemental's mass concentration and the concentration of reference element in the upper continental crust, respectively [6, 25].

To identify the potential origins of various elements species quantified by WD-XRF in PM_{10} particles in Nainital, PCA was employed. PCA employs orthogonal decomposition to separate distinct components, which are subsequently correlated with variables through loading factors [27, 31]. In PCA, the elemental results of PM_{10} are modified into a dimensionless standardized format [Eq. (2)].

$$Z_{ij} = \frac{C_{ij} - C_j}{\sigma_j} \tag{2}$$

where i is 1 to n samples, j is 1 to n elements, C_{ij} is the concentration of element j in sample i, C_j is the arithmetic mean concentration, and σ_j is the standard deviation for element j.

The PCA model is expressed in Eq. (3):

$$Z_{ij} = \sum_{k=1}^{p} g_{ik} h_{kj} \tag{3}$$

where k is an index ranging from 1 to p sources, g_{ik} is the factor loading, and h_{kj} is the factor sources [24, 27].

2.4 Health Risk Assessment (HRA)

The calculation of the HRA involved the application of criteria recommended by the United States Environmental Protection Agency (USEPA). Additionally, data from the Integrated Risk Information System (IRIS) database was utilized for assessing the carcinogenic and non-carcinogenic risk associated with inhalation exposure to PM_{10} for elements Al, Cr, Ni, Pb, Mn, Cu, and Zn [6].

To quantify these health risks, exposure concentration (EC) is measured in terms of lifetime average daily dose in mg kg^{-1} day^{-1}, and Hazard Quotient (HQ) and Carcinogenic Risk (CR) were determined using Eqs. (4, 5, and 6).

$$\mathrm{EC} = \frac{C \times \mathrm{IR} \times \mathrm{CF} \times \mathrm{EF} \times \mathrm{ED}}{\mathrm{BW} \times \mathrm{AT}_n} \tag{4}$$

where C is the concentration of the species (μg m^{-3}), IR is the air inhalation rate (for children = 10 m^3 day^{-1} and for adults = 20 m^3 day^{-1}), CF is the correction factor unit (0.001), EF is the relative exposure frequency (days $year^{-1}$), ED is the exposure duration (24 years), BW is the body weight (for children = 15 kg and adults = 70 kg), and AT_n is the average time (AT_n = 70 years × 365 days × 24 h day^{-1}).

For calculating non-carcinogenic risk, the Hazard Quotient (HQ) is quantified,

$$HQ = \frac{EC}{RFD}. \quad (5)$$

The reference exposure dose for the human population (RFD) (mg m^{-3}) was adopted from USEPA 2015. Hazard Quotient (HQ) exceeds 1, which signifies that the average daily dose surpasses the reference value indicating that any potential adverse effects on human health are attributable to inhalation exposure. Conversely, when HQ is less than 1, it signifies no expected negative impact on health.

The CR can be determined by evaluating the additional probability of developing cancer throughout a person's lifetime resulting from the overall exposure to potential carcinogens. This calculation follows the methodology outlined by the USEPA [29].

$$CR = \frac{C \times ET \times EF \times ED \times IUR}{AT}. \quad (6)$$

where ET is the exposure time (12 h day^{-1}), and IUR is the inhalation unit risk [$(\mu g\, m^{-3})^{-1}$] of the elements obtained from USEPA IRIS [7] database. USEPA [30] has set forth a recommended range for excess cancer risk concerning public health. Within this range, an acceptable level is defined as 10^{-6} for a high level of protection, while 10^{-4} is considered a level of risk that may be tolerated [6, 16, 32].

3 Results and Discussion

3.1 PM_{10} Concentration

At Nainital, the annual average mass concentration of PM_{10} was reported to be $64 \pm 6\ \mu g\ m^{-3}$ which was almost similar to the annual average concentration of National Ambient Air Quality Standards (NAAQS) for PM_{10} i.e., 60 $\mu g\ m^{-3}$ (annual) and 4 times higher than the World Health Organization (WHO) limits (15 $\mu g\ m^{-3}$).

Figure 2 shows the time series plot of PM_{10} concentration at the study site. Elevated levels of PM_{10} have been linked to adverse health effects, including impaired lung function and growth in children, as well as respiratory issues and asthma exacerbations. Additionally, adults exposed to high PM concentrations may be at risk for developing heart diseases, diabetes, and neurological problems [21]. Previous research conducted at Nainital indicated that the average concentration of PM_{10} in the air was $65 \pm 41\ \mu g\ m^{-3}$ [2], while at adjacent IGP locations, PM_{10} was very high, i.e., as follows: $135 \pm 64\ \mu g\ m^{-3}$ in Delhi, $99 \pm 33\ \mu g\ m^{-3}$ in Varanasi, and $116 \pm 38\ \mu g\ m^{-3}$ in Kolkata [9].

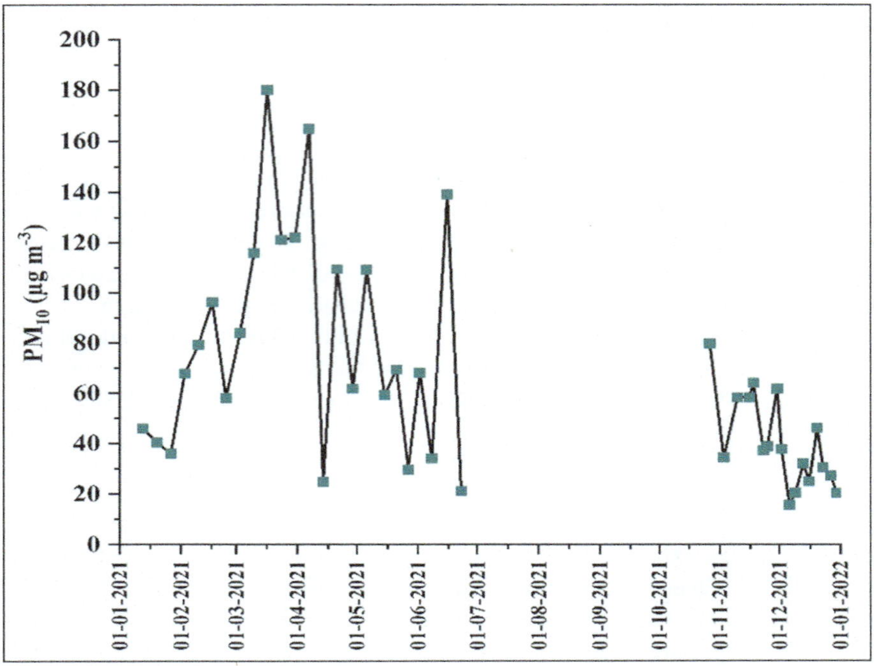

Fig. 2 Time series plot of PM_{10} concentration in Nainital, India

3.2 *Elemental Composition*

The annual average metal concentration is shown in Fig. 3. Among all the identified elements, Ca (1.789 ± 0.351 μg m^{-3}), Al (1.498 ± 0.272 μg m^{-3}), Fe (1.479 ± 0.207 μg m^{-3}), S (1.303 ± 0.142 μg m^{-3}), and K (1.175 ± 0.152 μg m^{-3}) were found to be contributed majorly in PM_{10} concentration.

The elements Mg (0.377 ± 0.081 μg m^{-3}), Pd (0.254 ± 0.091 μg m^{-3}), B (0.252 ± 0.024 μg m^{-3}), Mo (0.238 ± 0.174 μg m^{-3}), Zn (0.231 ± 0.034 μg m^{-3}), Ag (0.166 ± 0.076 μg m^{-3}), Nb (0.165 ± 0.111 μg m^{-3}), Ga (0.130 ± 0.032 μg m^{-3}), Cl (0.127 ± 0.033 μg m^{-3}), Ti (0.110 ± 0.018 μg m^{-3}), Zr (0.107 ± 0.033 μg m^{-3}), Na (0.106 ± 0.031 μg m^{-3}), and Cr (0.104 ± 0.010 μg m^{-3}) were found to be in trace level. The other elements such as Cu, Mn, Cs, P, Y, Sb, Ni, Sn, F, Sr, Br, Pm, U, Pb, Th, Fr, Rb, and V were quantified in the nanogram range (10 ng m^{-3} to 88 ng m^{-3}). The concentration of Ni and Pb is 23 ng m^{-3} and 5 ng m^{-3}, respectively, these values are in the permissible range as guidelines provided by NAAQS (Pb: 0.5 μg m^{-3} and Ni: 20 ng m^{-3}).

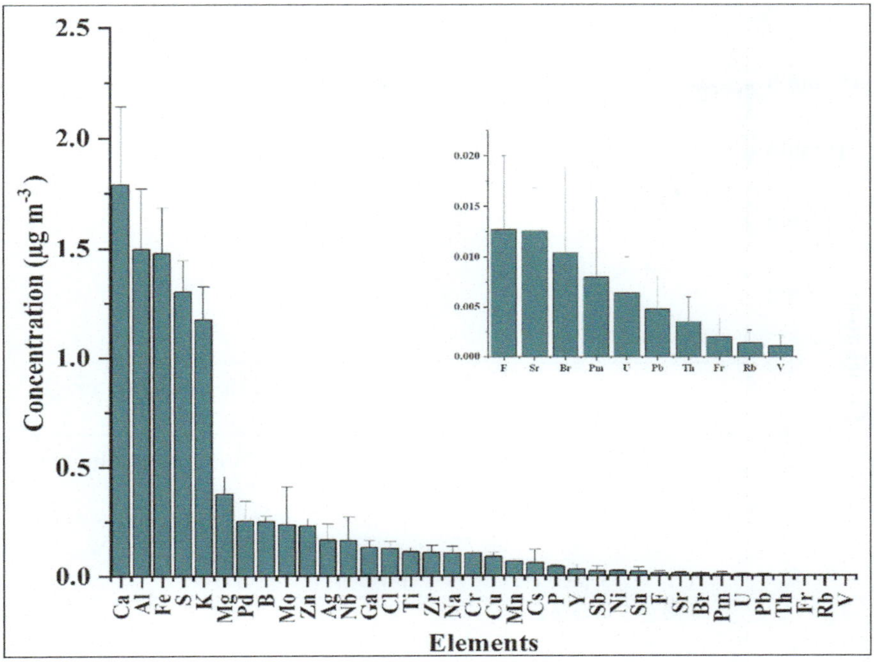

Fig. 3 Annual mean (± SE) concentration of the elemental content of PM_{10} in Nainital, India

3.3 Potential Source Identification

Enrichment Factor (EF) helps us distinguish between natural and anthropogenic sources of elements. When EF is less than 5, it suggests a predominantly natural crustal or soil particulate origin. An EF ranging from 5 to 10 indicates a mixture of natural and anthropogenic sources. EF values exceeding 10 points to a primarily anthropogenic origin, often associated with human activities [26]. The study reveals that metals like Rb, Na, V, Sr, Fe, K, Pb, Mg, Th, Ca, P, and Mn have EF values below 5, indicating that these elements likely originate from natural crustal sources. On the other hand, elements Zr and Ti have EF values between 5 and 10, suggesting that their sources are a combination of both natural processes and human activities, particularly combustion activities, construction or demolition activities, etc. Elements such as U, Y, Ni, Sn, Zn, Cu, Cr, Nb, Ga, Cs, Sb, B, Mo, Ag, and Pd have EF values greater than 10, strongly suggesting that their origin is predominantly anthropogenic, likely stemming from industrial emissions (Fig. 4).

The observed results were analyzed through PCA, which effectively identified the sources. PCA was applied to 22 quantified elements (Al, Mg, Ti, Ca, Na, Fe, K, P, Cl, Ga, Cu, Cr, Mn, Zn, Ni, Zr, Nb, Pd, Sr, B, S, and Mo) of PM10 for source identification and it pinpointed 5 sources. Based on the factor profile, PC1 showed a high loading of marker elements associated with crustal origin and combustion

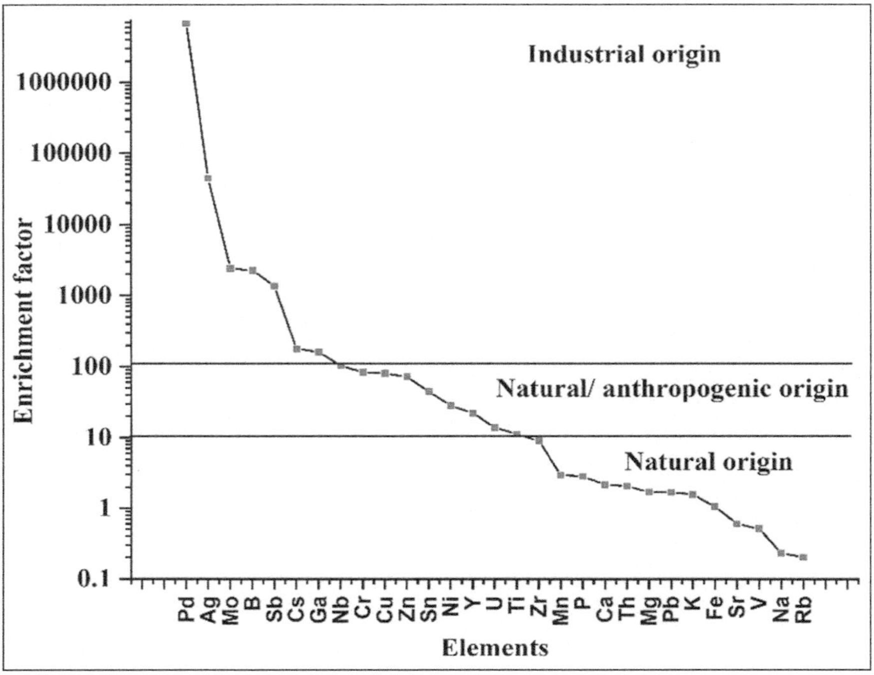

Fig. 4 Enrichment factor (EF) of metals in PM_{10}

activities, such as Al, Mg, Ti, Ca, Na, Fe, S, K, and Cl. The presence of elements Al, Mg, Ti, Ca, Na, and Fe and their observed EFs suggests their crustal origin [8, 26]. Al, Fe, Si, Ca, Ti, and Sr are commonly employed as tracers to investigate and identify the sources of dust-related emissions [10, 15, 18, 19, 25]. The presence of elements like S, K, and Cl indicated the combustion activities as K is a universal marker for biomass burning (BB), Cl is a used as a marker for coal combustion (CC), fireworks and waste incineration, and S is designated as a marker for BB and CC [1, 18, 19, 22]. Therefore, PC1 was identified as a mixed source (crustal/soil/road dust + combustion activities) with 45% contribution. PC2 was identified as vehicular emission showing the significant loadings of Ga, Cu, Mn, and Zn. The elements Cu, Zn, and Mn are used as a tracer of non-exhaust traffic emission, i.e., brake and tire wears, while Ga is used in petroleum refining and petrochemical manufacturing as a catalyst in fluidized-bed catalytic cracking (FCC) processes [17, 20, 26]. PC2 contributed 13% to the total identified elemental mass concentration. PC3 resolved as fossil fuel combustion with higher loadings of Zr, Nb, and Pd. The elements Zr and Nb are used in coal combustion in brick kilns, oil, and petroleum refineries, where coal is combusted [17] with a contribution of 9%. PC4 has a higher loading of B, and S indicates the BB source. As S is a marker for BB (crop residue, wood, dung cake, paper, etc.) [13, 22, 24], PC4 contributed 27%. PC5 was identified as an industrial emission with a contribution of 4% and a considerably high loading of Mo, primarily originating from steel industries [8, 14].

3.4 Carcinogenic and Non-carcinogenic Risk Assessment

The assessment of EC, HQ, and CR was conducted for seven heavy elements, i.e., Al, Pb, Cr, Mn, Cu, Zn, and Ni for both adults and children are given in Table 1. In the case of Cr and Mn, the HQ values for children are 2.33 and 2.99, respectively, which exceeded the acceptable limit of 1, indicating a non-carcinogenic health risk. Similarly, in adults, the HQ values for Mn are 1.28 which is higher than the safe limit, indicating a non-carcinogenic health risk as well. Additionally, Cr might pose a carcinogenic risk in adults, as it values surpass the permissible limit of 10^{-4}. Both Cr and Mn emitted from non-exhaust traffic emissions like tire or brake wears [5, 18, 19].

On the other hand, Al, Pb, Cu, Zn, and Ni have HQ values within the permissible limits suggested by USEPA, indicating no health hazard associated with these elements.

Table 1 Exposure concentration (EC), hazard quotient (HQ) and carcinogenic risk (CR) of PM_{10} heavy elements in Nainital

Elements	Adults			Children		
	EC (mg kg^{-1})	HQ	CR	EC (mg kg^{-1})	HQ	CR
Al	4.10×10^{-4}	2.16×10^{-1}	–	9.57×10^{-4}	5.03×10^{-1}	–
Pb	1.31×10^{-6}	3.72×10^{-4}	9.44×10^{-9}	3.06×10^{-6}	8.69×10^{-4}	2.36×10^{-9}
Cr	2.86×10^{-5}	9.99×10^{-1}	2.06×10^{-4}	6.67×10^{-5}	2.33×10^{0}	5.15×10^{-5}
Mn	1.84×10^{-5}	1.28×10^{0}	–	4.28×10^{-5}	2.99×10^{0}	–
Cu	2.41×10^{-5}	6.00×10^{-4}	–	5.63×10^{-5}	1.4×10^{-3}	–
Zn	6.33×10^{-5}	2.10×10^{-4}	–	1.48×10^{-4}	4.91×10^{-4}	–
Ni	6.27×10^{-6}	1.781×10^{-3}	9.03×10^{-7}	1.46×10^{-5}	4.16×10^{-3}	2.26×10^{-7}

4 Conclusion

The annual average concentration of PM_{10} was 64 ± 6 $\mu g\ m^{-3}$ during the study period in Nainital. The annual average concentration of 36 elements (Ca, Al, Fe, S, K, Mg, Pd, B, Mo, Zn, Ag, Nb, Ga, Cl, Ti, Zr, Na, Cr, Cu, Mn, Cs, P, Y, Sb, Ni, Sn, F, Sr, Br, Pm, U, Pb, Th, Br, Rb, and V) was quantified with higher concentration of elements like Ca, Al, Fe, S, and K followed by trace/ heavy elements. Based on the PCA analysis, 5 sources of the identified elements were determined, i.e., crustal/soil/ road dust, vehicle-generated emissions, combustion of fossil fuel, biomass burning, and industries-related emission. Furthermore, the analysis of HRA quantified the non-carcinogenic risks associated with Mn exposure in both adults and children, as well as the non-carcinogenic risks of Cr in children and the carcinogenic risks of Cr exposure in adults. Hence, this research can provide valuable insights to inform stakeholders and policymakers about the sources of PM_{10} particles based on elemental composition, including the interplay between regional and local sources. Furthermore, it can shed light on the implications of these sources on air quality and human health of the local population, offering critical information for decision-making.

Acknowledgements The authors are thankful to the Director, CSIR-NPL, New Delhi and Head, Environmental Sciences and Biomedical Metrology Division, CSIR-NPL, New Delhi, for their encouragement and support of this study. Department of Science and Technology, New Delhi, is also thankful for financial support.

References

1. Chang Y, Huang K, Xie M, Deng C, Zou Z, Liu S, Zhang Y (2018) First long-term and near real-time measurement of trace elements in China's urban atmosphere: temporal variability, source apportionment and precipitation effect. Atmos Chem Phys 18:11793–11812
2. Choudhary N, Srivastava P, Dutta M, Mukherjee S, Rai A, Kuniyal JC, Lata R, Chatterjee A, Naja M, Vijayan N, Mandal TK, Sharma SK (2022) Seasonal characteristics, sources and pollution pathways of PM_{10} at high altitudes Himalayas of India. Aerosol Air Qual Res 22(7):220092
3. Choudhary N, Rai A, Kuniyal JC, Srivastava P, Lata R, Dutta M, Ghosh A, Dey S, Sarkar S, Gupta S, Chaudhary S, Thakur I, Bawari A, Naja M, Vijayan N, Chatterjee A, Mandal TK, Sharma SK, Kotnala RK (2023) Chemical characterization and source apportionment of PM_{10} using receptor models over the Himalayan Region of India. Atmosphere 14:880
4. Gajananda K, Kuniyal JC, Momin GA, Rao PSP, Safai PD, Tiwari S, Ali K (2005) Trend of atmospheric aerosols over the north northwestern Himalayan region, India. Atmos Environ 39:4817–4825
5. Gianini MFD, Fischer A, Gehrig R, Ulrich A, Wichser A, Piot C, Besombes J-L, Hueglin C (2012) Comparative source apportionment of PM_{10} in Switzerland for 2008/2009 and 1998/ 1999 by positive matrix factorisation. Atmos Environ 54:149–158
6. Gupta S, Sharma SK, Mandal TK (2023) Elemental analysis and health risk assessment of $PM_{2.5}$ at an urban site of Delhi. In: Yadav S, Garg N, Aggarwal SG, Jaiswal SK, Kumar H, Achanta VG (eds) Chapter 16, Recent advances in metrology. Springer Science and Business Media

7. IRIS (Integrated Risk Assessment System) (1995) United States Environmental Protection Agency. www.epa.gov/IRIS/
8. Jain S, Sharma SK, Choudhary N, Masiwal R, Saxena M, Sharma A, Mandal TK, Gupta A, Gupta NC, Sharma C (2017) Chemical characteristics and source apportionment of $PM_{2.5}$ using PCA/ APCS, UNMIX and PMF at an urban site of Delhi, India. Environ Sci Pollut Res 24:14637–14656
9. Jain S, Sharma SK, Srivastava MK, Chatterjee A, Vijayan N, Tripathy SS, Sharma C (2021) Chemical characterization, source apportionment and transport pathways of $PM_{2.5}$ and PM_{10} over Indo Gangetic Plain of India. Urban climate 36:100805
10. Jain S, Sharma SK, Vijayan N, Mandal TK (2020) Seasonal characteristics of aerosols ($PM_{2.5}$ and PM_{10}) and their source apportionment using PMF: a four-year study over Delhi, India. Environ Pollut
11. Kaushal D, Kumar A, Yadav S, Tandon A, Attri AK (2018) Wintertime carbonaceous aerosols over Dhauladhar region of North-Western Himalayas. Environ Sci Poll Res 25:8044–8056
12. Kumar A, Attri AK (2016) Biomass combustion a dominant source of carbonaceous aerosols in the ambient environment of western Himalayas. Aerosol Air Qual Res 16(3):519–529
13. Kumar V, Yadav M, Behera SN (2022) Characterization of $PM_{2.5}$-bound trace elements, source apportionment, and assessment of associated human health risks during summer and winter in Greater Noida, the National Capital Region of India. Front Environ Sci 10:949913. https://doi.org/10.3389/fenvs.2022.949913
14. Mansha M, Ghauri B, Rahman S, Amman A (2012) Characterization and source apportionment of ambient air particulate matter ($PM_{2.5}$) in Karachi. Sci Total Environ 425:176–183
15. Pant P, Harrison RM (2012) Critical review of receptor modeling for particulate matter: a case study of India. Atmos Environ 49:1–12
16. Prakash J, Lohia T, Mandariya AK, Habib G, Gupta T, Gupta SK (2018) Chemical characterization and quantitative assessment of source-specific health risk of trace metals in $PM_{1.0}$ at a road site of Delhi, India. Environ Sci Pollut Res 25:8747–8764
17. Rabha S, Subramanayam KSV, Sawant SS, Saikia BK (2022) Rare-earth elements and heavy metals in atmospheric particulate matter in an urban area. ACS Earth Space Chem 6:1725–1732
18. Rai P, Furger M, El Haddad I, Kumar V, Wang L, Singh A, Dixit K, Bhattu D, Petit JE, Ganguly D, Rastogi N (2020) Real-time measurement and source apportionment of elements in Delhi's atmosphere. Sci Total Environ 742:140332
19. Rai P, Furger M, Slowik JG et al (2020) Source apportionment of highly time-resolved elements during a firework episode from a rural freeway site in Switzerland. Atmos Chem Phys 20:1657–1674
20. Roy A, Mandal M, Das S et al (2023) Non-exhaust particulate pollution in Asian countries: a comprehensive review of sources, composition, and health effects. Environ Eng Res
21. Sah D (2022) Characterization, seasonal variations, source apportionment and health risk assessment of heavy metals in $PM_{2.5}$ in the urban site of Agra, India. Research Square
22. Sharma SK, Mandal TK, Banoo R, Rai A, Rani M (2022) Long-term variation in carbonaceous components of $PM_{2.5}$ from 2012–2021 in Delhi. Bull Environ Contamin Toxicol 109:502–510
23. Sharma SK, Choudhary N, Srivastava P, Naja M, Vijayan N, Kotnala G, Mandal TK (2020) Variation of carbonaceous species and trace elements in PM_{10} at a mountain site in the central Himalayan region of India. J Atmos Chem 77:1–14
24. Sharma SK, Mandal TK (2023) Elemental composition and sources of fine particulate matter ($PM_{2.5}$) in Delhi, India. Bull Environ Contam Toxicol 110(3):60
25. Sharma SK, Mandal TK, Jain S, Saraswati, Sharma A, Saxena M (2016) Source apportionment of $PM_{2.5}$ in Delhi, India using PMF model. Bull Environ Contam Toxicol 97:286–293
26. Sharma SK, Mandal TK, Saxena M, Sharma A, Gautam R (2014) Source apportionment of PM_{10} by using positive matrix factorization at an urban site of Delhi, India. Urban Clim 10:656–670
27. Song Y, Zhang Y, Xie S, Zeng L, Zheng M, Salmon L, Shao M, Slanina S (2006) Source apportionment of $PM_{2.5}$ in Beijing by positive matrix factorization. Atmos Environ 40:1526–1537

28. Taylor SR, McLennan SH (1985) The continental crust: its composition and evolution. Blackwell, Oxford, p 312
29. USEPA (2011) Exposure factors Handbook 2011 Edition (final report). U.S. Environmental Protection Agency, Washington DC (EPA/600/R-09/052F)
30. USEPA (United States Environmental Protection Agency) (2015) User's guide/technical background document for US EPA region 9's RSL (Regional Screening Levels) tables. http://www.epa.gov/region9/superfund/prg/
31. Viana M, Kuhlbusch TAJ, Querol X, Alastuey A, Harrison RM, Hopke PK, Winiwarter W, Vallius M (2008) Source apportionment of particulate matter in Europe: a review of methods and results. J Aerosol Sci 39:827–849
32. Zheng N, Liu J, Wang Q, Liang Z (2010) Science of the total environment health risk assessment of heavy metal exposure to street dust in the zinc smelting district, Northeast of China. Sci Total Environ 408(4):726–733

Monitoring of Personal Exposure Due to Particulate Matter

Madhumita Chakraborty, Smaranika Panda, and Robin A. Christian

Abstract The study is aimed to monitor the personal exposure of traffic policemen and workers of industry to particulate matter in the industrial area of Surat City, India. The objective was to estimate the personal exposure concentration to PM_5 and $PM_{2.5}$. The study conducted experiments to estimate, specifically $PM_{2.5}$ and PM_5, during the working hours of the selected group. The results demonstrate that traffic policemen standing on road junctions (103.87 ± 20.71 μg/m^3) and workers of the textile industry (345.31 ± 113.48 μg/m^3) were exposed to significantly higher mean $PM_{2.5}$ levels. It was observed that workers working in a closed textile industry are exposed to more particulate matter than the traffic policemen standing on open road junctions with good air circulation. The major source of particulate pollution to traffic police was vehicular emissions and road dust resuspension.

Keywords Traffic policemen · Exposure · Industrial exposure · Workers

1 Introduction

'Air' is a mixture of gases that surrounds us everywhere and at every time. The concentration of Particulate Matter (PM) can provide us with insight into air pollution. PM enters the air as a result of different natural as well as anthropogenic activities. While these particles are always present in the air, a concentration greater than the prescribed safe limit by the World Health Organization (WHO) poses detrimental impacts on human health [13]. The impacts include heart rate variability, asthma,

M. Chakraborty (✉) · S. Panda · R. A. Christian
Department of Civil Engineering, Sardar Vallabhbhai National Institute of Technology, Surat, India
e-mail: madhumita.chakraborty19@gmail.com; d21ce003@ced.svnit.ac.in

S. Panda
e-mail: smp@ced.svnit.ac.in

R. A. Christian
e-mail: rac@ced.svnit.ac.in

S. M. Shiva Nagendra et al. (eds.), *Select Proceedings of the 8th Indian International Conference on Air Quality Management (IICAQM 2023)*, Lecture Notes in Civil Engineering 582, https://doi.org/10.1007/978-981-96-2359-4_17

chronic obstructive pulmonary disease, stroke, premature deaths, and leukemia [3, 12, 16]. Air pollution is one of the leading causes of high mortality (Dockery et al., 1993; [4]).

In developing economies like India, the rate of industrialization is very high, which results in making air pollution a more serious problem for human health [8]. There is a constant shift of human settlements in urban centers in search of livelihood. This changes the pattern of land use and increases vehicle density, along with meteorological conditions, which have resulted in more concentration of pollutants at these urban centers [9, 10]. The people residing in such locations are exposed to high PM concentrations, and several epidemiological studies have revealed that exposure to air pollutants has adverse impacts on human health [11]. Coarse particles can go deep into the respiration tract, but finer particles travel much deeper and have a high tendency to settle in the lungs, causing more serious health problems [2, 5]. This makes monitoring of exposure due to PM_5 more important, along with $PM_{2.5}$.

The exposure to PM varies in quantity according to the exposure duration, indoor and outdoor environment, household cooking, choice of fuel for cooking, ventilation conditions in the household, traffic microenvironments, dust, and workplace conditions [1]. Even distance from the source of smoke affects the health implications due to air pollution. Knowing the exposure concentration in all possible microenvironments is very important to study this pattern. This is the reason why fixed site monitors cannot provide accurate data that can be used to study personal exposure and, eventually, heath risk [14]. Personal exposure studies help to identify the possible exposure concentration and source of pollution, which will include the effect of the microenvironment, which can be used to assess the associated risk to human health [7].

2 Methodology

2.1 Study Area

The study area is selected in the Pandesera Gujarat Industrial Development Corporation (GIDC), the industrial area of Surat city of Gujrat state of India. The city of Surat has a total area of 472.149 square kilometers with the latitude of $21^0 10' 12.8640''$ N and $72049' 51.8232''$ E longitude. It is one of India's most dynamic cities, with the fastest growth rate. The textile industries of the city are well-acclaimed in the country. The hot season lasts for two months, from March to May, with an average daily high temperature above 35.55^0C. The hottest month of the year in Surat is May, with an average high of 44.44 °C and a low of 3 °C. The winter lasts for two months, from December to February, with an average daily high temperature below 30.56 °C. The coldest month of the year in Surat is January, with an average low of 17.22 °C and a high of 29 °C.

This research studies the personal exposure to particulate matter of a group of people in different working areas in the Pandesera Industrial Area. Currently, two groups are selected which are more exposed to air pollution. One group was identified as traffic policemen standing on road junctions (mostly without masks) in the Pandesera GIDC area, and the other group was the workers working in the textile industry.

The traffic policemen standing at road junctions are mostly exposed to vehicular emissions and roadside dust resuspension. Since the junctions selected are in the GIDC area of Surat, there are high chances of getting PM concentration due to industrial emissions.

The other group of people for study are workers in one of the textile industries in Pandesera GIDC Surat. The textile industry has different sections performing different operations to make the final product. The operations include bleaching, dying, printing, cutting, washing, and drying processes, and workers inside the industry keep moving from one section to another according to their work requirements.

2.2 *Instrumentation*

Personal Exposure (PE) to $PM_{2.5}$ was measured with a personal $PM_{2.5}$ sampler loaded with 37-mm 1-μm pore size glass fiber filters. The subject wore this sampler, while the subject did all the routine work during the sampling period. The airflow rate was maintained at 2.0 lpm for the sampling duration.

Personal exposure to PM_5 was a low-cost sampler with an attached cyclone. The airflow rate was maintained at 2.5 lpm for the entire sampling duration. The sampler was loaded with 37-mm 1-μm pore size glass fiber filter papers. All the filter papers were analyzed using the gravimetric method to find out the final particulate matter concentration.

2.3 *Data Collection*

Sampling for personal exposure was done using the instruments stated above for the month of April, May, and June 2023. SKC universal pump and an Envirotech APM 821 sampler were used to collect $PM_{2.5}$ and PM_5, respectively. Personal exposure of traffic policemen was estimated by sampling the air for 8 h at an inflow rate of 2.0 lpm for both instruments [6, 15]. The traffic policeman wore the SKC universal sampler and stood at his designated place on the road junction, which was pre-identified in the GIDC area. The Envirotech APM 821 was kept on the platform in the stand for traffic police at the road junction, keeping the inlet piece near the exhaling level.

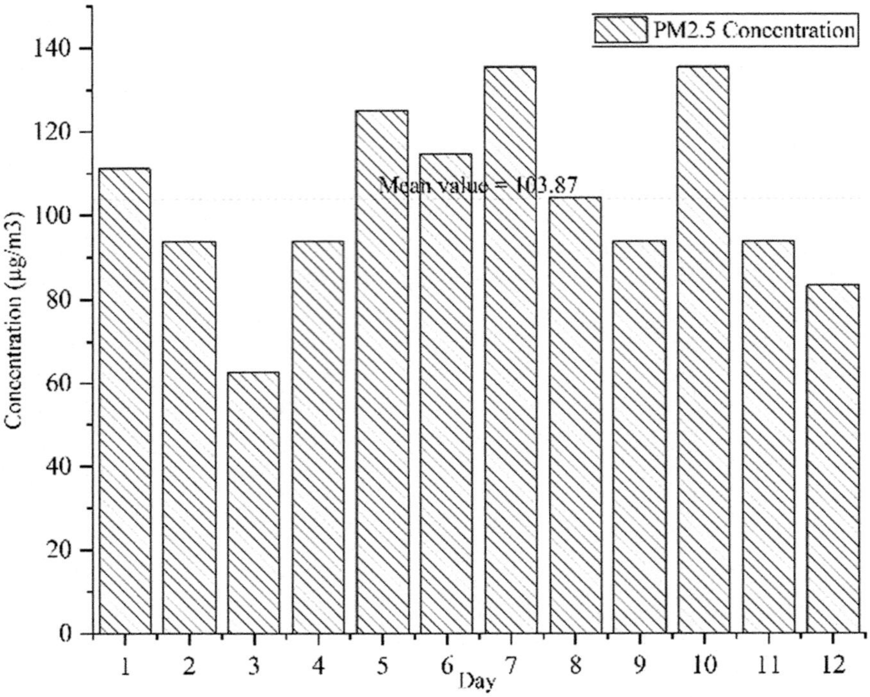

Fig. 1 Variation of PM2.5 exposure of TP

3 Results

3.1 *Personal Exposure Concentration of Traffic Policemen (TP) in Industrial Area*

The sampling has been done for 12 days, from 9 am to 3 pm (6 h). The maximum $PM_{2.5}$ and PM_5 concentration obtained during the study on traffic police was 135.42 µg/m^3 and 244.44 µg/m^3, respectively, whereas the minimum concentration for $PM_{2.5}$ and PM_5 was 62.50 µg/m^3 and 104.17 µg/m^3. Figures 1 and 2 show the variation in exposure concentration of $PM_{2.5}$ and PM_5 of TP.

3.2 *Personal Exposure Concentration of Workers in Industry*

Personal exposure to particulate matter of workers working in the textile industry was measured for 16 days during their shift timings. The maximum and minimum values of $PM_{2.5}$ obtained during the study were 583.33 µg/m^3 and 187.50 µg/m^3, respectively; variation is shown in Fig. 3. And the maximum PM_5 concentration

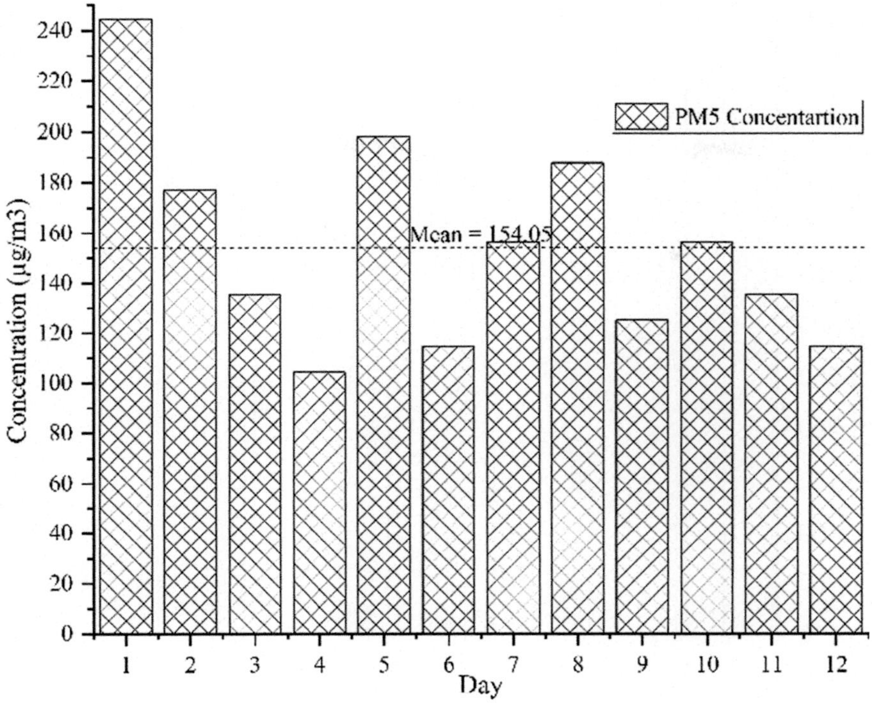

Fig. 2 Variation of PM5 exposure of TP

obtained during the study was 604.17 µg/m^3, whereas the minimum concentration was 250 µg/m^3, shown in Fig. 4.

4 Conclusions

The study involves monitoring the personal exposure of two groups of people working in different microenvironments. One group is traffic policemen standing on the major road junctions in the industrial area of Surat city, whereas the other group selected consists of workers working in the textile industry. The exposure to particulate matter, namely $PM_{2.5}$ and PM_5, is measured using personal samplers.

The mean $PM_{2.5}$ concentration for traffic policemen was observed to be 103.87 ± 20.71 µg/m^3 when monitored for eight hours for 12 days. The maximum and minimum concentrations obtained were 135.41 and 62.5 µg/m^3, respectively. The mean PM_5 concentration for traffic policemen was observed to be 154.05 ± 39.68 µg/m^3 when monitored for six hours for 12 days. The maximum and minimum concentrations obtained were 244.44 and 104.17 µg/m^3, respectively. The exposure concentration is high for workers as they are working inside an industry at different machines.

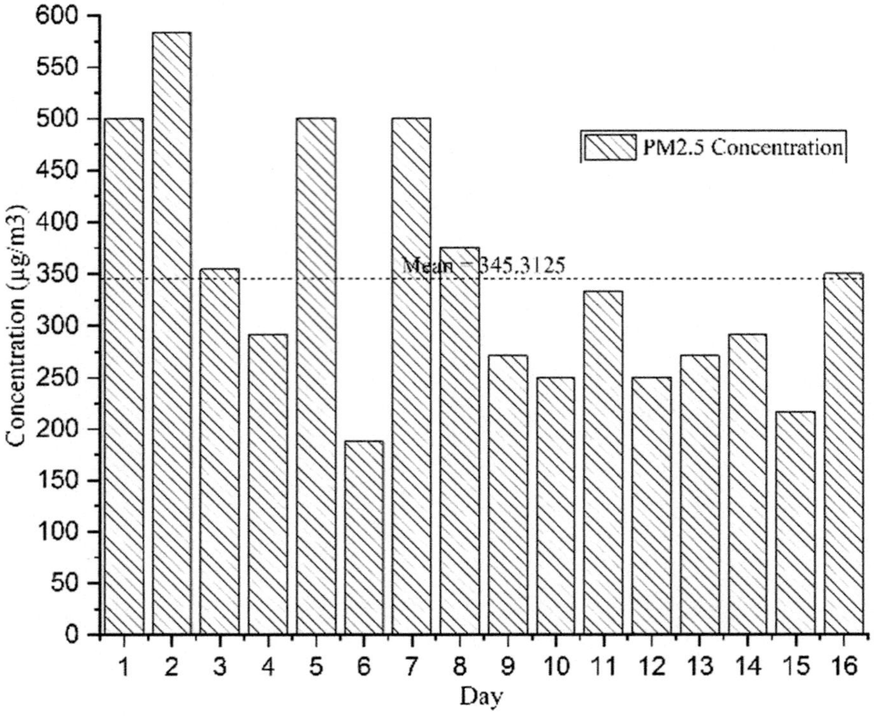

Fig. 3 Variation of PM2.5 exposure of workers

Workers work on machines that perform bleaching, dying, washing, printing, cutting, movement of loading vehicles, exhaust emission from vehicles inside industry premises, dust from fabrics, and lack of proper ventilation (air circulation) contribute to a higher concentration of personal exposure. This study provided insight into the exposure levels of two sensitive groups, which are more likely to be affected by air pollution. This can be helpful in the assessment of associated health risks.

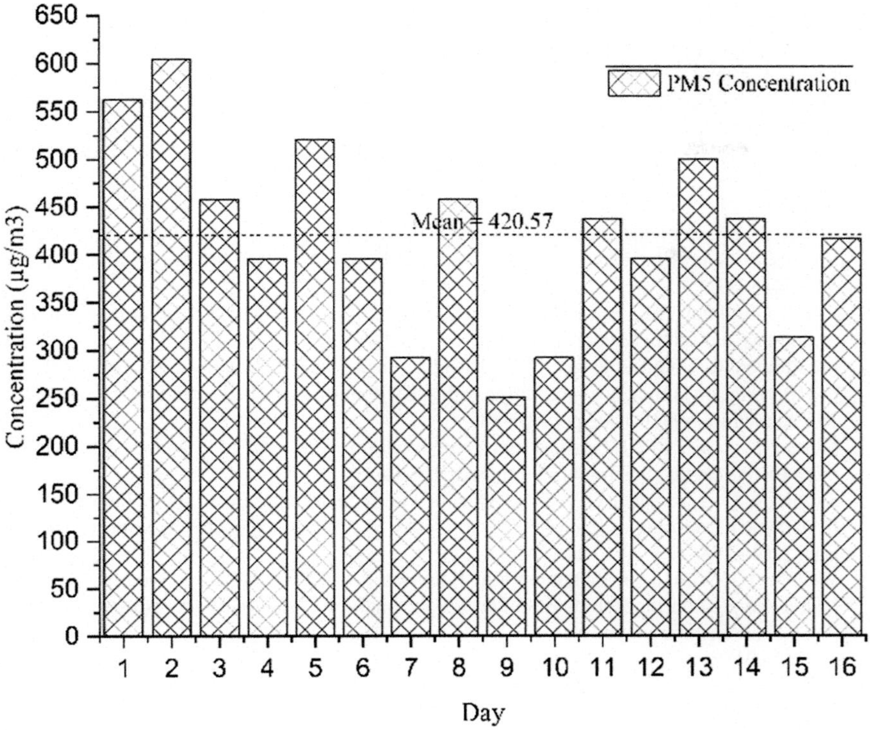

Fig. 4 Variation of PM5 exposure of workers

References

1. Ashok V, Gupta T, Dubey S, Jat R (2014) Personal exposure measurement of students to various microenvironments inside and outside the college campus. Environ Monit Assess 186(2):735–750. https://doi.org/10.1007/s10661-013-3413-y
2. Bodor K, Szép R, Bodor Z (2022) The human health risk assessment of particulate air pollution (PM2.5 and PM10) in Romania. Toxicology Reports 9:556–562. https://doi.org/10.1016/j.toxrep.2022.03.022
3. Cohen AJ, Brauer M, Burnett R, Anderson HR, Frostad J, Estep K, Balakrishnan K, Brunekreef B, Dandona L, Dandona R, Feigin V, Freedman G, Hubbell B, Jobling A, Kan H, Knibbs L, Liu Y, Martin R, Morawska L, ... Forouzanfar MH (2017) Estimates and 25-year trends of the global burden of disease attributable to ambient air pollution: an analysis of data from the Global Burden of Diseases Study 2015. The Lancet 389(10082):1907–1918. https://doi.org/10.1016/S0140-6736(17)30505-6
4. Fan Z, Pun VC, Chen XC, Hong Q, Tian L, Ho SSH, Lee SC, Tse LA, Ho KF (2018) Personal exposure to fine particles (PM 2.5) and respiratory inflammation of common residents in Hong Kong. Environ Res 164:24–31. https://doi.org/10.1016/j.envres.2018.02.009
5. Ghosh N et al (2021) Particulate matter and adverse respiratory health outcome: exposure of street vendors in Kolkata City in India. Pollution 7(4):769–85
6. Koehler KA, Peters TM (2015) New methods for personal exposure monitoring for airborne particles. Curr Environ Health Rep 2(4):399–411

7. Kolluru SSR, Patra AK (2020) Personal exposures to pm during short distance highway travel in India. Transp Res Part D Transp Environ 81.
8. Kumar P, Gupta NC (2016) Commuter exposure to inhalable, thoracic and alveolic particles in various transportation modes in Delhi. Sci Total Environ 541:535–541. https://doi.org/10.1016/j.scitotenv.2015.09.076
9. Lv H et al (2021) Assessment of pedestrian exposure and deposition of PM10, PM2.5 and ultrafine particles at an urban roadside: a case study of Xi'an, China. Atmos Pollut Res 12(4):112–21
10. Menon JS, Shiva Nagendra SM (2018) Personal exposure to fine particulate matter concentrations in central business district of a tropical coastal city. J Air Waste Manag Assoc 68(5):415–29
11. Pant P, Guttikunda SK, Peltier RE (2016) Exposure to particulate matter in India: a synthesis of findings and future directions. Environ Res 147:480–96
12. Preutthipan A, Udomsubpayakul U, Chaisupamongkollarp T, Pentamwa P (2004) Effect of PM10, pollution in Bangkok on children with and without asthma. Pediatr Pulmonol 37(3):187–192. https://doi.org/10.1002/ppul.10398
13. Steinle S et al (2015) Personal exposure monitoring of pm2.5 in indoor and outdoor microenvironments. Sci Total Environ 508:383–94
14. Steinle S, Reis S, Sabel CE (2013) Quantifying human exposure to air pollution-moving from static monitoring to spatio-temporally resolved personal exposure assessment. Sci Total Environ 443:184–93
15. Wheeler AJ, Williams I, Beaumont RA, Hamilton RS (2000) Characterisation of particulate matter sampled during a study of children's personal exposure to airborne particulate Matter in a UK urban environment
16. Wilhelm M, Ritz B (2003) Residential proximity to traffic and adverse birth outcomes in Los Angeles County, California, 1994–1996. Environ Health Perspect 111(2):207–216. https://doi.org/10.1289/ehp.5688

Probable Health Risk of Pollutants and Noise Found Inside Bottling Beverage Plant of Himachal Pradesh, India

Indramani Dhada, Prince Vats, and Brajesh Singh

Abstract Rapid industrialization worldwide has led to a noteworthy rise in air pollution, affecting the indoor air quality of urban areas and posing health risks. This research precisely explored the impact of air pollutants and noise found within 12 bottling beverage industries dealing with alcohol, situated in Himachal Pradesh, India. The research focused on (TVOCs), noise, various sizes of particulate matter ($PM_{2.5}$, PM_{10}, PM_1), inhalable, thoracic, and alveolic PM, along with Black Carbon concentrations during the monsoon season. It also assessed the potential harmful impacts linked with these pollutants. The highest values of TVOC, PM_{10}, $PM_{2.5}$, PM_1, BC, and noise were found to be 142 ppm, 359 $\mu g/m^3$, 102 $\mu g/m^3$, 65 $\mu g/m^3$, 26 $\mu g/m^3$, and 87 dB, respectively. Non-carcinogenic health risk is found to be in the range of 4.7E-05–8.9E-05, and the age group 16–21 years males and females are more prone to non-carcinogenic risk. Carcinogenic risk due to BC is estimated to be in the range of 5.8E-04–6.8E-04, which is not significant. Due to ethical reasons and limited resources, the monitoring is restricted to one season, so the health actual risk will be less than the reported value here if annual average concentrations are considered.

Keywords Industrial pollution · TVOC · PM · Black carbon · Sound levels

1 Introduction

The escalating levels of air pollution have emerged as one of the most significant threats to environmental health in recent years. Air pollution has been linked to higher rates of respiratory deaths and diseases in various studies [48, 31]. In 2016, it was estimated that air pollution levels exceeded the acceptable standards in areas where 91% of the global population resided [49, 2]. Ambient and indoor air is calming 6.7 million premature deaths annually throughout the world. However, in 2020, indoor

I. Dhada (✉) · P. Vats · B. Singh
Department of Civil Engineering, Indian Institute of Technology Ropar, Rupnagar, India
e-mail: idhada@iitrpr.ac.in

S. M. Shiva Nagendra et al. (eds.), *Select Proceedings of the 8th Indian International Conference on Air Quality Management (IICAQM 2023)*, Lecture Notes in Civil Engineering 582, https://doi.org/10.1007/978-981-96-2359-4_18

air pollution claims 3.2 million deaths (WHO). Approximately 80% of people's time is spent in enclosed spaces such as homes, offices, institutions, industries, or other indoor areas [1, 4], and exposed to increased concentration of air pollutants in indoor than outdoor [39], which attracts the concern of researchers to study various indoor micro-environment. The primary pollutants found in indoor air include compounds that easily evaporate, fine particles suspended in the air, soot particles, and various gaseous pollutants [15]. Exposure to particulate matter is associated with several serious health effects, especially cardiovascular disease, pulmonary disorder, and cancer [49, 10, 35, 38].

Particulate matter (PM) includes complex mixture of particles such as metals, acids, organics, crustal material, and liquid droplets covering a range of different sizes [22]. The surface area, composition, and number of particles, which are significant parameters in determining health effects, can absorb and distribute a wide variety of pollutants [44, 20]. Black Carbon (BC), a smaller particle, is a component of $PM_{2.5}$ that is released during the burning of fossil fuels [17], and ultrafine particles (UFP, PM $\leq$ 100 nm) are of recent health interest [31]. These tiny particles, due to their small size, have the ability to penetrate deep into the lung alveoli [19]. By various mechanisms, such as causing oxidative damage, altering cell communication, and triggering the release of inflammatory mediators, they can harm both the heart and the lungs [10, 19, 33].

BC is a constituent of aerosols that originates from the incomplete burning of biomass and fossil fuels [18]. BC is a short-lived air pollutant with an average atmospheric lifespan of 4–12 days. It has a global warming potential that is 460–1,500 times greater than that of CO_2 [8]. BC not only has harmful effects on human health [7], but it also has the potential to contaminate the indoor atmosphere in urban areas [18], affect the cryosphere (regions of snow and ice), and impact agriculture [8]. Outdoor air pollutants have a significant contribution to indoor air pollutants and continuous exposure to those can be attributed to sick building syndrome (SBS) [20]. Despite the limited number of studies on BC in indoor environments, its role in contributing to respiratory diseases cannot be overlooked [30]. Toxicological studies have shown that BC can cause progressive tissue damage. This damage primarily occurs through mechanisms such as redox imbalance and inflammatory responses [6, 34]. Additionally, BC has been linked to gene mutations [14, 29, 34], and has been found to have detrimental effects on the nervous system. [32, 33]. Epidemiological studies have found a correlation between the concentration of atmospheric BC and the incidence of cardiovascular [5] and respiratory diseases [27]. These diseases are increasingly being linked to elevated levels of BC and PM in the air. In fact, in 2020, these pollutants were estimated to contribute to approximately 45% of premature deaths caused by indoor air pollution ("WHO"). Due to their nano size, irregular shape, and large surface area,

BC can also act as a carrier for carcinogenic compounds such as polycyclic aromatic hydrocarbons and volatile organic compounds. These compounds can penetrate the bronchial tree because of their small size [40, 28]. This penetration into the human respiratory system can lead to serious health effects [23]. Bottling plants producing liquor are using sprit like ethyl alcohol as raw material which is the major source of total volatile organic compounds (TVOCs) inside those industries. VOCs can irritate the nose, throat, and eyes, make breathing difficult and cause nausea, and harm the central nervous system and other organs when inhaled. Some VOCs are even known to be carcinogenic, meaning they can cause cancer [12]. Apart from the health effects, VOC has potential to create ozone in the atmosphere in presence of sunlight and ozone in troposphere can create greenhouse effect. Also bottling industries are associated with process starting from handling of empty bottles and cans for washing till the packaging of final product and loading which creates noise. As per the noise pollution regulations established by the Central Pollution Control Board (CPCB) in India, the permissible noise level ranges from 68 to 75 dB for different types of sources [9]. Exceeding the acceptable noise level limit can potentially harm our hearing and trigger a variety of reactions, including mental and bodily reactions, and possibly even disease-related conditions.

Considering the above pollutants and their health effect, this research focuses the monitoring of level of such pollutants and their probable health risk on the workers inside the bottling industries of Himachal Pradesh.

2 Methodology

2.1 Study Area: Himachal Pradesh

Himachal Pradesh, situated in the northern part of India, serves as the backdrop for this comprehensive indoor air quality study provides a unique opportunity to examine the interface between industrial development and environmental conservation in such geographically diverse and ecologically sensitive state.

Geographic Diversity: The state covering low lying Shivalik foothills to the towering Himalayan peaks, resulting in a wide range of climatic conditions and ecosystems, and its varying altitudes and microclimates contribute to the complexity of its air quality dynamics, making it an intriguing location for this study.

Industrial Heterogeneity: The study encompasses thirteen bottling plants, each representing a distinct facet of the state's industrial landscape. These bottling plants are scattered across Himachal Pradesh, with variations in production processes, operational scales, and proximity to urban centers. This variety enables a comprehensive assessment of how industrial activities affect air quality. The provided Fig. 1 presents a map of India, with the state of Himachal Pradesh emphasized, and it also indicates the locations of the industries that are being examined in this study.

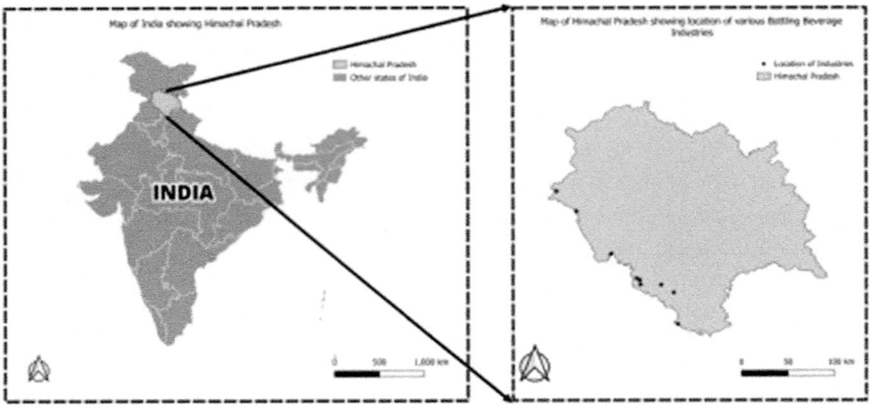

Fig. 1 Schematic diagram of sampling location in Himachal Pradesh

2.2 Air Quality Data and Sampling Methodology

Air quality data in terms of PM and BC in indoor environments was assessed with the help of USEPA certified a portable battery-operated laser photometer (Aerosol spectrometer, model 1.109, make: GRIMM, USA) and optical sensor-based handy instrument (Black Carbon Analyzer, make: Distributed Sensing Technologies, California), respectively. The TVOC is measured through photoionization detection (PID) technology-based TVOC analyzer (Ion science make). The noise is measured through sound level meter (Lutron make).

The pollutants measured consisted of TVOC, fine PM ranging from 0.25 μm to 32 μm (in 31 channels), BC, and sound levels in the indoor environments of industries. The instruments were placed near the breathing zone as much as possible during the study period (approximately 1.5 m in height). This was done to accurately record the concentration of inhalable particles. However, care was taken to avoid any obstructions near the inlet nozzle of the sampler within a 3-m radius.

2.3 Estimation of Health Risk

Health risks associated with PM and BC have been assessed by calculating a hazard index (HI) to evaluate non-carcinogenic risks and estimating a lifetime cancer risk (LCR) specifically for BC, addressing carcinogenic risks. These assessments are conducted through a deterministic approach to understand the potential health impacts. To account for variations in physiology and behavior based on age, weight, and gender, the population in the study has been segmented into five distinct age groups: 16–21 years, 21–30 years, 30–40 years, 40–50 years, and 50–60 years, separately for males and females. The average parameter values for each of these age categories have been sourced from the exposure handbook of USEPA [43]. Subsequently, health risk assessments have been conducted for all these age groups.

2.4 Deterministic Health Risk

The health risk assessment model developed by the USEPA in 1989 was used to carry out the health risk assessments in this study. Since the industrial employees belong to various age groups, the health risk calculations consider the variability of parameters, as detailed in Table 1. This methodology aligns with similar guidelines followed in published literature [11] for assessing health risks related to $PM_{2.5}$, [25] used it for polycyclic aromatic hydrocarbons (PAHs) and employed a comparable method for assessing health risks associated with hydrogen sulfide (H_2S) gas [3].

The process of assessing health risks in this study involves calculating the daily intake of pollutants, as outlined in (1). Then, the reference dose for these pollutants is determined using (2), and the associated hazard quotient is calculated using (3). These steps collectively form the basis for estimating the health risks associated with exposure to the studied pollutants.

Table 1 Mean parameter values considered in estimation of health risk, suggested by USEPA [44] and an on-spot survey (from a similar site)

	Males					Females				
Age (years)	16–21	21–30	30–40	40–50	50–60	16–21	21–30	30–40	40–50	50–60
Mean inhalation rate (m^3/day)	17.21	18.82	20.29	20.94	20.91	13.59	14.57	14.98	16.2	16.19
Mean body weight (kg)	77.3	84.9	87	90.5	89.5	65.9	71.9	74.8	77.1	77.5
Exposure time (hours per day)	8	8	8	8	8	8	8	8	8	8
Exposure frequency (days per year)	313	313	313	313	313	313	313	313	313	313
Exposure duration (years)	35	35	35	35	35	35	35	35	35	35
Averaging time (years)	70	70	70	70	70	70	70	70	70	70

$$\text{CDI} = \frac{\text{CA} \times \text{IR} \times \text{ET} \times \text{EF} \times \text{ED}}{\text{BW} \times \text{AT}} \tag{1}$$

The terms are defined as follows:

- CDI: This stands for Chronic Daily Intake of pollutants, measured in milligrams per kilogram of body weight per day.
- CA: This is the Average Concentration of the inhaled pollutant, measured in milligrams per cubic meter.
- IR: This is the Inhalation Rate, which signifies the volume of air an individual inhales, measured in cubic meters per hour.
- ET: This is the Exposure Time, measured in hours per day, during which an individual is exposed to the pollutant.
- EF: This is the Exposure Frequency, indicating the frequency of exposure, measured in days per year.
- ED: This is the Exposure Duration, representing the number of years a person is exposed to the pollutant.
- BW: This is the Mean Body Weight, measured in kilograms.
- AT: This is the Averaging Time, which can vary depending on whether the risk is associated with non-carcinogenic or carcinogenic effects.

AT is taken equal to ED for non-carcinogenic risk, while for carcinogenic risk, it is set at 70 years.

For this study, the values of IR and BW were taken as the average values suggested by the manual, ensuring a consistent approach. However, CA, ET, EF, and ED were specifically tailored based on the primary data collected at a similar site of a bottling beverage plant. These values can be found in Table 1, providing a customized and site-specific foundation for the risk assessment estimation.

$$\text{RfD} = \text{RfC} \times \text{IR} \times \text{BW} \tag{2}$$

The terms are defined as follows:

RfD (mg/kg/day): This stands for the Reference Dose, which is the acceptable daily intake of a pollutant.

RfC (μg/m^3): This is the Reference Concentration, which is the permissible concentration of a pollutant in the environment.

This study uses Reference RfD and RfC values recommended by CPCB and WHO (detailed in Table 2), to assess health risks. When 8-h average data is unavailable, the study incorporates 24-h RfC values from CPCB and WHO guidelines for a comprehensive risk assessment.

Further, the following equations show the formulae used to estimate the HQ and HI:

$$\text{HO} = \frac{\text{CDI}}{\text{RfD}} \tag{3}$$

Table 2 RfC of PM set by different regulatory bodies in μg/m^3

Pollutants	NAAQS-CPCB (Daily average)	NAAQS-CPCB (Annual average)	WHO [50] (Daily average)	WHO [50] (Annual average)	OSHA-NIOSH (8-h average)
$PM_{2.5}$	60	40	15	5	–
PM_{10}	100	60	45	15	–
BC	–	–	–	–	3500

$$\mathrm{HI} = \sum \mathrm{HQ} \tag{4}$$

BC is hazardous due to its characteristics and nano-sized particles and is also recognized as carcinogenic if it is emitted from diesel soot [24]. Carcinogenic risk due to BC is estimated in terms of LCR using (5). Slope factor (SF), taken as 1.1 mg/kg-day by considering the risk of BC, equivalent to risk due to diesel engines exhaust [13, 21].

$$\mathrm{LCR} = \mathrm{CDI} \times \mathrm{SF} \tag{5}$$

The limits for HI and LCR beyond which the pollutants may cause significant non-carcinogenic and carcinogenic risks are 1 and 1×10^{-6}, respectively [42]. The EPA deems the risk of cancer to be insignificant if there is less than one instance of cancer in a population of one million people (1×10^{-6}). However, if the risk exceeds one in ten thousand people (1×10^{-4}), it is considered substantial and necessitates medical intervention.

3 Results and Discussion

3.1 Tvoc

The TVOC concentration data obtained from the 12 industries provide critical insights into the levels of VOCs within these environments. For ethical reasons, actual names of industries have labeled as I-1 to I-12. Notably, I-5 exhibits the highest TVOC concentration (141.6 ppm), while I-1 has the lowest concentration (0.1 ppm). This might be due to high leakage of volatile sprits from I-5 using more raw materials and having more production capacity, whereas for I-1, the capacity is minimum. The standard deviations in TVOC concentrations indicate the degree of dispersion or variability within each industrial setting. Higher standard deviations (48.253 and 16.585), as seen in I-3 and I-8, respectively, suggest greater fluctuations

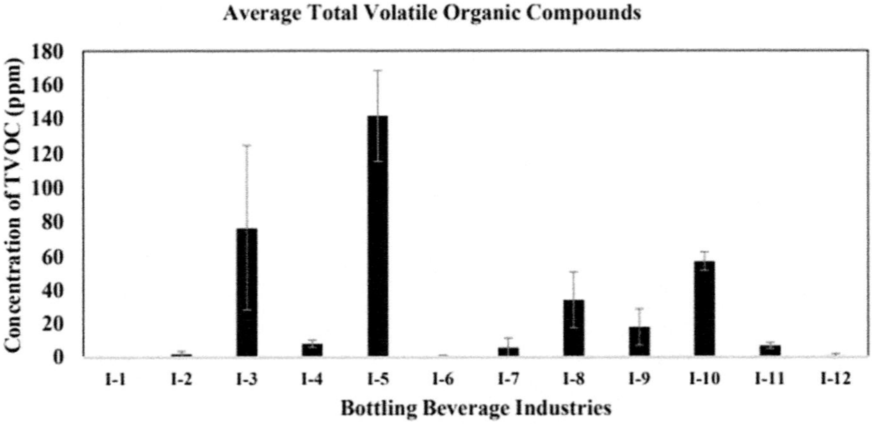

Fig. 2 Graph showing variation in average values of TVOC

in TVOC levels over time, possibly due to dynamic production processes or emissions. These emissions can also affect outdoor air quality and the health of workers inside (Fig. 2).

3.2 *Sound Levels*

I-5 records the highest average sound level (87 $\pm$ 0.725 dB), while I-7 has the lowest sound level (67 $\pm$ 6.144 dB). This significant range highlights the diverse noise profiles associated with different industrial processes.

Industries such as I-2, I-5, and I-9 exhibit relatively higher sound levels compared to others. These industries likely involve machinery or processes that generate elevated noise levels. Conversely, industries like I-4 and I-8 maintain lower sound levels, suggesting quieter working environments. Elevated sound levels within industrial settings can have significant occupational health implications, including the risk of noise-induced hearing loss and other adverse health effects. Industries with higher sound levels should prioritize hearing protection measures for employees exposed to loud environments. The variations in sound levels underscore the importance of adherence to noise pollution regulations and guidelines within industrial facilities. Noise control measures, such as soundproofing and equipment maintenance, may be necessary to ensure compliance with acceptable noise levels. The sound levels within industrial settings also influence the overall quality of the work environment. A quieter workplace can contribute to improved concentration, communication, and employee well-being. The nature of industrial processes within each facility can significantly impact sound levels. Industries with dynamic machinery or frequent equipment operation tend to have higher sound levels, while those with less noisy processes maintain quieter environments (Fig. 3).

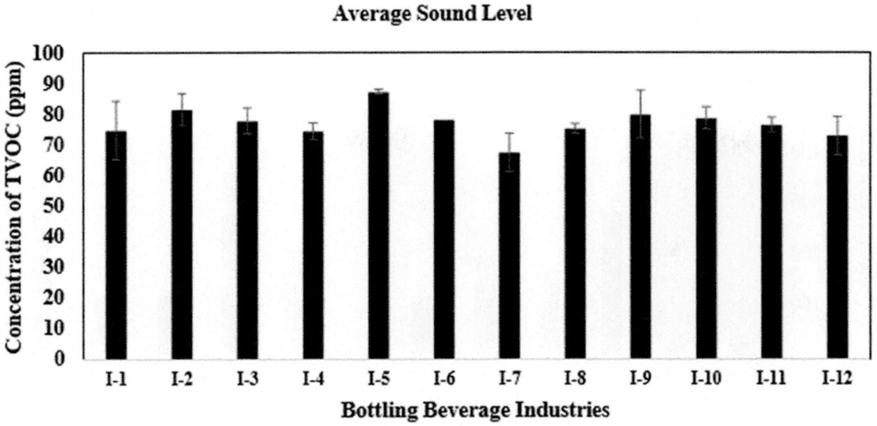

Fig. 3 Graph showing variation in average values of sound levels

3.3 *Particulate Matter (PM)*

These monitored PM concentrations are represented in Fig. 4, 5, 6, 7, 8 and 9. There is significant variability in air quality parameters among the industrial settings, reflecting differences in industrial processes, emissions sources, and pollution control measures. Industries such as I-3, I-4, I-8, and I-9 show elevated PM concentrations across all size fractions (PM10, PM2.5, and PM1). This suggests that these industries may generate substantial particulate matter emissions, which can have implications for air quality and health. Inhalable and thoracic particulate matter concentrations generally mirror the trends observed in PM10 and PM2.5. Industries with high PM concentrations also exhibit elevated levels of inhalable and thoracic particulate matter. These fractions are important for assessing the potential health impact of particulate matter exposure, as they can penetrate deeper into the respiratory system. Alveolic particulate matter, which represents the finest particles that can reach the deepest portions of the lungs and possibly enter in the bloodstream, shows elevated concentrations in I-3 and I-9. Elevated particulate matter concentrations, especially in I-3 and I-9, can have adverse health effects on workers and nearby communities. These effects may include respiratory problems, cardiovascular issues, and other health risks. Industries that have high levels of particulate matter should put in place strong measures to control pollution. This will help in reducing emissions and preserving the quality of air. Such measures may include advanced filtration systems, improved industrial processes, and regular maintenance. Elevated particulate matter levels also have environmental implications, including potential harm to ecosystems and air quality degradation in the surrounding areas.

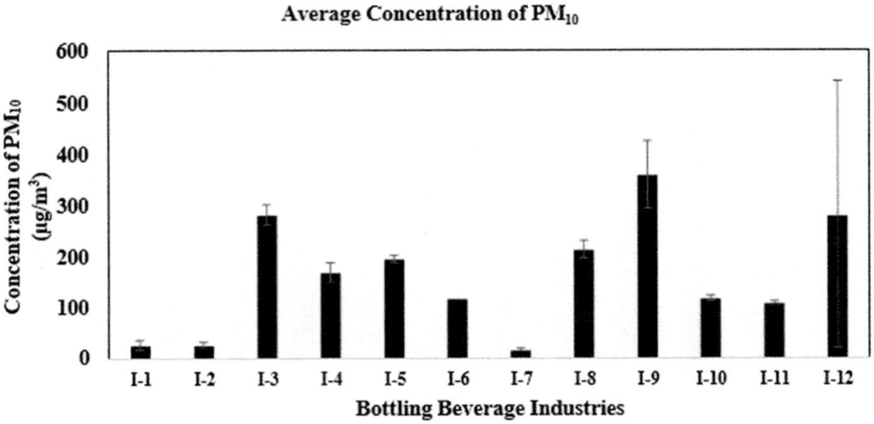

Fig. 4 Graph showing variation in average values of PM10

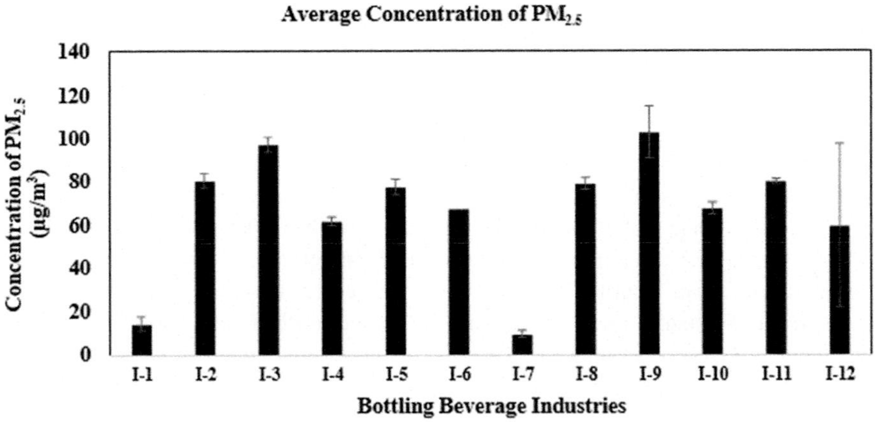

Fig. 5 Graph showing variation in average values of PM2.5

3.4 Black Carbon (BC)

The sampling was performed in the studied bottling industries in the working shift for a fixed time and the average of BC concentrations of the sampling time was taken for analysis. Since the particle size of BC contributes to a major range of PM2.5, it shows health effects that are similar to PM2.5 [37, 46]. Studies have shown that BC can have impacts on respiratory, cardiovascular, and eye diseases, with an average concentration of 5.2 $\mu g/m^3$ [26]. Furthermore, it has been observed that when the concentration of BC is equal to or greater than 1.7 $\mu g/m^3$, there is a correlated increase in hospitalizations due to emergency myocardial infarctions [5]. Likewise, it has been observed that a rise in the levels of PM2.5 and BC correlates with an increase

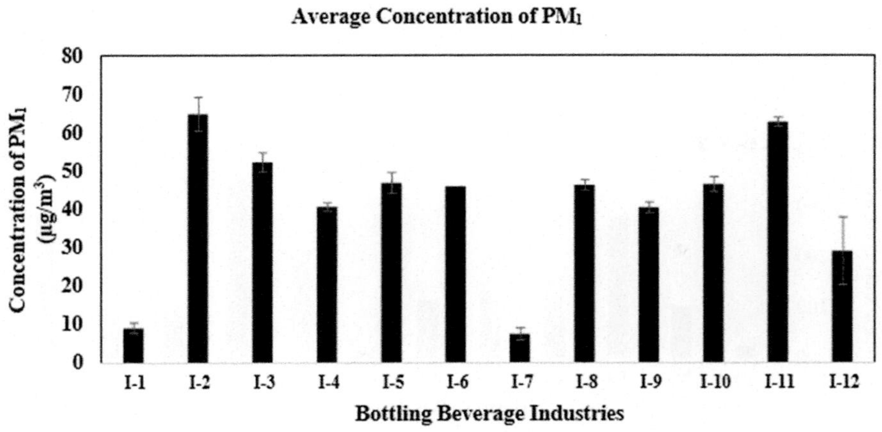

Fig. 6 Graph showing variation in average values of PM1

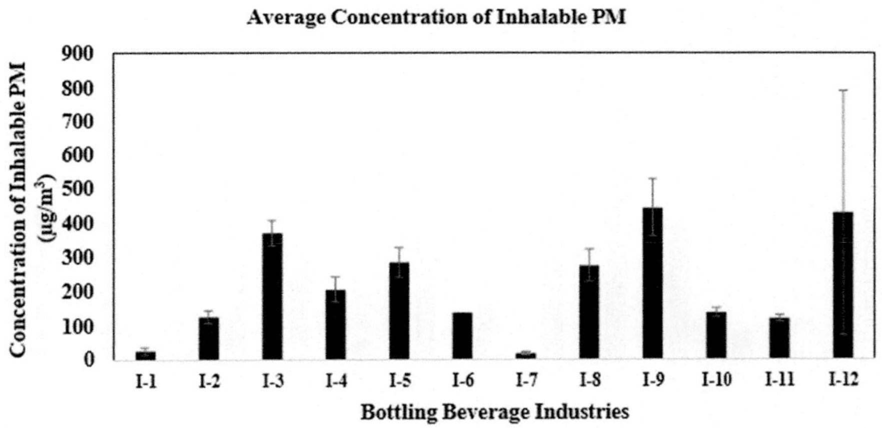

Fig. 7 Graph showing variation in average values of Inhalable PM

in the number of emergency room visits, as reported by hospitals in Shanghai [47]. Additionally, research indicates that the components of PM related to combustion processes are more detrimental to health compared to the non-combustible fractions [51]. Figure 10 shows a column chart illustrating the average concentration of BC found in those industries. The collected data indicates a significant variation in the concentrations of BC across the industrial sites that were surveyed. This suggests that different sites may have different levels of emissions or environmental conditions affecting BC concentrations. I-4 has one of the highest BC concentrations, indicating potentially significant emissions of Black Carbon at this site, this might be because of heavy vehicular traffic near to this industry. I-9 has a relatively low BC concentration, which suggests that it might have lower emissions or better environmental controls in place. I-1, I-2, and I-5 all have moderate BC concentrations. These sites may warrant

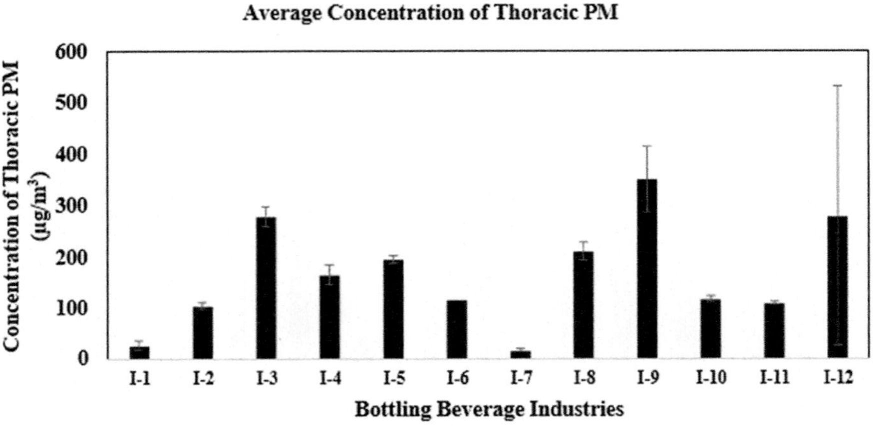

Fig. 8 Graph showing variation in average values of Thoracic PM

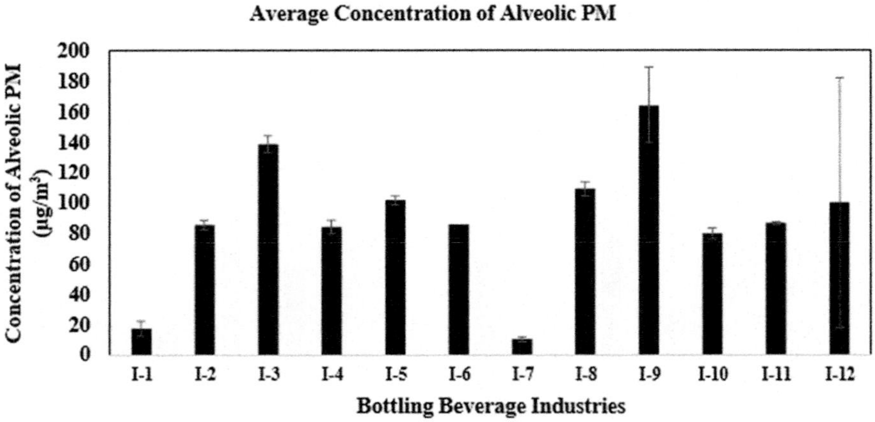

Fig. 9 Graph showing variation in average values of Alveolic PM

further investigation to determine if there are any trends or factors influencing these concentrations.

3.5 *Correlation between PM and BC*

The correlation between PM and BC varies depending on location and pollution sources. In areas with high combustion-related emissions, such as urban centers with heavy traffic, there tends to be a strong correlation between BC and fine PM (PM2.5), while the correlation may be weaker for larger PM fractions (PM10) or in regions with different pollution sources. Both PM and BC pose health risks, with BC

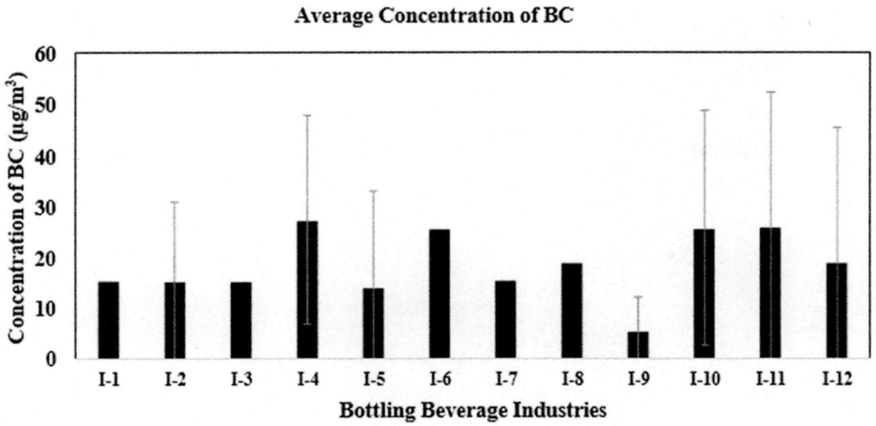

Fig. 10 Graph showing variation in average values of BC concentrations

also contributing to climate change, making it crucial to monitor and control both pollutants to address air quality, human health, and environmental concerns.

The Pearson correlations between mass concentration of various ranges of PM and BC are depicted in Fig. 11.

Our study findings align with previous research by [16, 36, 41], which indicate that the primary size range of indoor BC particles falls within the PM2.5 and PM1 categories. This suggests that vehicular exhaust and diesel generator emissions are significant contributors to BC indoors, as BC exhibits a strong correlation with these smaller particulate matter sizes, a pattern consistent with the findings of [36] for vehicular emissions.

3.6 *Health Risk (HR)*

The non-carcinogenic HR for PM2.5, PM10 & BC, and carcinogenic HR (LCR) for BC were quantified for both males and females as per the methodology given in the previous section. The HI estimated over the study period for PM2.5, PM10 and BC were found to be in the safe range (HI < 1) for different age groups for males and females when CPCB and OSHA standards were considered. Figures 12 and 13 show a comparative analysis of differences in values of HI and LCR (BC), respectively, for different age groups and for males and females. The small magnitude of the HI suggests that, in general, the risk of non-carcinogenic air pollution-related problems is relatively low. However, even low levels of exposure can be of concern, especially if the population is large or vulnerable. The consistently higher HI for females across all age groups suggests that females may have a slightly higher estimated risk compared to males. This could be due to differences in exposure patterns, susceptibility, or other factors related to gender.

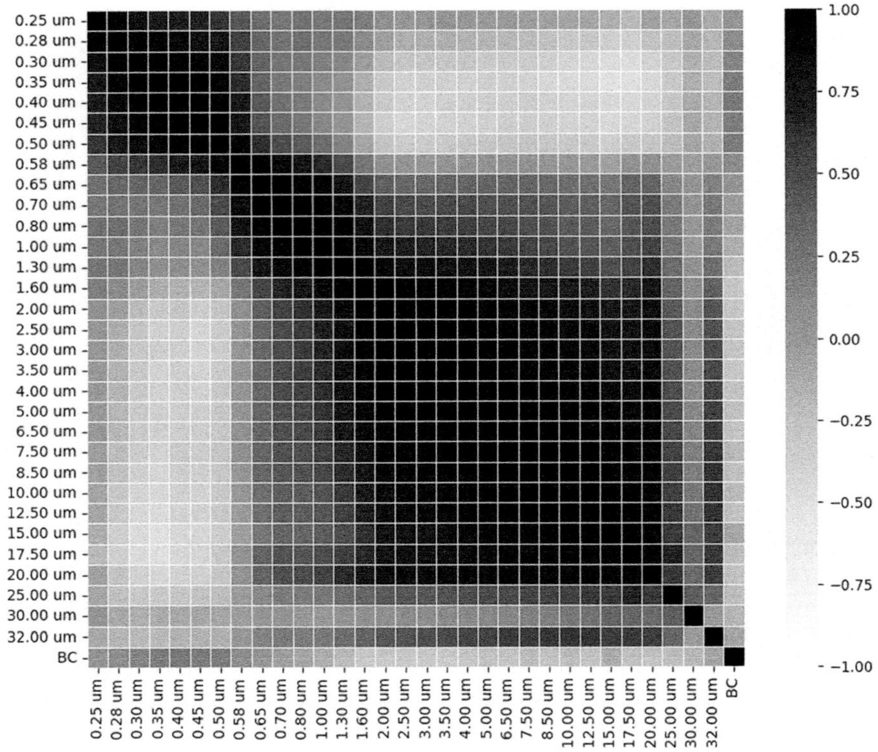

Fig. 11 Correlation between mass concentration of PM and BC concentrations for all the dataset

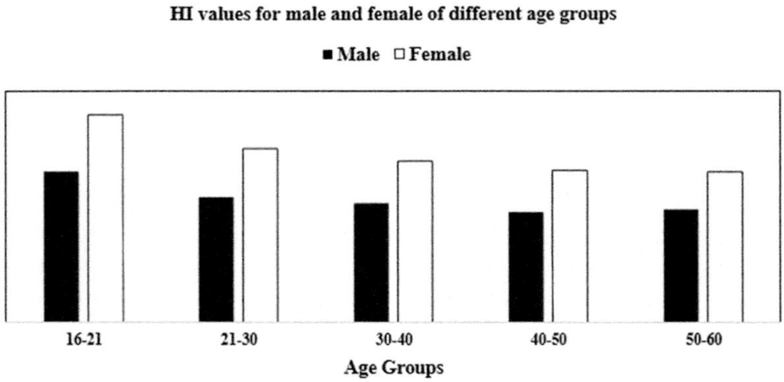

Fig. 12 Comparative analysis of HI values for different age groups, segregated by gender

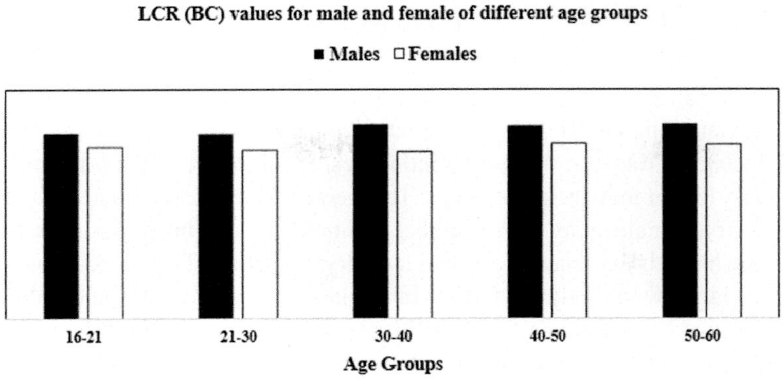

Fig. 13 Comparative analysis of LCR (BC) values for different age groups, segregated by gender

The decreasing HI with increasing age still indicates that younger individuals (16–21 age group) are at a slightly higher estimated risk for non-carcinogenic air pollution-related problems compared to older age groups. This might be due to lifestyle factors, activity patterns, or physiological differences between age groups.

The highest Hazard Index (HI) is observed in the age group of 16–21 years for both males and females. So, to focus on this particular age group, a detailed industry-wise analysis is performed on all the available datasets. This will help in understanding better the vulnerability of this age group. Figure 14 shows variation of HI for males and females of 16–21 years age group.

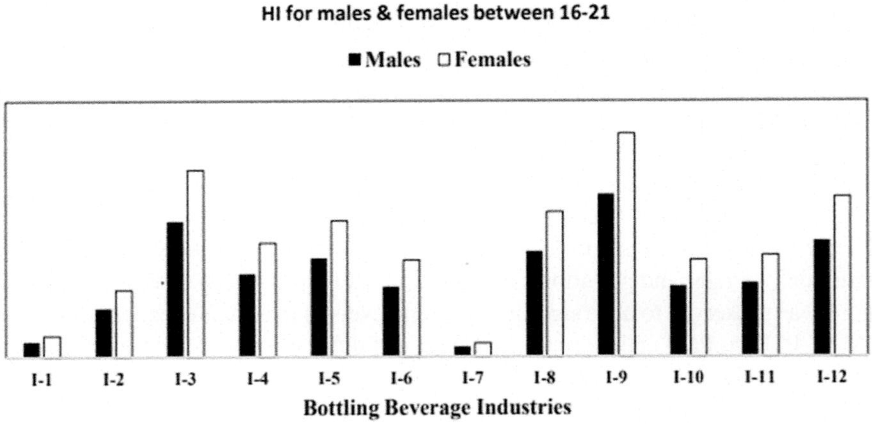

Fig. 14 Comparison of HI values for both males and females in the 16–21 years age group

3.7 HI for Males and Females between 16 and 21

The hazard indices vary significantly between different industries (I-1 to I-12) for both males and females. This suggests that the potential health risks associated with each industry are distinct. Across all industries, the hazard indices for females are consistently higher than those for males. This gender difference indicates that, within each industry, females may have a higher estimated risk of health issues related to the specific hazards associated with that industry. Understanding the reasons for this gender disparity would require further investigation. To fully interpret and discuss these results, it is crucial to understand the specific hazards and conditions within each industry. Different industries may have varying levels and types of exposures, which can lead to differences in risk. The interpretation and response should be tailored to address the unique challenges of each industry.

4 Conclusions

In conclusion, the diverse range of TVOC concentrations observed across these 12 industries underscores the need for tailored approaches to monitor and manage volatile organic compound emissions within industrial environments. This data can inform regulatory decisions, industrial practices, and efforts to improve both indoor air quality and environmental sustainability within these sectors. Further research may explore the specific sources and dynamics of TVOC emissions in each industrial context to develop targeted mitigation strategies.

This study highlights the diverse sound levels prevalent in industrial settings and their potential implications for occupational health and workplace quality. The data emphasize the need for industries to implement noise control measures, ensure regulatory compliance, and prioritize employee well-being. Further research may delve into the specific sources of noise emissions within each industrial context to develop targeted mitigation strategies and enhance the overall work environment.

The presented data revealed a wide spectrum of air quality parameters across diverse industrial settings, reflecting the intricate interplay of industrial processes, emissions sources, and pollution control measures. Industries labeled as I-3, I-4, I-8, and I-9 are observed to have significantly high levels of particulate matter (including PM10, PM2.5, PM1), along with inhalable and thoracic particles. This indicates a high potential for considerable particulate emissions, posing increased risks to health and the environment.

Alveolic particulate matter, representing the finest particles with deeper health implications, is notably elevated in I-3 and I-9. The variability within each industrial setting, as indicated by standard deviations, underscores the complexity of air quality dynamics. Notably, data gaps in I-5 and I-6 hint at unmonitored or under-reported air quality conditions. These findings emphasize the importance of continuous monitoring and robust pollution control measures, particularly in industries with elevated particulate matter levels, to protect public health and the environment while promoting sustainable industrial practices.

The dataset provides the average concentrations of BC, measured in micrograms per cubic meter ($\mu g/m^3$), at various industrial locations. This data is crucial for understanding the extent of BC emissions across different industries. Notably, some sites exhibit high BC levels, possibly indicating significant BC emissions requiring closer scrutiny. Conversely, some demonstrated a lower BC concentration, suggesting better emission controls or environmental conditions. High BC concentrations can have adverse health and environmental impacts, underscoring the need for rigorous monitoring, regulatory compliance, and potential mitigation efforts at these industrial locations. Our study confirms that indoor BC predominantly falls within the PM2.5 and PM1 size ranges, likely originating from vehicular exhaust and diesel generators. Outdoors, BC correlations with specific particle size ranges indicate a mixture of sources, with finer BC particles linked to vehicular emissions and coarser BC particles potentially arising from industrial fossil fuel combustion. Indoor BC levels vary widely with particle size due to industrial activities, with smaller BC particles being the dominant fraction.

The hazard indices suggest variations in estimated health risks across different categories, with females generally having higher estimated risks than males. The hazard indices for different industries suggest variations in estimated health risks, with females generally having higher estimated risks than males within each industry. The significance of these results depends on the specific hazards associated with each industry and should be used as a basis for targeted risk management strategies and regulatory actions to protect the health of workers in these industries. Further investigation into the underlying causes of gender differences in risk is also warranted.

Acknowledgements The authors would like to express their heartfelt gratitude to the proprietors of the industries under study for allowing access to their facilities during operational hours. This cooperative endeavor has greatly contributed to the research and highlighted the significance of collaboration in furthering our comprehension of the effects of industrial activities on air quality.

References

1. Ali MU, Yu Y, Yousaf B, Munir MAM, Ullah S, Zheng C, Kuang X, Wong MH (2021) Health impacts of indoor air pollution from household solid fuel on children and women. J Hazard Mater 416. https://doi.org/10.1016/j.jhazmat.2021.126127
2. Anwar MN, Shabbir M, Tahir E, Iftikhar M, Saif H, Tahir A, Murtaza MA, Khokhar MF, Rehan M, Aghbashlo M, Tabatabaei M, Nizami AS (2021) Emerging challenges of air pollution and particulate matter in China, India, and Pakistan and mitigating solutions. J Hazard Mater 416:125851. https://doi.org/10.1016/J.JHAZMAT.2021.125851
3. Bakhtiari AR, Zakaria MP, Yaziz MI, Lajis MNHL, Bi X (2014) Environment Asia. Environ Asia 7:104–111
4. Baldelli A (2021) Evaluation of a low-cost multi-channel monitor for indoor air quality through a novel, low-cost, and reproducible platform. Meas Sens 17. https://doi.org/10.1016/j.measen.2021.100059

5. Bell ML, Zanobetti A, Dominici F (2013) Evidence on vulnerability and susceptibility to health risks associated with short-term exposure to particulate matter: a systematic review and meta-analysis. Am J Epidemiol 178:865–876. https://doi.org/10.1093/AJE/KWT090
6. Bhat AH, Dar KB, Anees S, Zargar MA, Masood A, Sofi MA, Ganie SA (2015) Oxidative stress, mitochondrial dysfunction and neurodegenerative diseases; a mechanistic insight. Biomed Pharmacother 74:101–110. https://doi.org/10.1016/j.biopha.2015.07.025
7. Bond TC, Doherty SJ, Fahey DW, Forster PM, Berntsen T, DeAngelo BJ, Zender CS (2013) Bounding the role of black carbon in the climate system: A scientific assessment. Journal of Geophysical Research: Atmospheres 118(11):5380–5552. https://doi.org/10.1002/jgrd.50171
8. CCA (2022) Black carbon. Climate & Clean Air Coalition [WWW Document]. Climate and Clean Air Coalition. URL https://www.ccacoalition.org/en/slcps/black-carbon. Accessed 21 Mar 23
9. CPCB (2001) Noise pollution regulations in India
10. Cohen AJ, Brauer M, Burnett R, Anderson HR, Frostad J, Estep K, Balakrishnan K, Brunekreef B, Dandona L, Dandona R, Feigin V, Freedman G, Hubbell B, Jobling A, Kan H, Knibbs L, Liu Y, Martin R, Morawska L, Pope CA, Shin H, Straif K, Shaddick G, Thomas M, van Dingenen R, van Donkelaar A, Vos T, Murray CJL, Forouzanfar MH (2017) Estimates and 25-year trends of the global burden of disease attributable to ambient air pollution: an analysis of data from the global burden of diseases Study 2015. The Lancet 389:1907–1918. https://doi.org/10.1016/S0140-6736(17)30505-6
11. Deng WJ, Zheng HL, Tsui AKY, Chen XW (2016) Measurement and health risk assessment of PM2.5, flame retardants, carbonyls and black carbon in indoor and outdoor air in kindergartens in Hong Kong. Environ Int 96:65–74. https://doi.org/10.1016/J.ENVINT.2016.08.013
12. Dhada I, Nagar PK, Sharma M, Gupta N (2016) Photo-catalytic oxidation technology for VOC control: evaluation and risk characterization of intermediates of benzene degradation adsorbed on catalyst. Environ Eng Sci 33:970–977. https://doi.org/10.1089/ees.2016.0109
13. EPA, 2000. Science Policy Council Handbook: Risk characterization, vol 31
14. Gong T, Sun Z, Zhang X, Zhang Y, Wang S, Han L, Zhao D, Ding D, Zheng C (2019) Associations of black carbon and PM2.5 with daily cardiovascular mortality in Beijing, China. Atmos Environ 214. https://doi.org/10.1016/j.atmosenv.2019.116876
15. González-Martín J, Kraakman NJR, Pérez C, Lebrero R, Muñoz R (2021) A state–of–the- art review on indoor air pollution and strategies for indoor air pollution control. Chemosphere 262. https://doi.org/10.1016/j.chemosphere.2020.128376
16. Gramsch E, Reyes F, Oyola P, Rubio MA, López G, Pérez P, Martínez R (2014) Particle size distribution and its relationship to black carbon in two urban and one rural site in Santiago de Chile. J Air Waste Manage Assoc 64:785–796. https://doi.org/10.1080/10962247.2014.890141
17. Janssen NAH, Hoek G, Simic-Lawson M, Fischer P, van Bree L, Brink HT, Keuken M, Atkinson RW, Ross Anderson H, Brunekreef B, Cassee FR (2011) Black carbon as an additional indicator of the adverse health effects of airborne particles compared with PM10 and PM2.5. Environ Health Perspect 119:1691–1699. https://doi.org/10.1289/ehp.1003369
18. Kahnert M, Kanngießer F (2020) Modelling optical properties of atmospheric black carbon aerosols. J Quant Spectrosc Radiat Transf 244:106849. https://doi.org/10.1016/J.JQSRT.2020.106849
19. Kampa M, Castanas E (2008) Human health effects of air pollution. Environ Pollut 151:362–367. https://doi.org/10.1016/j.envpol.2007.06.012
20. Kankaria A, Nongkynrih B, Gupta SK (2014) Indoor air pollution in India: implications on health and its control. Indian J Community Med 39:203–207. https://doi.org/10.4103/0970-0218.143019
21. Khan MDH, Sarkar MS, Haque SS, Hossain MA (2021) Health risk assessment of black carbon emission from fossil fuel. J Eng Sci 12:23–28. https://doi.org/10.3329/jes.v12i2.54628
22. Kirrane EF, Luben TJ, Benson A, Owens EO, Sacks JD, Dutton SJ, Madden M, Nichols JL (2019) A systematic review of cardiovascular responses associated with ambient black carbon and fine particulate matter. Environ Int. https://doi.org/10.1016/j.envint.2019.02.027

23. Kreyling WG, Semmler M, Erbe F, Mayer P, Takenaka S, Schulz H, Oberdörster G, Ziesenis A (2002) Translocation of ultrafine insoluble iridium particles from lung epithelium to Extrapulmonary organs is size dependent but very low. J Toxicol Environ Health A 65:1513–1530. https://doi.org/10.1080/00984100290071649
24. Lequy E, Siemiatycki J, de Hoogh K, Vienneau D, Dupuy JF, Garès V, Hertel O, Christensen JH, Zhivin S, Goldberg M, Zins M, Jacquemin B (2021) Contribution of long-term exposure to outdoor black carbon to the carcinogenicity of air pollution: evidence regarding risk of cancer in the Gazel cohort. Environ Health Perspect 129. https://doi.org/10.1289/EHP8719
25. Li G, Wu S, Wang L, Akoh CC (2016) Concentration, dietary exposure and health risk estimation of polycyclic aromatic hydrocarbons (PAHs) in youtiao, a Chinese traditional fried food. Food Control 59:328–336. https://doi.org/10.1016/J.FOODCONT.2015.06.003
26. Liang F, Tian L, Guo Q, Westerdahl D, Liu Y, Jin X, Li G, Pan X (2017) Associations of PM2.5 and black carbon with hospital emergency room visits during heavy haze events: a case study in Beijing, China. Int J Environ Res Public Health 14. https://doi.org/10.3390/ijerph14070725
27. Liu J, Yang N, Liu Y (2014) High-risk factors of respiratory distress syndrome in term neonates: a retrospective case-control study. Balkan Med J 2014:64–68. https://doi.org/10.5152/BALKANMEDJ.2014.8733
28. Van Der Meer AJ, Veldt BJ, Feld JJ, Wedemeyer H, Dufour JF, Lammert F, Duarte-Rojo A, Heathcote EJ, Manns MP, Kuske L, Zeuzem S, Hofmann WP, De Knegt RJ, Hansen BE, Janssen HLA (2012) Association between sustained virological response and all-cause mortality among patients with chronic hepatitis C and advanced hepatic fibrosis. JAMA 308:2584–2593. https://doi.org/10.1001/JAMA.2012.144878
29. Mordukhovich I, Coull B, Kloog I, Koutrakis P, Vokonas P, Schwartz J (2015) Exposure to subchronic and long-term particulate air pollution and heart rate variability in an elderly cohort: the normative aging study. Environ Health 14:1–10. https://doi.org/10.1186/S12940-015-0074-Z/TABLES/4
30. Niranjan R, Thakur AK (2017) The toxicological mechanisms of environmental soot (black carbon) and carbon black: focus on oxidative stress and inflammatory pathways. Front Immunol 8:763. https://doi.org/10.3389/fimmu.2017.00763
31. Nordeide Kuiper I, Svanes C, Markevych I, Accordini S, Bertelsen RJ, Bråbäck L, Heile Christensen J, Forsberg B, Halvorsen T, Heinrich J, Hertel O, Hoek G, Holm M, de Hoogh K, Janson C, Malinovschi A, Marcon A, Miodini Nilsen R, Sigsgaard T, Johannessen A (2021) Lifelong exposure to air pollution and greenness in relation to asthma, rhinitis and lung function in adulthood. Environ Int 146. https://doi.org/10.1016/j.envint.2020.106219
32. Onoda A, Umezawa M, Takeda K, Ihara T, Sugamata M (2014) Effects of maternal exposure to ultrafine carbon black on brain perivascular macrophages and surrounding astrocytes in offspring mice. PLoS One 9. https://doi.org/10.1371/journal.pone.0094336
33. Paunescu AC, Casas M, Ferrero A, Pañella P, Bougas N, Beydon N, Just J, Lezmi G, Sunyer J, Ballester F, Momas I (2019) Associations of black carbon with lung function and airway inflammation in schoolchildren. Environ Int 131. https://doi.org/10.1016/j.envint.2019.104984
34. Peng C, Bind MAC, Colicino E, Kloog I, Byun HM, Cantone L, Trevisi L, Zhong J, Brennan K, Dereix AE, Vokonas PS, Coull BA, Schwartz JD, Baccarelli AA (2016) Particulate air pollution and fasting blood glucose in nondiabetic individuals: associations and epigenetic mediation in the normative aging study, 2000–2011. Environ Health Perspect 124:1715–1721. https://doi.org/10.1289/EHP183
35. Pöschl U (2005) Atmospheric chemistry atmospheric aerosols: composition, transformation, climate and health effects. Angew Chem Int Ed 44. https://doi.org/10.1002/anie.200501122
36. Raparthi N, Phuleria HC (2021) Real-world vehicular emissions in the Indian megacity: carbonaceous, metal and morphological characterization, and the emission factors. Urban Clim 39:100955. https://doi.org/10.1016/J.UCLIM.2021.100955
37. Sarkar C, Roy A, Chatterjee A, Ghosh SK, Raha S (2019) Factors controlling the long-term (2009–2015) trend of PM2.5 and black carbon aerosols at eastern Himalaya India. Sci Total Environ 656:280–296. https://doi.org/10.1016/J.SCITOTENV.2018.11.367

38. Singh AA, Fatima A, Mishra AK, Chaudhary N, Mukherjee A, Agrawal M, Agrawal SB (2018) Assessment of ozone toxicity among 14 Indian wheat cultivars under field conditions: growth and productivity. Environ Monit Assess 190:1–14. https://doi.org/10.1007/S10661-018-6563-0/FIGURES/5
39. Singh D, Kumar A, Kumar K, Singh B, Mina U, Singh BB, Jain VK (2016) Statistical modeling of O3, NOx, CO, PM2.5, VOCs and noise levels in commercial complex and associated health risk assessment in an academic institution. Sci Total Environ 572:586–594. https://doi.org/10.1016/J.SCITOTENV.2016.08.086
40. Suglia SF, Gryparis A, Schwartz J, Wright RJ (2008) Association between traffic-related black carbon exposure and lung function among urban women. Environ Health Perspect 116:1333–1337. https://doi.org/10.1289/ehp.11223
41. Tiwari S, Pipal AS, Srivastava AK, Bisht DS, Pandithurai G (2015) Determination of wood burning and fossil fuel contribution of black carbon at Delhi, India using aerosol light absorption technique. Environ Sci Pollut Res 22:2846–2855. https://doi.org/10.1007/s11356-014-3531-2
42. USEPA (2015) Regional screening levels (RSLs)—generic tables (November 2015) [WWW Document]. United States Environmental Protection Agency. URL https://www.epa.gov/risk/regional-screening-levels-rsls-generic-tables-november-2015. Accessed 26 Mar 23
43. USEPA (1989) Risk management guidance for superfund, vol 2. Environmental Evaluation Manual. Office of Emergency and Remedial Response Washington, DC, USEPA. EPA/540/1-89/001. I
44. USEPA (2011) Exposure Factors Handbook: 2011 Edition (Final Report) (EPA/600/R-09/052F). Washington, DC: National Center for Environmental Assessment
45. Valavanidis A, Fiotakis K, Vlachogianni T (2008) Airborne particulate matter and human health: toxicological assessment and importance of size and composition of particles for oxidative damage and carcinogenic mechanisms. J Environ Sci Health C Environ Carcinog Ecotoxicol Rev 26:339–362. https://doi.org/10.1080/10590500802494538
46. Viidanoja J, Sillanpää M, Laakia J, Kerminen VM, Hillamo R, Aarnio P, Koskentalo T (2002) Organic and black carbon in PM2.5 and PM10: 1 year of data from an urban site in Helsinki. Finland. Atmos Environ 36:3183–3193. https://doi.org/10.1016/S1352-2310(02)00205-4
47. Wang X, Chen R, Meng X, Geng F, Wang C, Kan H (2013) Associations between fine particle, coarse particle, black carbon and hospital visits in a Chinese city. Sci Total Environ 458–460:1–6. https://doi.org/10.1016/J.SCITOTENV.2013.04.008
48. World Health Organization (2006) Air quality guidelines: Global update 2005. Particulate matter, ozone, nitrogen dioxide and sulfur dioxide. WHO Regional Office for Europe ISBN 92 890 2192 6
49. World Health Organization (2018) WHO global ambient air quality database (update 2018). Retrieved from WHO Air Quality Database 2018
50. World Health Organization (2021) World health statistics 2021: Monitoring health for the SDGs, sustainable development goals. Geneva: World Health Organ ISBN: 978-92-4-002705-3
51. World Health Organization (2005) Health effects of transport-related air pollution—World Health Organization—Google books [WWW Document]. URL https://books.google.co.in/books?hl=en&lr=&id=b2G3k51rd0oC&oi=fnd&pg=PR1&dq=13.%09World+Health+Organization.+Health+effects+of+transport-related+air+pollution.+Copenhagen:+WHO+Regional+Office+for+Europe%3B+2005.+p.+125-65&ots=O86sbBCm3w&sig=0rBpc2IgVGhiLx. Accessed 24 Mar 2023

Assessment of Benzene Exposure Through Urinary S-Phenylmercapturic Acid (SPMA) in the Adult Population of West Bengal Emphasizing on Sociodemographic and Lifestyle Factor

Anupa Yadav, Aniruddha Mukhopadhayay, Amit Chakrabarti, Asim Saha, and Pritha Bhattacharjee

Abstract Evaluated benzene exposure in West Bengal's adult population through urinary S-phenylmercapturic acid (SPMA), aiming to identify sociodemographic and lifestyle factors influencing environmental benzene exposure. SPMA ($N = 258$) quantified by high-pressure liquid chromatography (HPLC). Sociodemographic and lifestyle data collected through questionnaire. Obtained mean SPMA 4.60 μg/g, with 95th percentile value 12.62 μg/g. Urban dwellers exhibited higher SPMA than rural ($p < 0.01$). Females had elevated SPMA than males. Highest SPMA (5.11 μg/g) recorded in < 35 years age group. Low-income groups (LIG), urban LIG, smoker males and those females exposed to passive tobacco smoke had higher SPMA. Urban females engaged in cooking had higher SPMA ($p = 0.05$). Smoker male travellers had 2.5 times higher SPMA than their counterparts. Traffic emissions and tobacco smoke were primary sources of benzene exposure in urban dwellers, while biomass fuel burning in rural. Urgent action is needed, emphasizing implementation of targeted government initiatives like clean fuel and efficient cook stoves to reduce benzene exposure in study areas.

Keywords Environmental benzene · Urinary SPMA · Urban · Rural · Biomass fuel burning

A. Yadav · A. Chakrabarti · A. Saha
Centre for Ageing and Mental Health (I-CAM), Indian Council of Medical Research (ICMR), Kolkata, India
e-mail: anupayadav08@gmail.com

A. Yadav · A. Mukhopadhayay · P. Bhattacharjee (✉)
Department of Environmental Science, University of Calcutta, Kolkata, India
e-mail: pbenvs@caluniv.ac.in

S. M. Shiva Nagendra et al. (eds.), *Select Proceedings of the 8th Indian International Conference on Air Quality Management (IICAQM 2023)*, Lecture Notes in Civil Engineering 582, https://doi.org/10.1007/978-981-96-2359-4_19

1 Introduction

Benzene, a volatile organic compound, serves as a raw material in various products such as gasoline, plastic, rubber, and detergents [22, 55]. Natural and anthropogenic sources, including industrial processes, gasoline vapours, vehicular emissions, wood burning, and tobacco smoking, contribute to airborne benzene [1, 17, 52, 67]. Due to its widespread presence and carcinogenic nature, benzene warrants significant attention. The World Health Organization (WHO) designates it as one of the ten priority chemicals for study and regulation [3]. European Commission regulations (Directive 98/70/EC) limit gasoline benzene content to 1% by volume (Directive 98/70/CE). Indian gasoline currently contains 1% benzene, adhering to regulatory standards with an ambient air permissible limit of 5 $\mu g/m^3$ for benzene (annual average) [18]. Factors influencing benzene exposure among non-occupationally exposed individuals include traffic density, benzene content in petrol, wood burning, and smoking. Research in developing nations reveals that the indoor use of simple, inefficient cooking stoves with solid biomass fuels leads to elevated VOC emissions, including benzene, with emission levels affected by factors like burning temperature and oxygen availability [30, 36]. Majumdar and Srivastava [65] reported high benzene levels (103.19 $\mu g/m^3$) in Indian kitchens.

Study from Kolkata city recorded a peak ambient benzene concentration of 214.8 $\mu g/m^3$ during traffic hours [43], with mean levels ranging from 13.8–72.0 $\mu g/m^3$ in residential and commercial areas [44]. These benzene levels exceed the permissible NAASQ limit [13, 43, 67]. More studies from Kolkata indicate a continuous rise in air benzene-1.6 $\mu g/m^3$ (2003), 2.5 $\mu g/m^3$ (2005), and 5.19 $\mu g/m^3$ (2018) [29, 48]. Majumdar and Mukherjee [44] found an association between environmental air VOCs to cancer risk in Kolkata residents, surpassing the cancer factor threshold of 1×10^{-6}. Smoking is a significant benzene source for both active and passive exposure [26, 27, 53]. In smokers' rooms, high benzene levels (48 $\mu g/m^3$) were measured (Majumdar and Srivastava 2011). Johnson et al. [36] noted that smokers receive 90% of their benzene exposure from cigarette smoke, with indoor smoking contributing to about half of estimated benzene exposure [2, 36, 72].

Benzene, an IARC group 1 human carcinogen [10, 33], poses health risks with any exposure due to its non-threshold nature [61]. Acute exposure can result in symptoms like headaches, dizziness, fatigue, convulsions, mucosal irritation, excitement, depression, and may lead to death from respiratory failure [70, 4]. Chronic exposure to benzene is linked to non-cancerous diseases, including an increased risk of non-Hodgkin's lymphoma, hematotoxicity, acute myeloid, multiple myeloma, and other hematopoietic disorders [25, 28, 34, 55, 59, 66]. Benzene exposure could cause metabolic syndrome [6, 58, 74], diabetes, cardiovascular disorders [20, 40, 74], immunological issues [15, 37], respiratory problems [15, 37], and reproductive disorders. In males, it can cause alterations in sperm characteristics, while in females, it is associated with preterm birth, pregnancy complications, and reduced fertility [32, 54, 56].

Exposure to environmental benzene is monitored through its biomarker of exposure S-phenylmercapturic acid [7]. Biomarker concentration in unexposed subjects reflects biological susceptibility and lifestyle variability [8, 10]. Only 0.11% of absorbed benzene transforms into SPMA [16]. With a 10-h biological half-life, SPMA is a specific biomarker for recent benzene exposure monitoring [10, 59]. This biomarker is not affected by diet, but smoking, a strong confounding factor, can influence its concentration in smokers and those exposed to environmental tobacco smoke (ETS) [26, 27, 53]. The American Conference of Governmental Industrial Hygienists (ACGIH) sets a biological exposure limit of 25 μg/g [5]. Silva et al. [59] found SPMA associated (0.80–8.01 μg/g) with odds ratios (OR) of 0.78, 1.29, 2.30, and 1.60 for changes in leukocytes, mean corpuscular volume (MCV), neutrophils, and haematological changes, respectively.

Monitoring benzene exposure in occupational groups is simplified by well-defined work environments. Conversely, quantifying exposure in the general public is challenging due to diverse lifestyles and varied work environments [10]. Limited Indian data exist on personal-level benzene exposure for risk assessment, and no reference values for urinary SPMA are available to orient public health policies for the reduction in exposure to environmental benzene as well. This study aimed to measure benzene exposure in occupationally unexposed adults in urban and rural West Bengal, India, identifying socioeconomic and lifestyle factors influencing environmental benzene exposure.

2 Material and Methods

2.1 *Study Population*

The study was conducted in West Bengal, India, between November 2021 and February 2022 to evaluate environmental exposure to benzene in adult individuals aged 22 to 80 years (mean age 50 years). It included 258 healthy individuals (129 males and 129 females) selected by random sampling from both rural and urban areas who had been residing in the study area for more than one year and did not have any medical history or health problems. Both working and non-working individuals were included, and they did not have occupational exposure to benzene. The criteria of exclusion were residing in an industrial zone and staying at their current address for less than one year.

2.2 Ethical Approval and Data Collection

The study, approved by the Institutional Ethical Committee (approval number 15thIEC/CNCD/5.5 dated 15.07.2021), ensured participant understanding and confidentiality. Face-to-face interviews with a structured questionnaire covered family details, employment, and lifestyle, encompassing sociodemographics, anthropometrics, personal/family history, employment/residential history, and lifestyle factors like smoking, travelling, and cooking practices.

2.3 Variables Information

Individuals were grouped by age (<35 years, 35–50 years, > 50–65 years, and > 65 years), food cooking (NA and Yes), smoking (Yes and No), and travelling (Yes and No, the former implying a daily commute for occupational or other household purpose), study area (rural and urban). Study individuals were also grouped based on their monthly family income into low-income group (LIG) earning < 20 K/Month, middle-income group (MIG) earning up to 50 K/Month, and high-income group (HIG) earning > 50 K/Month. Females were further divided into two groups based on their (1) employment status-(a) homemakers as well as working and (b) homemakers only; and (2) exposure to tobacco smoke—(a) those who were passively exposed and (b) those who were not. Males were also divided into two groups based on their smoking and travelling activity (1) smoker traveller and (2) non-smoker non-traveller. These classifications were done to better understand benzene exposure from the ambient environment via tobacco smoke and traffic pollutants.

2.4 Urine Collection and SPMA Analysis

Individuals provided 100 mL morning mid-stream urine in sterilized containers (Tarson) the day after the interview. Samples, labelled for identification, were refrigerated during transport and stored at − 20°C for subsequent analysis.

SPMA was extracted from the urine sample as described by Osamu Inoue et al. [35], briefly explained here, the urine sample was stored at − 20°C were taken out and kept to attain room temperature. 500 μL of urine was mixed with 500 μL of ultrapure water and 200μL sulfuric acid (9.9N). After 10 min of sulfuric acid addition, this strong acid mixture was made weakly acidic by the addition of 200μL of potassium hydroxide (7.8N), being a final pH of about 3. The mixture was extracted with 4 mL of methanol:ether mixture (1:9); and homogenized by mechanical shaking for 3 min. Then the whole content was centrifuged for 10 min at 1600 × g, to separate out the organic phase. The upper organic layer (3 mL) was taken into another test tube, and

500 μL of mobile phase was mixed into it. Its volume was reduced to 500μL under nitrogen stream, and 10 μL of it was injected into the HPLC system for analysis.

HPLC (Shimadzu, Japan) coupled with system controller (Model: SLC-10 A), connected to an automated liquid sampler (Model: LC-10AT), detector (Model: SPD—M 10A), column oven, and a monitor. The chromatographic condition was a flow rate of 0.8 mL/min; reverse phase column ODS-2 Hypersil (250 mm, 4.6 mm, and 5 μm) and oven temperature was 40 °C. SPMA Chromatogram was detected at λ_{max} 253 nm. The run time of linear gradient programme was 20 min for each sample run, retention time (RT) was at around 3.8 min [35]. Quality control measures such as recovery study and inter or intra-day precision measurement were well implemented. The intra-day and inter-day precision were 0.5–1.2% and 1.5–2%, respectively, for concentrations1 μg/L to 25 μg/L. Study recovery ranged between 98 and 114% at spiked concentration 3 times of LOQ. The limit of quantification was 15 μg/L. The limit of detection was 5 μg/L.

2.5 *Creatinine Estimation*

Urine creatinine was measured using the Jaffe method with a reagent kit from Arkray (ARKRAY, India) on a UV–visible spectrophotometer (Model: Lambda 45, Perkin Elmer, USA) at 520 nm [50, 71].

2.6 *Statistical Analyses*

Data analysis performed in Microsoft Excel. Urinary SPMA levels (μg/g creatinine) were assessed by nonparametric tests due to skewed distribution, verified by the Kolmogorov–Smirnov test. Descriptive statistics were applied. Mann–Whitney U-test analysed SPMA differences among individuals. SPSS software used for statistical calculations. Samples below the LOD (about 5%) used instrument-measured concentrations, following Gyllenhammar et al. [31].

3 Results

3.1 *Demographic Characteristics of the Study Population*

Table 1 illustrates sociodemographic characteristics (area-wise) by gender, age group, education, family income, cooking, travel, and smoking status for study individuals. Initially, 341 individuals were involved, however 83 were excluded due to questionnaire non-response, inadequate urine samples, damaged containers, and

Table 1 Demographic characteristics of study population ($N = 258$)

Variables	Categories	Urban ($N = 186$)	Rural ($N = 72$)	Total ($N = 258$)
Gender N (%)	Male	102 (55%)	27 (38%)	129 (50%)
	Female	84 (45%)	45 (62%)	129 (50%)
Age group (years) N (%)	< 35	29 (16%)	08 (11%)	37 (14%)
	35–50	92 (49%)	27 (38%)	119 (46%)
	> 50–65	43 (23%)	21 (29%)	64 (25%)
	> 65	22 (12%)	16 (22%)	38 (15%)
Education N (%)	No Formal Edu	34 (39%)	21 (29%)	55 (21%)
	Primary	32 (35%)	30 (42%)	62 (24%)
	High School	54 (58%)	14 (19%)	68 (26%)
	S Sec. School	27 (27%)	0 (0%)	27 (11%)
	College	39 (40%)	7 (10%)	46 (18%)
Family income N (%)	LIG	111 (60%)	64 (89%)	175 (68%)
	MIG	65 (35%)	8 (11%)	73 (28%)
	HIG	10 (5%)	0 (0%)	10 (04%)
Cooking N (%)	NA	113 (61%)	33 (46%)	146 (57%)
	Yes	73 (39%)	39(54%)	112(43%)
Travelling N (%)	No	100 (54%)	63 (88%)	163 (63%)
	Yes	86 (46%)	9(12%)	95(37%)
Smoking N (%)	No	157 (84%)	67 (93%)	224 (87%)
	Yes	29 (16%)	5 (7%)	34 (13%)

Abbreviations: LIG, low-income group; MIG, middle-income group; HIG, high-income group; NA, not applicable

urine creatinine levels beyond ACGIH recommendations (<0.3 g/L or > 3.0 g/L) [8]. Consequently, final sample comprised 258 individuals meeting inclusion criteria, with a balanced gender ratio (1:1). Most were aged 35–50 years (46%), had up to High School education (26%), and fell into the low-income group (68%). Individuals involved in cooking 43%, daily commuters were 37%, and smokers constituted ~ 13% of the total. Urban participants were from specific coordinates (N 22°29′40″ E 88°23′01″ residential and N 22°34′56″ E 88°26′21″ commercial). The rural control site was Debipur village (N 22°01′23″ E 87°56′45″), Fig. 1.

3.2 *SPMA Levels Distribution with Demographic and Lifestyle Factors*

Table 2 displays SPMA results in μg/g, following ACGIH preferences. Creatinine-adjusted values were used for comparison with literature. Mean SPMA was 4.60 μg/

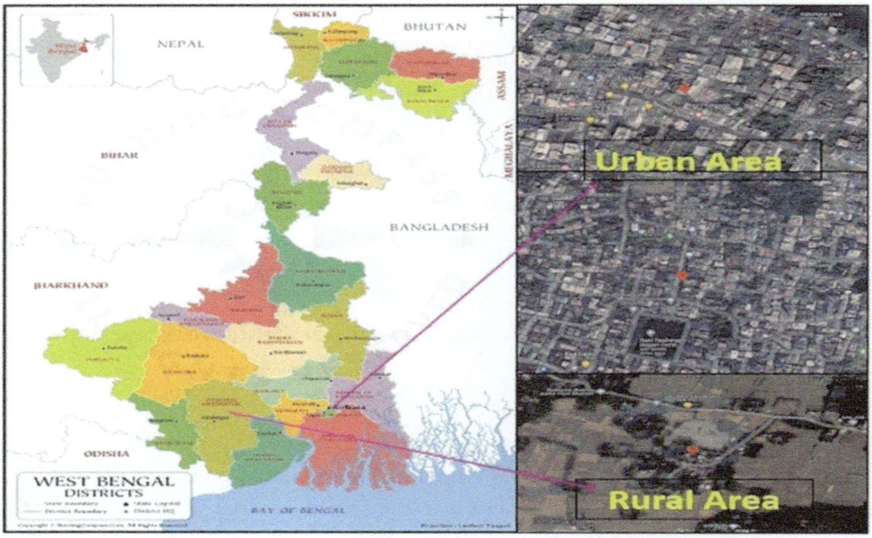

Fig. 1 Study sites in Kolkata city and Debipur Village, West Bengal, India

Table 2 Urinary concentration of SPMA (μg/g) in study participants ($N = 258$)

	N (%)	Mean	Range	*p*-value	Median	Percentile		
						25th	75th	95th
Total	258 (100%)	4.60	0.51–21.02		3.23	1.46	6.65	12.62
Male	129(50%)	3.98	0.51–21.02	0.3	2.54	1.45	6.46	15.25
Female	129(50%)	4.37	0.35–19.83		3.42	1.51	6.65	11.39
Urban	186(72%)	4.77	0.61–21.02	0.00003	3.00	1.54	6.53	13.2
Rural	72(28%)	4.15	0.20–10.99		3.74	1.27	6.75	9.83

g (median 3.23 μg/g, range 0.51–21.02 μg/g). Interquartile range (i.e. 25th and 75th percentile) details are also provided in Table 2. SPMA was higher in females (4.37 μg/g) but not significantly different from males (3.98 μg/g). However, urban individuals had significantly higher SPMA levels ($p = 0.00003$) than rural ones. The 95th percentile of SPMA was 12.62 μg/g, 15.25 μg/g, 11.39 μg/g, 13.2 μg/g, and 9.83 μg/g for total, males, females, urban, and rural individuals, respectively.

3.3 Age

Figure 2a shows the urinary SPMA variation trend with age. Individuals < 35 years had the highest SPMA levels (5.11 μg/g, 5.17 μg/g, 4.9 μg/g in total, urban, and rural, respectively). Conversely, those > 65 years had the lowest SPMA levels (0.88 μg/g,

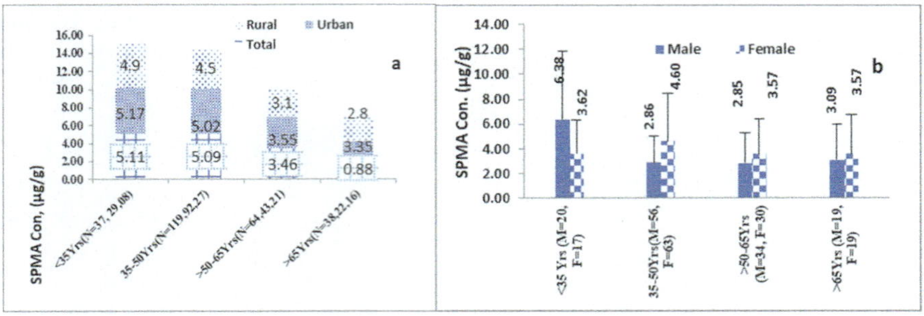

Fig. 2 Urinary SPMA level (µg/g) with age **a** in total study individuals, and **b** in males and females of same age group

3.35 µg/g, 2.8 µg/g in total, urban, and rural, respectively). This reveals a decreasing SPMA trend with age in both areas. Figure 2b depicts gender-specific SPMA differences across age groups. After 35 years age, females had higher SPMA levels, while males had higher levels before 35 years age. Statistically significant gender differences were found in the < 35 years and 35–50 years age groups ($p = 0.01$ and $p = 0.00001$, respectively).

3.4 *Family Income*

Family income significantly influences environmental benzene exposure, impacting SPMA excretion. Individuals of HIG and MIG were merged for this analysis. Low-income group (LIG) individuals exhibited higher SPMA (5.2 µg/g) compared to the merged middle and high-income groups (MIG and HIG, 3.34 µg/g), with statistically significant difference ($p < 0.01$, Fig. 3a. We further categorized LIG individuals based on their study area. Urban LIG individuals had significantly ($p < 0.01$) higher SPMA than rural LIG individuals (Fig. 3b).

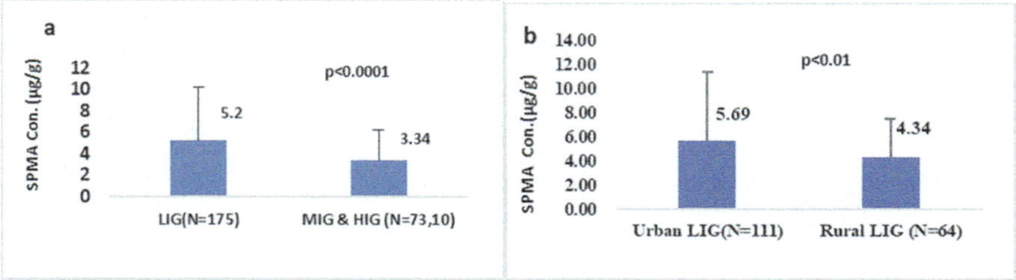

Fig. 3 Levels of urinary SPMA (µg/g) in study individuals **a** LIG, merged MIG and HIG; **b** urban LIG and rural LIG

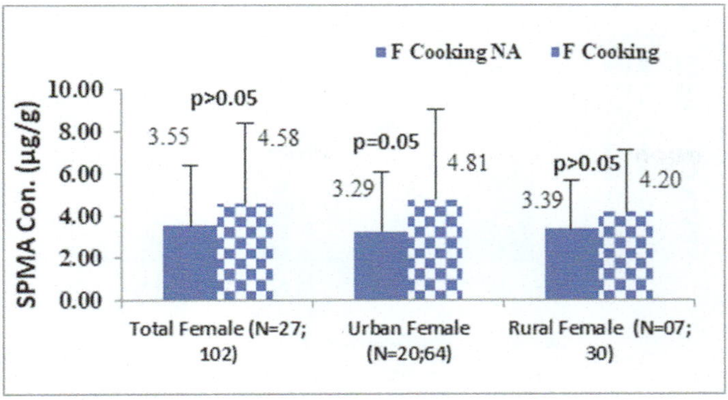

Fig. 4 Levels of SPMA (µg/g) in females based on cooking practice

3.5 Cooking Practices

Only female individuals were predominantly involved in household cooking. The smoke released during solid biomass fuel burning in the cooking process is a crucial source of benzene exposure via inhalation in females. Study results found that females who were involved in cooking practice had higher SPMA (4.58 µg/g, 4.81 µg/g, and 4.20 µg/g) for the total, urban, and rural female participants, respectively, than their counterparts those who were not involved in any kind of food cooking, as presented in Fig. 4. This difference was significant ($p = 0.05$) only in the urban area females. However, there was no significant difference observed in total and rural area females.

3.6 Smoking

The study investigated SPMA association with smoking (active or passive) in urban participants. Smoking was absent in rural individuals. To avoid confounding factors, if any, from vehicular-emitted benzene exposure for this analysis, we have included only non-traveller males and were categorized into smokers ($N = 12$) and non-smokers ($N = 39$). Females were grouped based on passive tobacco smoke exposure—those exposed ($N = 20$) and those without exposure ($N = 38$). Results in Fig. 5a revealed significantly higher SPMA levels in smoking males (4.21 µg/g, $p = 0.03$) and females with passive smoke exposure (4.94 µg/g, $p = 0.02$) compared to their counterparts. The study implies tobacco smoke as a potential benzene exposure source, not only for smokers but also those exposed to second-hand smoke.

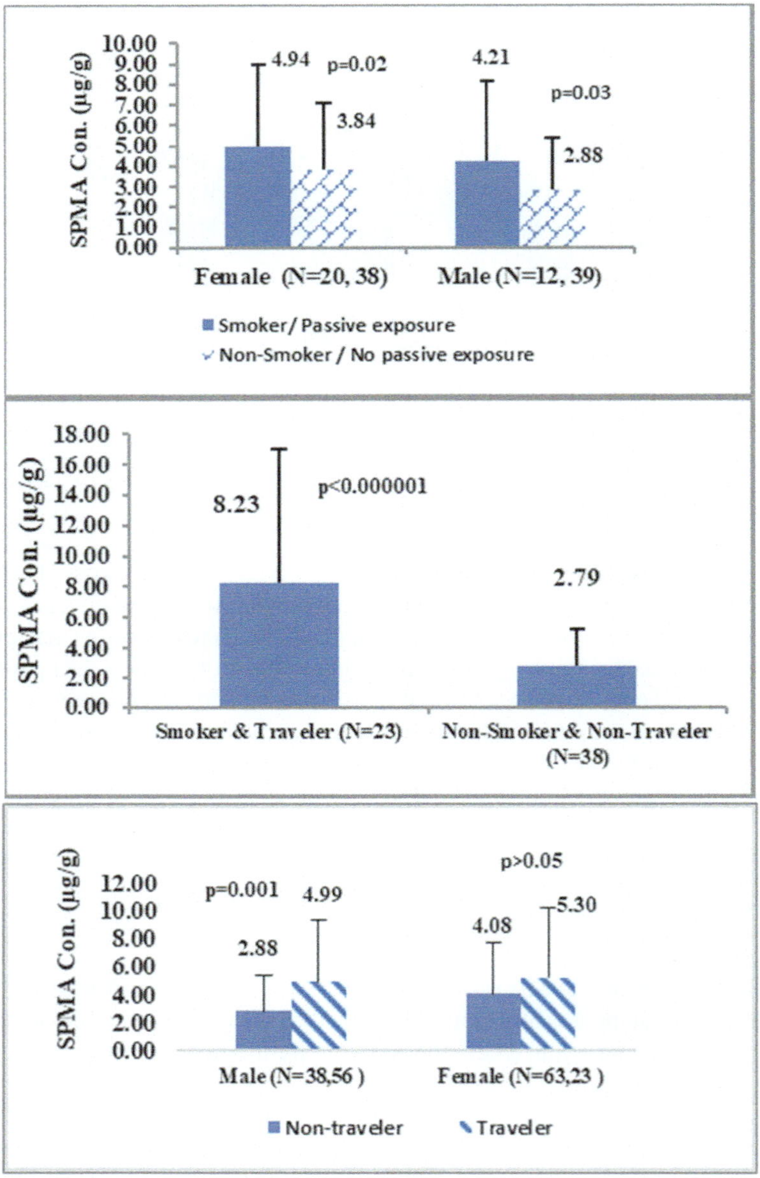

Fig. 5 **a** Variation in SPMA (µg/g) concentration with tobacco smoke among urban male and female; **b** Variation in SPMA (µg/g) concentration with traffic pollution in urban male and female; **c** Combined impact of tobacco smoke and traffic pollutants on levels of SPMA (µg/g) in male individuals

3.7 Travelling

For those individuals without direct benzene exposure, vehicular emissions posed a notable risk in urban areas. Urban individuals, both males and females, were categorized as travellers (56male; 23female) and non-travellers (38male; 63female) to explore traffic-related benzene exposure. Rural areas lacked daily commuters for traffic pollution assessment. To minimize tobacco smoke effects on SPMA in males, only non-smoker males were included, categorized as travellers and non-travellers. All females were non-smokers and also categorized similarly. Mean SPMA levels were higher in travellers (male 4.99 μg/g; female 5.30 μg/g) than non-travellers (male 2.88 μg/g; female 4.08 μg/g, Fig. 5b). Traveller males showed significantly ($p = 0.001$) higher SPMA than non-traveller males, indicating traffic emissions as a significant contributor of benzene exposure in urban individuals. However, in females, this difference was not significant.

This study highlighted the synergistic impact of tobacco smoke and traffic emissions on SPMA levels in urban males. Daily smokers who travelled exhibited significantly higher ($p < 0.000001$) SPMA levels (8.23 μg/g) compared to non-smokers who didn't travel daily (2.79 μg/g, Fig. 5c).

3.8 Work Environment

To assess the impact of environmental benzene exposure, urban and rural individuals were categorized as indoor and outdoor employed. SPMA was higher in indoor rural workers than urban workers, but not significantly different. However, outdoor urban workers had significantly higher SPMA levels (8.35 μg/g) than outdoor rural workers (3.55 μg/g, $p = 0.02$, Fig. 6).

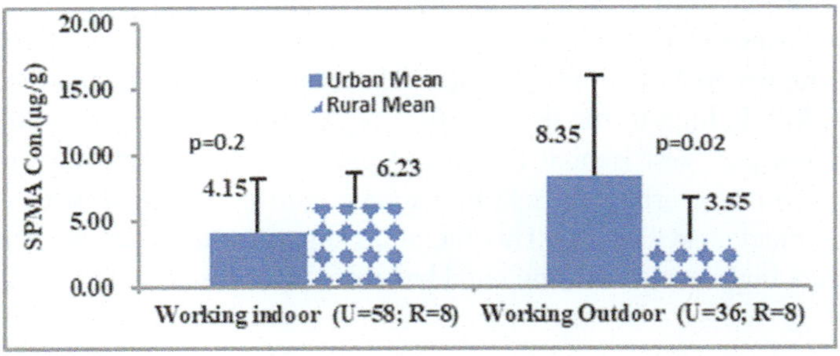

Fig. 6 Variation in SPMA (μg/g) among rural and urban individuals working indoor and outdoor

4 Discussion

To our knowledge, this is the first study examining benzene exposure in West Bengal's adult population, encompassing both urban and rural areas in Eastern India. SPMA biomonitoring assessed integrated benzene exposure in study individuals with no occupational exposure with smoking prevalence of 13% only. The mean urinary SPMA (4.60 μg/g; median 3.23 μg/g) was compared with values from other studies. Results showed that it was 3 times, 1.8 times, and 2 times higher than that reported in Italy (0.14 μg/g), Germany (0.083 μg/g), and Brazil (1.90 μg/g), respectively [57, 59, 68]. However, the obtained median value of 3.23 μg/g was similar to that reported by Fan et al. [23] in non-occupationally exposed subjects (median, 3.35 μg/g). The range of SPMA measured in our study is consistent with the SPMA range of 1.2–16 μg/g reported by Johan et al. (2007) among unexposed populations. In India, benzene exposure was mainly monitored among occupational workers in a few studies [47]. Our study's 95th percentile SPMA values were 12.62 μg/g, 15.25 μg/g, 11.39 μg/g, 13.2 μg/g, and 9.83 μg/g for total, males, females, urban, and rural individuals, surpassing values in Germany (0.38 μg/g), Italy (3.40 μg/g), the UK (7.0 μg/g), and Brazil (3.26 μg/g) [14, 57, 59, 68]. Our results indicate higher benzene exposure in our population compared to some other countries.

4.1 SPMA Variation in Urban and Rural Individuals

The study evident benzene exposure in both urban and rural individuals, with significantly higher SPMA levels in urban areas ($p = 0.00003$). This difference may stem from urban individuals' smoking habits and exposure to environmental benzene from traffic and industrial emissions. Several studies have reported high levels of benzene in urban ambient environments due to traffic emissions [13, 29, 43, 67, 64]. Majumdar et al. [44] reported high air benzene in residential and commercial areas of Kolkata city ranging from 13.8 to 72.0 μg/m^3 [44]. Our findings align with Protano et al. [53], who noted higher SPMA in urban than rural areas. Schwedler et al. [57] observed elevated SPMA in high-traffic density (0.22 μg/L) compared to low-traffic density (0.18 μg/L) areas. West Bengal Pollution Control Board reports identified traffic emissions as a major urban air pollution source, noting increased benzene content with CNG vehicle use [38, 73]. This literature supports our observation that urban residents face high levels environmental benzene exposure.

4.2 Impact of Various Sociodemographic and Lifestyle Factors on SPMA

Specified the significance of gender in monitoring benzene exposure [24, 45, 51], our study, maintaining a gender ratio of 1:1, observed higher SPMA levels in females compared to males. Our results align with Tranfo et al. [68], who found higher SPMA in women (1.12 μg/g) than men (0.69 μg/g), even after creatinine adjustment. Tuakulia [69] also reported higher SPMA in females (1.4 μg/L) than males (1.1 μg/L), without creatinine correction. Poli et al. [51] identified a 23–26% higher risk of hematic disorder in women than men at a similar benzene exposure level. Gender differences, influenced by body mass index (BMI) and hormonal changes, can lead to varied health outcomes [9, 41]. Females, with higher adipose tissue, may be more prone to lipophilic compound deposition like benzene [9, 21]. In our study, this gender-specific SPMA difference could be attributed to females' predominant involvement in cooking practices and extended indoor time, where indoor benzene levels are higher [49, 65]. About 80% ($N = 102$) of females involved in cooking food, possibly explaining their elevated SPMA levels. These findings emphasize the need to consider gender-specific differences in interpreting benzene exposure biomarkers.

Our findings revealed a consistent decline in urinary SPMA levels with age across all study areas. Middle-aged individuals, regardless of gender, engaged actively in both indoor and outdoor household activities, potentially increasing exposure to outdoor air pollution and higher urinary SPMA levels. Gender differences were observed among individuals aged > 35 years, with females having higher SPMA levels than males. This trend persisted until the age group of > 65 years, possibly attributed to prolonged cooking practices in females beyond the age of 50 years, while changes in male smoking and travel patterns led to reduced outdoor benzene exposure. These results imply environmental benzene exposure discrepancies between genders in specific age brackets. Age significantly influences urinary SPMA levels, highlighting its potential as a valuable biomarker for environmental benzene exposure in the general public. This exposure could contribute to reproductive disorders in both genders. In women, various adverse outcomes, such as decreased birth weight [19], preterm birth (PTB) [54], diminished fertility [56], and pregnancy complications [32], have been documented. Similarly, in men, an alteration in sperm characteristics was observed [15]. In our study, 54% of females ($N = 71$) and 53% of males ($N = 69$) were in the reproductive age group (< 45 years), raising significant concerns about benzene exposure. Unlike other studies, we found no age-specific gender differences in SPMA levels, and no consistent trend in SPMA levels with age has been reported by previous researchers.

Our study found a significant link between family income and SPMA excretion. Among the 175 individuals in LIG, 63% ($N = 111$) were urban dwellers, residing close to high-traffic areas. Occupations included housekeepers, labourers, street hawkers, and tea sellers. Rural LIG individuals (37%, $N = 64$) primarily faced benzene exposure through biomass fuel use. These factors indicate a higher risk of environmental benzene exposure in LIG compared to MIG and HIG groups.

Mukherjee et al. [47] found that low family income might increase SPMA levels in petrol pump workers, consistent with Schwedler et al. [57] findings among low-economic-status individuals. Our study aligns, indicating higher benzene exposure risk for LIG individuals compared to MIG and HIG individuals. These results underscore the impact of socioeconomic factors on environmental health disparities, emphasizing the necessity for targeted policies and interventions to mitigate exposures in vulnerable populations.

Females those were involved in cooking, particularly in urban areas, exhibited higher SPMA levels than non-cooking counterparts. This disparity was absent in rural females, potentially attributed to non-clean biomass solid fuel usage in poorly ventilated kitchens with traditional stoves. Both involved in cooking and non-cooking rural females faced benzene exposure from household air due to emission from incomplete fuel combustion. Majumdar and Srivastava (2011) found up to 103.19 μg/m^3 of benzene in West Bengal households using biomass fuel. Other Indian studies reported levels ranging from 23–50 μg/m^3 to 103 μg/m^3 [49, 60]. International studies also showed elevated benzene levels (25–40 μg/m^3) indoors and outdoors with biomass fuel use [30, 63]. Existing literature consistently highlights heightened benzene concentrations in households utilizing biomass fuel. Exposure to benzene levels below 1 ppm (3.25 mg/m^3) can cause heamatotoxicity [25], while chronic exposure (1–5 ppm) over 40 years is linked to increased acute myeloid risk [36]. In our study, females using biomass fuels, potentially exposed to high benzene levels, showed increased SPMA excretion. Considering existing literature, females engaged in cooking activities may be at risk of hematologic and reproductive disorders. Factors like ventilation, cooking frequency, and duration also influence benzene exposure. Prenatal benzene exposure can affect offspring differently, with males developing hepatic tumours and females developing hematopoietic tumours [12, 39]. Among 102 females in our study who cooked, 50% ($N = 53$) were of reproductive age (< 45 years), raising concerns. Government schemes like Pradhan Mantri Ujjwala Yojana [75] and National Biomass Cookstoves Initiative [46] aim to reduce harmful cooking smoke exposure, but our findings emphasise the need for more targeted implementation in studied areas.

Our findings indicate a positive association between smoking, passive tobacco smoke exposure, and elevated SPMA levels in urban males and females. These results were consistent with previous studies [57, 65, 69]. Cigarette smoking is a primary source of benzene exposure, with approximately 90% of benzene intake attributed to smoking alone [36]. High concentrations of benzene (48 μg/m^3) were measured in smokers' rooms by Majumdar and Srivastava (2011). Schwedler et al. [57] found higher SPMA in active smokers (8.09 μg/L) and passively exposed individuals than non-smokers (5.42 μg/L). Tuakulia [69] reported about a 3.5-fold higher SPMA in smokers than non-smokers. Other studies also link passive cigarette smoke exposure to increased SPMA [26, 53]. Our results emphasize tobacco smoke (active or passive exposure) as a primary source of benzene exposure in smokers and those passively exposed, potentially leading to higher SPMA levels. Silva et al. [59] linked SPMA (0.80–8.01 μg/g) to altered neutrophil count and haematological variations. In our study, SPMA ranged from 0.51 to 21.02 μg/g, raising concerns about unidentified health risks for participants. These observations highlight the crucial public health need to minimize exposure to tobacco smoke to mitigate associated health risks.

The study identified significantly higher mean SPMA levels in male travellers (non-smoker) compared to non-travellers, highlighting traffic-related benzene exposure. This trend was not significant for female travellers. In Kolkata, motorbikes and three-wheeler autos were prevalent mode of conveyance, contributing to benzene emissions in the city due to slow, congested traffic, and the use of old automobiles. Leong et al. [42] identified two-stroke engine vehicles as a significant source of elevated urban benzene. Studies found in large cities, mean benzene levels ranged from 15 to 120 $\mu g/m^3$ [11, 42]. Kolkata's ambient air recorded benzene 5.19 $\mu g/m^3$, exceeding the recommended benzene limit ($< 5\ \mu g/m^3$) by the European Commission and CPCB, India [48]. At present, only 1% of benzene by volume is allowed in Indian gasoline, even though the literature from the city revealed a continuous increase of benzene in Kolkata's ambient environment [29, 48]. Khillare et al. [38] noted a 60% increase in urban air benzene due to compressed natural gas (CNG) vehicle use. Som et al. [62] reported high benzene levels (112.4 $\mu g/m^3$) inside commuter vehicles in Kolkata, highlighting substantial exposure to vehicular exhaust benzene during travel. Our study aligns with others highlighting traffic emissions as a major contributor to Kolkata's ambient benzene [29, 43, 73, 48]. Daily urban commuting poses a significant risk of benzene exposure, suggesting a higher health risk for urban residents. Therefore, strategic reduction in traffic pollution is crucial to mitigate potential health risks from benzene exposure.

Our study found higher urinary SPMA levels in rural individuals employed indoors (6.23 $\mu g/g$) compared to urban counterparts (4.15 $\mu g/g$), though not statistically significant. Significantly higher SPMA levels were observed in urban individuals employed outdoors compared to their rural counterparts ($p = 0.02$). This suggests elevated benzene levels in rural indoor air than urban indoor air and higher benzene levels in urban outdoor air compared to rural outdoor air. This disparity may be due to non-clean solid biomass fuel use in poorly ventilated rural kitchens, while clean fuels like LPG are prevalent in urban households. Gustafson et al. [30] noted a 29% higher benzene presence in households using wood for various purposes compared to those using clean fuel. Several studies have consistently linked biomass fuel usage to increased indoor benzene levels (Majumdar and Srivastava, 2011, [30]. Conversely, in urban areas, elevated outdoor benzene levels are often associated with vehicular and industrial emissions [29, 43, 67, 73, 48]. Tranfo et al. [68] reported high SPMA levels in housewives staying indoors (1.53 $\mu g/g$), aligning with our study results. Silva et al. [59] found participants with detectable urinary SPMA levels had odds ratios (OR) of 0.78, 1.29, 2.30, and 1.60 for changes in leukocytes, mean corpuscular volume (MCV), neutrophils, and overall haematological alterations. This indicates a risk of haematological disorders due to elevated benzene exposure in our study population. The results underscore significant public health implications, emphasizing the necessity for targeted interventions to reduce benzene exposure in both urban and rural settings.

5 Limitations

Acknowledging limitations is crucial. The sample size may not fully represent West Bengal's population. Using HPLC, rather than more sensitive methods like LC/MS, could limit result applicability. Nevertheless, the study indicates urinary SPMA as a valuable benzene exposure biomarker, identifying high-exposure groups.

6 Conclusion

High benzene exposure was observed among West Bengal residents, both urban and rural. Females engaged in cooking faced increased risk. Urban exposure linked to smoking, passive smoking, and traffic emissions; rural exposure tied to biomass fuel burning. Smoker travellers males exhibited 2.5 times higher SPMA levels. Evaluating SPMA in non-smokers and unexposed individuals was crucial, serving as a useful biomarker. Findings underscored the necessity for targeted government schemes to reduce environmental benzene emissions, emphasizing the urgent need for awareness and regulatory solutions to minimize benzene exposure.

Acknowledgements Gratitude is extended to director in charge and officer in charge of the Indian Council of Medical Research-Centre for Ageing and Mental Health, Kolkata for funding and support. Appreciation goes to the University of Calcutta's Department of Environmental Science for facilitating the study. Colleagues, scholars, and office staff are acknowledged for their valuable contributions.

References

1. Agency for Toxic Substances and Disease Registry (2007) ToxGuideTM for benzene C6H6 CAS# 71-43- 2. Department of Health and Human Services Public Health Service, U.S. https://www.atsdr.cdc.gov/toxguides/toxguide-3.pdf. . Accessed 26 Feb 2020
2. Agency for Toxic Substances and Disease Registry (2007) Toxicological profile for benzene. Agency for Toxic Substances and Disease Registry, Atlanta. https://www.atsdr.cdc.gov/toxprofiles/tp.asp?id=40&tid=15
3. American Conference of Governmental Industrial Hygienists (2001) Benzene CAS number: 71-43-2. American Conference of Governmental Industrial Hygienists, Cincinnati
4. American Conference of Governmental Industrial Hygienists (2013) Benzene (CAS number: 71-43-2). Cincinnati, OH
5. American Conference of Governmental Industrial Hygienists (2019) Threshold limit values (TLVs) for chemical substances and physical agents biological exposure indices for 2019. American Conference of Governmental Industrial Hygienists, Cincinnati
6. Amin MM, Rafiei N, Poursafa P, Ebrahimpour K, Mozafarian N, Shoshtari-Yeganeh B, Hashemi M, Kelishadi R (2018) Association of benzene exposure with insulin resistance, SOD, and MDA as markers of oxidative stress in children and adolescents. Environ Sci Pollut Res Int 25(34):34046–34052

7. Amorim LCA (2003) O uso dos biomarcadoresnaavaliação da exposiçãoocupacional a substânciasquímicas [The use of biomarkers in the assessment of occupational exposure to chemical substances]. Rev Bras Med Trab. 1(2):124–132
8. Aprea C, Sciarra G, Bozzi N, Pagliantini M, Perico A, Bavazzano P, Leandri A, Carrieri M, Scapellato ML, Bettinelli M et al (2008) Reference values of urinary trans, trans-muconic acid: Italian multicentric study. Arch Environ Contam Toxicol 55(3):329–340
9. Arbuckle TE (2006) Are there sex and gender differences in acute exposure to chemicals in the same settings? Environ Res 101(2):195–204
10. Arnold SM, Angerer J, Boogaard PJ, Hughes MF, O'Lone RB, Robison SH, Schnatter AR (2013) The use of biomonitoring data in exposure and human health risk assessment: benzene case study. Crit Rev Toxicol 43(2):119–153
11. Avogbe PH, Ayi-Fanou L, Autrup H, Loft S, Fayomi B, Sanni A, Vinzents P, Moller P (2005) Ultrafine particulate matter and high-level benzene urban air pollution in relation to oxidative DNA damage. Carcinogenesis 26(3):613–620
12. Badham HJ, LeBrun DP, Rutter A, Winn LM (2010) Transplacental benzene exposure increases tumor incidence in mouse offspring: possible role of fetal benzene metabolism. Carcinogenesis 31(6):1142–1148
13. Batterman S, Su FC, Li S, Mukherjee B, Jia C (2014) Personal exposure to mixtures of volatile organic compounds: modeling and further analysis of the RIOPA data. Res Rep Health Eff Inst 181:3–63
14. Bevan R, Jones K, Cocker J, Assem FL, Levy LS (2013) Reference ranges for key biomarkers of chemical exposure within the UK population. Int J Hyg Environ Health 216(2):170–174
15. Bolden AL, Kwiatkowski CF, Colborn T (2015) New look at BTEX: are ambient levels a problem? Environ Sci Technol 49(9):5261–5276
16. Carbonari D, Chiarella P, Mansi A, Pigini D, Iavicoli S, Tranfo G (2016) Biomarkers of susceptibility following benzene exposure: influence of genetic polymorphisms on benzene metabolism and health effects. Biomark Med 10(2):145–163
17. Center for Disease Control and Prevention (2017) Biomonitoring summary benzene CAS No. 71-43-2. https://www.cdc.gov/biomonitoring/Benzene_BiomonitoringSummary.html
18. Central Pollution Control Board (CPCB) (2020) National ambient air quality monitoring. NAAQMS/45/2019-2020. Ministry of Environment, Forest and Climate Change, Government of India
19. Chen D, Cho SI, Chen C, Wang X, Damokosh AI, Ryan L, Smith TJ (2000) Exposure to benzene, occupational stress, and reduced birth weight. Occup Environ Med 57(10):661–667
20. Choi YH, Kim JH, Lee BE, Hong YC (2014) Urinary benzene metabolite and insulin resistance in elderly adults. Sci Total Environ 482:260–268
21. Clewell H, Teeguarden J, McDonald T, Sarangapani R, Lawrence G, Covington T, Gentry R, Shipp A (2002) Review and evaluation of the potential impact of age and gender specific pharmacokinetic differences on tissue dosimetry. Crit Rev Toxicol 32(4):329–389
22. Cottica D, Grignani E (2013) Evolution of technology and occupational exposures in petrochemical industry and in petroleum refining. GiornaleItaliano di Medicina del Lavoro ed Ergonomia 35(3):236–243
23. Fan R, Wang D, She J (2015) Method development for the simultaneous analysis of trans, trans-muconic acid, 1,2-dihydroxybenzene, S-phenylmercapturic acid and S-benzylmercapturic acid in human urine by liquid chromatography/tandem mass spectrometry. Anal Methods 7:573–580
24. Franconi F, Brunelleschi S, Steardo L, Cuomo V (2007) Gender differences in drug responses. Pharmacol Res 55:81–95
25. Fustinoni S, Campo L, Mercadante R, Manini P (2010) Methodological issues in the biological monitoring of urinary benzene and S-phenyl-mercapturic acid at low exposure levels. J Chromatogr B 878:2534–2540
26. Fustinoni S, Campo L, Satta G, Campagna M, Ibba A, Tocco MG, Meloni M (2012) Environmental and lifestyle factors affect benzene uptake biomonitoring of residents near a petrochemical plant. Environ Int 39:2–7

27. Fustinoni S, Consonni D, Campo L, Buratti M, Colombi A, Pesatori AC, Garte S (2005) Monitoring low benzene exposure: comparative evaluation of urinary biomarkers, influence of cigarette smoking, and genetic polymorphisms. Cancer Epidemiol Biomark Prev 14:2237–2244
28. Galbraith D, Gross SA, Paustenbach D (2010) Benzene and human health: a historical review and appraisal of associations with various diseases. Crit Rev Toxicol 40:1–46
29. Gaur M, Singh R, Shukla A (2016) Volatile organic compounds in India: concentration and sources. J Civ Environ Eng 6(5):1–7
30. Gustafson P, Barregard L, Strandberg B, Sallsten G (2007) The impact of domestic wood burning on personal, indoor and outdoor levels of 1,3-butadiene, benzene, formaldehyde and acetaldehyde. J Environ Monit 9(1):23–32
31. Gyllenhammar I, Johansson C, Jonsson BA, Lindh CH, Lind L (2017) Associations between circulating levels of polycyclic aromatic hydrocarbons and markers of inflammation in middle-aged and elderly men. Sci Total Environ 575:645–651
32. Hannigan JH, Bowen SE (2010) Reproductive toxicology and teratology of abused toluene. Syst Biol Reprod Med 56:184–200
33. IARC (1982) Some industrial chemicals and dystuffs. IARC (Int Agency Res Cancer) Monogr Eval Carcinog Risks Hum 29:93–149. https://publications.iarc.fr/47
34. Infante PF (2006) Benzene exposure and multiple myeloma: A detailed meta-analysis of benzene cohort studies. Ann N Y Acad Sci 1076:90–109
35. Inoue O, Kanno E, Kakizaki M, Watanabe T, Higashikawa K, Ikeda M (2000) Urinary phenylmercapturic acid as a marker of occupational exposure to benzene. Ind Health 38:195–204
36. Johnson ES, Langard S, Lin YS (2007) A critique of benzene exposure in the general population. Sci Total Environ 374(2–3):183–198
37. Kelishadi R, Mirghaffari N, Poursafa P, Gidding SS (2009) Lifestyle and environmental factors associated with inflammation, oxidative stress and insulin resistance in children. Atherosclerosis 203:311–319
38. Khillare PS, Hoque RR, Shridhar V, Agarwal T, Balachandran S (2008) Temporal variability of benzene concentration in the ambient air of Delhi: a comparative assessment of pre-and post-CNG periods. J Hazard Mater 154:1013–1018
39. Koshko L, Debarba LK, Sacla M, de Lima JBM, Didyuk O, Fakhoury P, Sadagurski M (2021) In utero maternal benzene exposure predisposes to the metabolic imbalance in the offspring. Toxicol Sci 180:252–261
40. Kramer U, Herder C, Sugiri D, Strassburger K, Schikowski T, Ranft U, Rathmann W (2010) Traffic-related air pollution and incident type 2 diabetes: results from the SALIA cohort study. Environ Health Perspect 118:1273–1279
41. Legato M (2010) Principles of gender-specific medicine. Elsevier Academic Press, London, UK
42. Leong ST, Muttamara S, Laortanakul P (2002) Influence of benzene emission from motorcycles on Bangkok air quality. Atmos Environ 36:651–661
43. Majumdar D, Mukherjee AK, Sen S (2009) Apportionment of sources to determine vehicular emission factors of BTEX in Kolkata India. Water Air Soil Pollut 201:379–388
44. Majumdar D, Mukherjee AK, Sen S (2011) BTEX in ambient air of a metropolitan city. J Environ Prot 2(1):11–20
45. Messing K, Silverstein BA (2009) Gender and occupational health. Scand J Work Environ Health 35(1):81–83
46. MNRE [Ministry of New and Renewable Energy] (2009) Government of India. National biomass cook-stove initiative, Available from: http://www.mnre.gov.in
47. Mukherjee AK, Chattopadhyay BP, Roy SK, Das S, Mazumdar D, Roy M, Yadav A (2016) Work-exposure to PM10 and aromatic volatile organic compounds, excretion of urinary biomarkers and effect on the pulmonary function and heme-metabolism: a study of petrol pump workers and traffic police personnel in Kolkata City, India. J Environ Sci Health Part A 51(2):135–149

48. Namrata D, Jain JB, Ratna T (2022) A comparative study focusing the effect of criteria pollutants, volatile organic compounds and meteorological parameters on ambient air quality of various cities of India. Appl Ecol Environ Sci 10(3):131–146
49. Pandit G, Srivastava P, Rao A (2001) Monitoring of indoor volatile organic compounds and polycyclic aromatic hydrocarbons arising from kerosene cooking fuel. Sci Total Environ 279:159–165
50. Peake M, Whiting M (2006) Measurement of serum creatinine–current status and future goals. Clin Biochem Rev 27(4):173
51. Poli D, Mozzoni P, Pinelli S, Cavallo D, Papaleo B, Caporossi L (2022) Sex difference and benzene exposure: does it matter? Int J Environ Res Public Health 19(4):2339
52. Protano C, Andreoli R, Manini P, Vitali M (2012) Urinary trans, trans-muconic acid and S-phenylmercapturic acid are indicative of exposure to urban benzene pollution during childhood. Sci Total Environ 435–436:115–123
53. Protano C, Guidotti M, Manini P, Petyx M, La Torre G, Vitali M (2010) Benzene exposure in childhood: role of living environments and assessment of available tools. Environ Int 36(8):779–787
54. Protano C, Scalise T, Orsi GB, Vitali M (2012) A systematic review of benzene exposure during pregnancy and adverse outcomes on intrauterine development and birth: still far from scientific evidence. Annali di igiene, medicinapreventiva e di comunita 24:451–463
55. Sahmel J, Devlin K, Burns A, Ferracini T, Ground M, Paustenbach D (2013) An analysis of workplace exposures to benzene over four decades at a petrochemical processing and manufacturing facility (1962–1999). J Toxicol Environ Health A 76(12):723–746
56. Sallmen M, Neto M, Mayan ON (2008) Reduced fertility among shoe manufacturing workers. Occup Environ Med 65(8):518–524
57. Schwedler G, Schulz C, Conrad A, Rucic E, Reiber L, Peisker J, Kolossa-Gehring M (2021) The German environmental survey for children and adolescents 2014–2017 (GerES V)–study population, response rates and representativeness. Int J Hyg Environ Health 237:113821
58. Shim YH, Ock JW, Kim YJ, Kim Y, Kim SY, Kang D (2019) Association between heavy metals, bisphenol a, volatile organic compounds and phthalates and metabolic syndrome. Int J Environ Res Public Health 16(5):671
59. Silva CB, Mota CDL, Almeida YR, Emidio V, Fonseca ASA, Mitri S, Moreira JC (2019) Environmental exposure to benzene: Evaluation of urinary SPMA and polymorphism (CYP2E1-1293G>C and NQO1 609C>T) in Campos Elíseos residents, Duque de Caxias, Rio de Janeiro State Brazil. Cadernos de SaúdePública 35:e00198618
60. Sinha SN, Kulkarni PK, Shah SH, Desai NM, Patel GM, Mansuri MM, Saiyed HN (2006) Environmental monitoring of benzene and toluene produced in indoor air due to combustion of solid biomass fuels. Sci Total Environ 357(1–3):280–287
61. Smith MT (2010) Advances in understanding benzene health effects and susceptibility. Annu Rev Public Health 31:133–148
62. Som D, Dutta C, Chatterjee A, Mallick D, Jana TK et al (2007) Studies on commuters' exposure to BTEX in passenger cars in Kolkata India. Sci Total Environ 372(2–3):426–432
63. Son B, Breysse P, Yang W (2003) Volatile organic compounds concentrations in residential indoor and outdoor and its personal exposure in Korea. Environ Int 29(1):79–85
64. Srivastava A, Joseph AE, More A, Patil S (2005) Emissions of VOCs at urban petrol retail distribution centres in India (Delhi and Mumbai). Environ Monit Assess 109:227–242
65. Srivastava A, Majumdar D (2011) Monitoring and reporting VOCs in ambient air. Air Qual Monit Assess Manage 8(2):181–202
66. Steinmaus C, Smith AH, Jones RM, Smith MT (2008) Meta-analysis of benzene exposure and non-Hodgkin lymphoma: biases could mask an important association. Occup Environ Med 65(6):371–378
67. Talapatra A, Srivastava A (2011) Ambient air non-methane volatile organic compound (NMVOC) study initiatives in India–a Review. J Environ Prot 2(1):21–36
68. Tranfo G, Pigini D, Paci E, Bauleo L, Forastiere F, Ancona C (2018) Biomonitoring of urinary benzene metabolite SPMA in the general population in Central Italy. Toxics 6(3):37

69. Tuakuila J (2013) S-phenylmercapturic acid (SPMA) levels in urine as an indicator of exposure to benzene in the Kinshasa population. Int J Hyg Environ Health 216(4):494–498
70. Tunsaringkarn T, Soogarun S, Palasuwan A (2013) Occupational exposure to benzene and changes in haematological parameters and urinary trans, trans-muconic acid. Int J Occup Med Environ Health 26(1):45–49
71. Weber JA, Van Zanten AP (1991) Interferences in current methods for measurements of creatinine. Clin Chem 37(5):695–700
72. Weisel CP (2010) Benzene exposure: an overview of monitoring methods and their findings. Chem Biol Interact 184(1–2):58–66
73. West Bengal Pollution Control Board (2005) Air quality management: final report. WBPCB in Collaboration with Asian Development Bank. Intercontinental Consultant and Technocrats Pvt. Ltd., New Delhi, India
74. Williams AD, Grantz KL, Zhang C, Nobles C, Sherman S, Mendola P (2019) Ambient volatile organic compounds and racial/ethnic disparities in gestational diabetes mellitus: are Asian/Pacific Islander women at greater risk? Am J Epidemiol 188(2):389–397
75. Yadav Y (2020) Women empowerment through Pradhan Mantri Ujjwala Yojana (PMUY) Scheme in Rajasthan: a study on ruralhouseholds in selected region. SSRN

Assessment of Source Effects and Action Plan for Controlling Air Pollution in Non-attainment Areas of Ribhoi District, Meghalaya: A Traffic Survey Report

Aishi Nath and Ganesh Chandra Dhal

Abstract Over the last decade, Meghalaya faced escalating threats from air pollution and climate change, prompting the Meghalaya State Pollution Control Board (MSPCB) to identify critical environmental stressors. Vehicular emissions, improper waste management, rapid deforestation, and industrialisation significantly contribute to the region's deteriorating environmental conditions, with vehicular emissions emerging as a pervasive issue. Despite existing regulations and awareness campaigns, grassroots reforms require amplification. To tackle these challenges, comprehensive studies are imperative to identify emission sources and accurately quantify contaminants. This study focuses on Ribhoi district, examining Meghalaya's industrial sector to explore remedial methods for mitigating environmental effects. Prioritising pollutants like $PM_{2.5}$, PM_{10}, SO_x, and NO_x demonstrates commitment to addressing specific air quality threats. The research aims to provide data-driven insights, raise awareness, and propose actionable strategies for sustainable environmental management in key areas like Shillong and Byrnihat, preserving Meghalaya's natural beauty and ecological balance.

Keywords Ambient air quality · Pollution sources · Pollution quantification · Action plans

1 Introduction

Environmental pollution is a persistent and serious global issue, contributing significantly to various harmful diseases and mortality rates. Human activities, such as urbanisation, mining, exploration, and industrialisation, play pivotal roles in exacerbating this problem. Both developed and developing countries bear the burden of

A. Nath · G. C. Dhal (✉)
Department of Civil Engineering, National Institute of Technology Meghalaya, Sohra, India
e-mail: ganeshdhal@nitm.ac.in

S. M. Shiva Nagendra et al. (eds.), *Select Proceedings of the 8th Indian International Conference on Air Quality Management (IICAQM 2023)*, Lecture Notes in Civil Engineering 582, https://doi.org/10.1007/978-981-96-2359-4_20

environmental pollution, with developed nations actively monitoring pollutants and implementing environmental regulations. The excessive presence of air pollutants has far-reaching and long-term implications for both the environment and human health and well-being [1].

Today, environmental pollution takes various forms, including air pollution, water pollution, and soil contamination, making it one of the most pressing global challenges. Air pollution remains a critical concern for developing nations like India. Inadequate air quality in a region has detrimental effects on ecosystems, agriculture, and the health of humans and other living organisms within the biosphere. Additionally, it impacts regional climate patterns, which, in turn, can influence global climate dynamics—a complex and challenging aspect to address [2].

The Environmental Protection Agency (EPA) has identified six of the most prevalent air pollutants as "criteria air pollutants" for assessing air quality in specific locations. These include particulate matter (PM), ground-level ozone (O_3), carbon monoxide (CO), lead, sulphur dioxide (SO_2), nitrogen dioxide (NO_2), and other toxic organic compounds. Among these criteria pollutants, particulate matter and aerosols are the most common [1]. This research aims to explore the intricate relationship between air quality, meteorological conditions, precipitation, and transportation, shedding light on the complexities of addressing and mitigating air pollution's adverse impacts on our environment and health.

In India, particulate matter (PM), hydrocarbons (HC), and nitrogen oxides (NOx) levels have already exceeded the national ambient air quality standard (NAAQS) limits in various parts of the country. Particularly, particulate matter with a diameter of 2.5 μm or less ($PM_{2.5}$) and hydrocarbons (including black carbon) pose serious health risks by obstructing respiratory pathways and contributing to premature death. These pollutants primarily result from various combustion processes, vehicular emissions, and other human activities [1, 15].

In 2019, the World Health Organization (WHO) reported a staggering 6.7 million premature deaths worldwide due to ambient air pollution. In India, nearly 600,000 premature deaths annually are attributed to ambient air pollution [21]. Studies, including those by [3, 20], have underscored the critical importance of improving poor air quality for public health.

Indian meteorological conditions, along with local pollutant emissions, exert a substantial influence on ambient air pollution concentrations. Factors such as wind direction and speed, among other climatic variables, play a pivotal role in regulating pollution levels in urban areas. Several studies have documented the impact of meteorological conditions on the levels of particulate matter (PM) and other gaseous pollutants in metropolitan regions of India, including research conducted by [10, 22]. Understanding these interactions is vital for effectively managing and mitigating air pollution in densely populated urban areas.

1.1 The Air Pollution Issue in the North-Eastern States of India

Northeast India, formally known as the North-eastern Region (NER), constitutes India's easternmost region and functions as a distinct geographical and political-administrative entity. The states that collectively form this region include Meghalaya, Manipur, Mizoram, Nagaland, Tripura, Sikkim, Assam, and Arunachal Pradesh.

In recent years, the air quality in the north-eastern part of India has been on a steady decline due to various factors affecting air quality. According to data from the Central Pollution Control Board (CPCB), the State Pollution Control Boards (SPCBs) of all the north-eastern states, the Ministry of Environment and Forest (MoEF), and the National Health Profile (NHP), Assam reported the highest number of deaths (225) attributed to Acute Respiratory Infection (ARI) in 2017. Following Assam, other north-eastern states also recorded ARI-related deaths: Meghalaya (7), Arunachal Pradesh (3), Manipur (23), Sikkim (8), Mizoram (36), and Tripura (21). In the year 2016, ARI claimed lives in various north-eastern states, with Assam reporting 200 deaths, Mizoram 68, Sikkim 43, Manipur 21, Tripura 7, and Arunachal Pradesh 7. However, there were no ARI-related deaths reported in Nagaland during that year [5].

Among the north-eastern states, Meghalaya had the highest number of reported ARI cases in 2017, totalling 3,33,772 cases, according to the National Health Profile (NHP). Tripura followed with 1,78,570 ARI cases, while Sikkim reported 93,270 cases. Manipur had 51,205 cases, Arunachal Pradesh 36,250, Nagaland 31,158, Assam 29,595, and Mizoram 29,595. The increasing prevalence of ARI cases and related deaths in the north-eastern states highlights the urgent need for measures to improve air quality, reduce pollution sources, and enhance public health and well-being in the region [19].

1.2 Shillong, Meghalaya, India

Shillong, often referred to as "The Abode of Clouds," is a picturesque hill station located at coordinates 25.57° N and 91.88° E. Nestled in the serene Shillong plateau in the north-eastern region of India, Shillong proudly serves as the capital of the state of Meghalaya. It also functions as the administrative hub for the East Khasi Hills district. As of the 2011 census, Shillong is recorded as India's 330th most populous city, with a population of approximately 143,229 residents. The city's breath-taking natural beauty, temperate climate, and unique cultural heritage makes it a popular destination for both tourists and those seeking a peaceful and scenic place to call home. With its lush landscapes, vibrant culture, and strategic location in the heart of Meghalaya, Shillong holds a special place in India's north-eastern region, drawing visitors and residents alike to its captivating charm [11].

1.3 Air Quality Index (AQI)

The air quality index (AQI) helps government agencies determine pollution concentrations and anticipate consequences. Distinct nations have distinct air quality indices and sub-indices that meet national norms. The Swachh Bharat Abhiyan 2014 is one of the Indian government's air quality initiatives. The CPCB and SPCB run the National Air Monitoring Programme (NAMP), which includes 342 monitoring sites in 240 cities. Tables 1 and 2 provide details on AQI categories and its health impact [4].

In this study, a survey is done to gather information on the fuel used by the cars entering Meghalaya through Byrnihat. The purpose of the study is to comprehend the connections between fuel preferences, gasoline usage, production year, registration year, and other factors. The survey is carried out in the field, where data is physically gathered by tallying the number of cars that use the petrol pump during the survey period and recording the type of fuel that each car uses. Since the Byrnihat A. Warjri petrol pump is the hub for cars entering Meghalaya, the field survey was carried out there.

2 Data and Methodology

Through a traffic survey, primary data was gathered. The traffic survey included information on the type of vehicle and motor fuel used. The Public Works Department (PWD), the district transport office in Shillong, the district transport office in Nongpoh, and other government sectors were visited to gather information about car registration, traffic volume, and a list of fuelling stations. The survey was conducted at the Byrnihat A. Warjri Fuel Stop on March 30, 2022, from 6 a.m. to 8 a.m., April 4, 2022, from 6 p.m. to 8 p.m., and April 6, 2022, from 12 p.m. to 3 p.m.

Byrnihat has been chosen for this research project as it is crucial for the evaluation of traffic data to choose a traffic-congested zone. According to several official reports, the area's rising air pollution rates are caused by an increase in the number of vehicles on the road and expanding industries.

2.1 Study Area (Byrnihat)

Byrnihat, a modest industrial town 100 m above sea level, is located 90 kms from Shillong and 14 kms from Guwahati on the Guwahati-Shillong route (NH 40). The Survey of India toposheet No. 78 0/12 covers the region on the Umtrew River in Ri-Bhoi, Meghalaya, between 26° 2′14″ North latitude and 91° 5′27″ East longitude. The northern section of Ri-Bhoi, which borders Assam, is included. The research region lies on the northern edge of the Meghalaya plateau and the Brahmaputra basin Fig. 1. The studied region lacks industrial development. Barapani Industrial is

Table 1 AQI category, pollutants and ambient concentrations

AQI Category (range)	$PM_{2.5}$ (24 h)	PM_{10} (24 h)	O_3 (8 h)	SO_2 (24 h)	NO_2 (24 h)	CO (8 h)	NH_3 (24 h)	Pb (24 h)
Good (0–50)	0–30	0–50	0–50	0–40	0–40	0–1.0	0–200	0–0.5
Satisfactory (51–100)	31–60	51–100	51–100	41–80	41–80	1.1–2.0	201–400	0.5–1.0
Moderately polluted (101–200)	61–90	101–250	101–168	81–380	81–180	2.1–10	401–800	1.1–2.0
Poor (201–300)	91–120	251–300	169–208	381–800	181–280	10–17	801–1200	2.1–3.0
Very poor (301–400)	121–250	351–430	209–748	801–1600	281–400	17–34	1200–1800	3.1–3.5
Severe (401–500)	250 +	430 +	748 +	1600 +	400 +	34 +	1800 +	3.5 +

Table 2 Effects on health at various AQI range

Air quality index	Consequences
Good (0–50)	Nominal repercussions
Satisfactory (51–100)	Unpleasantness and sensitivity to individuals
Moderately polluted (101–200)	An individual with persisting lung disease, such as asthma, heart disease, children, and the elderly, might face difficulties in breathing
Poor (201–300)	People with heart disease and those who have been exposed for a long time may experience discomfort
Very poor (301–400)	People who are exposed for an extended period may develop respiratory illnesses. People with lung and heart disease may experience a stronger effect
Severe (401–500)	Even healthy persons may experience respiratory effects, and people with lung/ heart illnesses may experience major health consequences. Even light physical activity might have negative health consequences

Ri-Bhoi's lone industrial estate [6]. Nongpoh, and Byrnihat, all in Meghalaya state jurisdiction, were visited and data was gathered to assess air quality. Some industrial clusters around Byrnihat and Jorabat are estates and Export Promotion Industrial Parks. A number of operating industries in the region are listed in Table 3, along with the type of fuel and technology involved. This detail demonstrated the region's increasing industrialisation and its contribution to air pollution. The Byrnihat region is largely a small-scale industry [12].

3 Results and Discussion

Figure 2a represents the growing trend in the number of automobiles over the course of 15 years. The region's pollution is mostly caused by the rise in the number of vehicles. Estimating the general state of traffic in this region is further made easier by Table 4's, total and average daily traffic counts for Shillong and Byrnihat. The Byrnihat industrial districts are extremely polluted, as seen by the findings made above, and they spew pollutants out of control, which raises the temperature and damages the surrounding surroundings. The ecology has deteriorated due to the lack of trees, which has led to an accumulation of dust particles that make breathing difficult. Serious policies and tactics need to be adjusted to keep the pollution level under control; they are covered in the section below. Figure 2b displays the findings from the three-day survey. The maximum fuel utilised by cars when there were more diesel-fuelled vehicles than petrol-fuelled vehicles.

The following observations are noted:

1. Compared to night time, more automobiles fuel up during the day.
2. There are more cars running on petrol than on diesel.

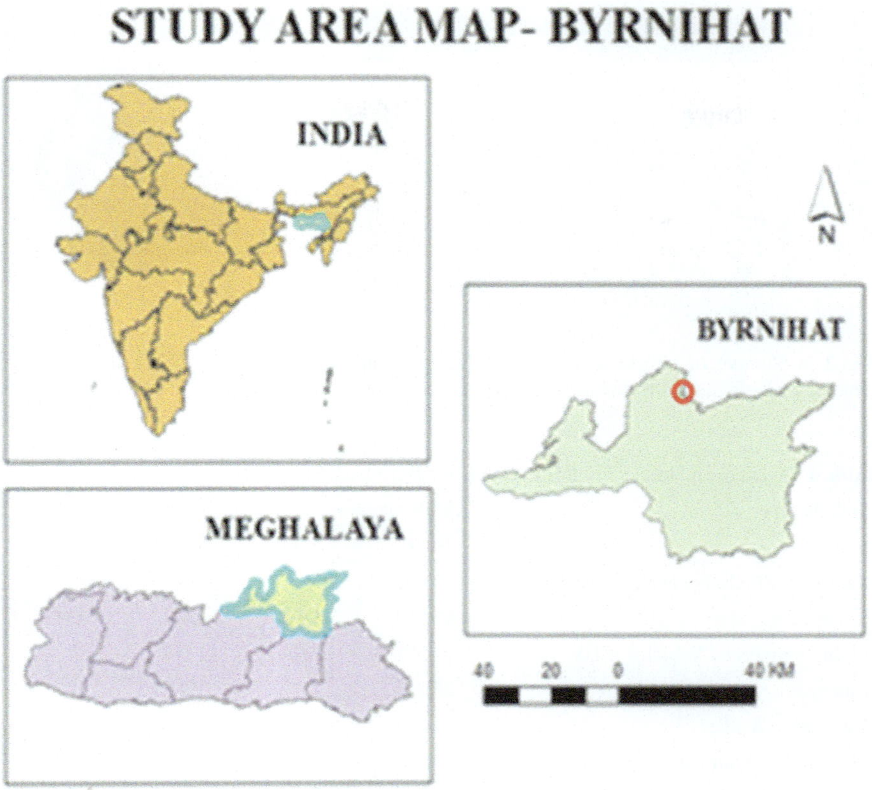

Fig. 1 Study area map

Table 3 List of operating industries in Byrnihat

Sl. No.	Technology involved	Fuel used	No. of operating industries
1	Furnace	Quartzite, Iron chips, Coke, Electrode paste, Electric energy Silicon yield	7
2	Gas engines	CNG and biomethane	2
3	Blast furnace	Coke	6
4	Roll stand	Electricity	3
5	Power plant	Coal	1

3. It is estimated that 20,000 L of diesel are consumed daily.

The above observations show that daytime pollution is higher than night time pollution. Most cars utilise petroleum engine oil, which accelerates fuel burning and produces NO_x, SO_x, and CO, the main vehicular pollutants.

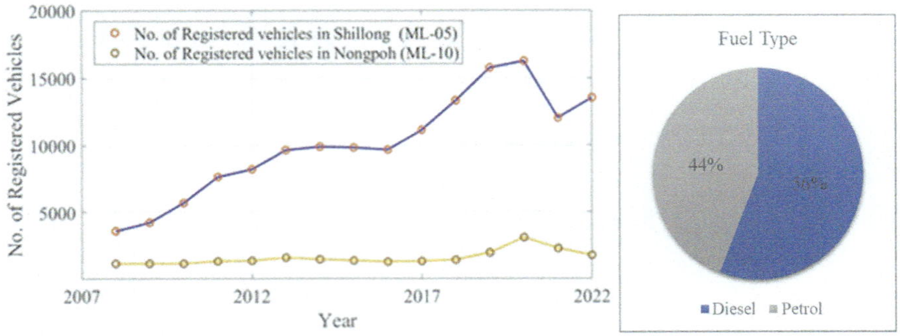

Fig. 2 **a** Number of registered vehicles in Shillong and Nongpoh (2007–2022), **b** Analysis of three-day survey data in Byrnihat A. Warjri fuel stop (March–April 2022)

Table 4 Average and total daily traffic counts for Byrnihat and Shillong

Vehicle type	Shillong		Byrnihat	
	Total	ADT	Total	ADT
Cars/Jeeps/Taxis/3-wheelers	435,522	21,776	5247	375
2-wheelers	85,881	4294	3342	3342
Light commercial Vehicles (LCV)	28,278	1414	1217	87
Bus	45,445	2272	1288	92
2 axle truck/tanker	81,238	4062	2670	191
Multi-axle truck/tanker	19,369	968	1477	106
Agricultural tanker	752	38	5	0
Cycle/rickshaws/Bullock cart/horse cart	3876	194	212	0
Others	640	32	11,799	0

Figure 2a displays further data that highlights the Meghalaya region's rising pollution rate. Measures should be taken in addition to legal policies, acts, and laws to limit pollution transmission at the source.

Days and nights in Shillong are foggy, so pollutants are more likely to stay in the air, creating respiratory discomfort and impaired eyesight for drivers. The policy controls and commonly used technologies are listed below.

3.1 Recommendations for Preventing Air Pollution

Mitigation: The rising incidence of air pollution caused by human activity requires strict emission control legislation.

Policies: Describes ways to reduce uncontrolled air pollution. Industrial expansion and car exhaust pollution have been issues for a decade. Technology has enabled strategic air pollution management system advancements. Advanced approaches and technology allow policy and ground implementation. The government has proposed many rules and regulations to regulate it [8, 15, 17].

Since 2000, India has enforced European pollution and fuel standards for four-wheeled light- and heavy-duty vehicles. The Indian government issued the National Auto Fuel Policy in 2003 to address vehicular emission norms, CNG/LPG fuel quality and standards, in-use vehicle emissions, automotive technology, air quality data, and R&D. The Auto Fuel Vision and Policy 2025, released in May 2014, tightened fuel, and pollution rules from 2003 [13].

Alternative Fuels: CNG, LPG, LNG, biodiesel, and biofuels are replacing fossil fuels to lower ground-level pollution. The Pradhan Mantri Ujwala Yojana (PMUY) was launched by MoPNG in 2016 to promote greener fuels. The Nation Clean Air Programme (NCAP) allows India to mix 20% ethanol in petrol and 5% biodiesel in diesel by 2030 under the National Policy on Biofuels, 2018 [14].

Thermal Power Plant Emission Policy: India has a large thermal power production market, comparable to industrialised nations, as per [18]. Air pollution in India is mostly caused by coal-fired thermal power units. The 2015 EPA guidelines require power facilities to comply by December 2022. Thermal power stations must reduce SO_2 emissions with Flue Gas Desulphurisation (FGD) and NO_x emissions with SCR/SNCR. Electrostatic precipitator (ESP) retrofitting and combustion modification are needed to meet MoEFCC emissions guidelines [16].

Faster Adoption and Manufacturing of Electric Vehicles (FAME): National and state governments provide incentives for EV purchases [23]. Toyota provides hybrid automobiles like the dynamic Corolla Hybrid, which gets 3.4 l/100 km. Both Maruti and Tata have developed hybrid automobiles in India since 2019 [7].

Emission Control Technologies: $PM_{2.5}$ and ultrafine particles contain dangerous heavy metals, black carbon, poisons, and carcinogens for humans and animals. Power and cement plants utilise cyclones, bag filters, granular-bed filters, scrubbers, and electrostatic precipitators. Although efficient at removing particles bigger than 2.5 μm, they become less effective as particle size decreases. Electrostatic precipitators for $PM_{2.5}$ particles can be improved by separating the charging and precipitation processes or agglomerating the particles before precipitation with a "two-stage electrostatic precipitator." The high electric field between the electrodes in the collecting stage and the high electric charge delivered to particles in specially designed rechargers boost the collection efficiency of two-stage electrostatic precipitators. Jaworek et al. [9] examined two-stage electrostatic precipitators' recharging and agglomeration stages. Some equipment used to clean diesel engine exhausts or coal-fired boilers had $PM_{2.5}$ particle mass collecting efficiency of above 95%. A two-stage electrostatic precipitator route diagram in Fig. 3 explains the equipment's operation [9].

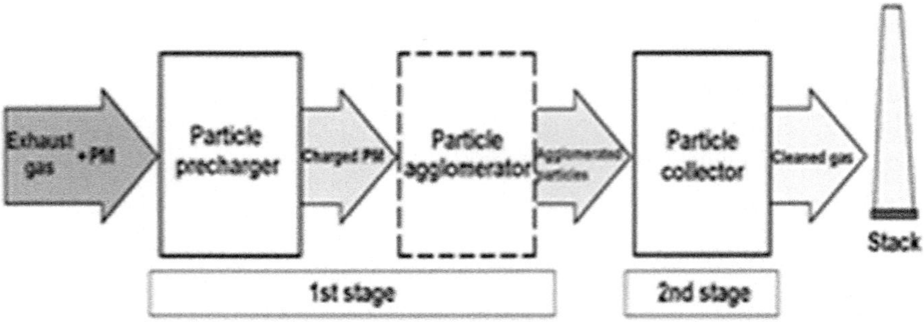

Fig. 3 Operational pathway of two-stage electrostatic precipitator [9]

4 Conclusions

Extensive data collection and site investigations in Byrnihat's industrial areas have revealed that elevated ambient air concentrations of respiratory suspended particulate matter (RSPM), total suspended particulate matter (TSPM), sulphur oxides (SO_x), nitrogen oxides (NO_x), and carbon monoxide (CO) are primarily caused by vehicular emissions. Additionally, air pollution in this region is found to be significantly contributed to by industrialisation, with the potential for commercial reduction. Solutions for industrial pollution, including electrostatic precipitators (ESP) and other technologies, have been identified.

The study highlights automotive emissions as the primary cause of tropospheric pollution, further exacerbated by heavy traffic, poor road infrastructure, and inadequate traffic management. Although decent air quality is generally maintained in Meghalaya, Byrnihat, designated a non-attainment city is found to fall below prescribed norms, posing a threat to local ecology. Addressing these challenges is deemed crucial to safeguard the environment and community health. The imperative for enhancing air quality in Byrnihat lies in the implementation of mitigation measures and pollution control, ensuring a healthier and more sustainable future for its residents.

References

1. Amann M, Purohit P, Bhanarkar AD, Bertok I, Borken-Kleefeld J, Cofala J, Heyes C, Kiesewetter G, Klimont Z, Liu J, Majumdar D (2017) Managing future air quality in megacities: a case study for Delhi. Atmos Environ 161:99–111
2. Amato F, Cassee FR, van der Gon HAD, Gehrig R, Gustafsson M, Hafner W, Harrison RM, Jozwicka M, Kelly FJ, Moreno T, Prevot AS (2014) Urban air quality: the challenge of traffic non-exhaust emissions. J Hazard Mater 275:31–36
3. Chowdhury S, Pozzer A, Dey S, Klingmueller K, Lelieveld J (2020) Changing risk factors that contribute to premature mortality from ambient air pollution between 2000 and 2015. Environ Res Lett 15(7):074010

4. CPCB (2023) Air quality index. cpcb.nic.in
5. CSE, 2021, CSE warns of growing air pollution in cities of India's northeast; Guwahati and Agartala top the list of most polluted cities in the region, https://www.cseindia.org/cse-warns-of-growing-air-pollution-in-cities-of-india-s-northeast-guwahati-and-agartala-top-the-list-of-most-polluted-cities-in-the-region-11093.
6. Datta S, Ray B, Vengatajalapathi G, Gogoi U, Radhapyari K (2017) Effects of solid wastes and industrial effluents on groundwater quality in Byrnihat industrial belt, Meghalaya. In: 7th International Ground Water Conference (IGWC-2017), New Delhi
7. Goel S, Sharma R, Rathore AK (2021) A review on barrier and challenges of electric vehicle in India and vehicle to grid optimisation. Transp Eng 4:100057
8. Gulia S, Shukla N, Padhi L, Bosu P, Goyal SK, Kumar R (2022) Evolution of air pollution management policies and related research in India. Environ Chall 6:100431
9. Jaworek A, Marchewicz A, Sobczyk AT, Krupa A, Czech T (2018) Two-stage electrostatic precipitators for the reduction of $PM_{2.5}$ particle emission. Prog Energy Combust Sci 67:206–233
10. Kumar P, Druckman A, Gallagher J, Gatersleben B, Allison S, Eisenman TS, Hoang U, Hama S, Tiwari A, Sharma A, Abhijith KV (2019) The nexus between air pollution, green infrastructure and human health. Environ Int 133:105181
11. Lamare RE, Chaturvedi SS (2014) Suspended particulate matter in ambient air of Shillong City, Meghalaya, India. Ind J Sci Res Technol 2:37–41
12. MSPCB (2019) Action plan for control of air pollution in non-attainment towns. Byrnihat including EPIP Industrial Area, Ri-Bhoi, Meghalaya (Revised). https://cpcb.nic.in/Actionplan/Byrnihat.pdf
13. NAP (2018) National auto policy. https://heavyindustries.gov.in/writereaddata/UploadFile/National%20Automotive%20Policy%20Draft%20v2.pdf
14. Othman MF, Adam A, Najafi G, Mamat R (2017) Green fuel as alternative fuel for diesel engine: a review. Renew Sustain Energy Rev 80:694–709
15. Bhave PP, Kulkarni N (2015) Air pollution and control legislation in India. J Inst Eng (India): Ser A 96:259–265
16. Pachouri R, Saxena A (2020) Emissions control in thermal power stations: issues, challenges and the way forward. Energy and Resources Institute, New Delhi, India
17. Roychowdhury A, Somvanshi A (2020) Breathing space; how to track and report air pollution under the national clean air programme. Center for Science and Environment. file:///C:/Users/Aishi/Downloads/http___cdn.cseindia.org_attachments_0.65306900_1581333991_breathing-space.pdf
18. TERI (2020) Emissions control in thermal power stations: issues, challenges, and the way forward. https://www.teriin.org/sites/default/files/2020-02/emissions-control-thermal-power.pdf
19. TNN (2022) Guwahati and Agartala most polluted cities in northeast this summer: study. http://timesofindia.indiatimes.com/articleshow/93763708.cms?utm_source=contentofinterest&utm_medium=text&utm_campaign=cppst
20. Venter ZS, Aunan K, Chowdhury S, Lelieveld J (2021) Air pollution declines during COVID-19 lockdowns mitigate the global health burden. Environ Res 192:110403
21. WHO (2022) Ambient (outdoor) air pollution. https://www.who.int/news-room/fact-sheets/detail/ambient-(outdoor)-air-quality-and-health#:~:text=In%202019%2C%2099%25%20of%20the,6.7%20million%20premature%20deaths%20annually
22. Yadav P, Usha K, Singh B (2022) Air pollution mitigation and global dimming: a challenge to agriculture under changing climate. In: Climate change and crop stress, pp 271–298
23. Zheng Y, Niu S, Shang Y, Shao Z, Jian L (2019) Integrating plug-in electric vehicles into power grids: a comprehensive review on power interaction mode, scheduling methodology and mathematical foundation. Renew Sustain Energy Rev 112:424–439

Interplay Between Urban Development Dynamics and Policy Frameworks in Shaping Air Quality in Delhi

Bhanu Chaudhary and Bhaven N. Tandel

Abstract Air quality and related policies have not received much attention from urban political ecology. The recently implemented air pollution policies in the capital of India are examined and analyzed in this paper, as well as the contribution of development to rising levels of air pollution. The paper considers the implications of these policies on change in pollution. Many recent studies found that urbanization is a major contributor to air pollution in Delhi. Construction activities, industrial emissions, and waste burning are other major sources of air pollution in the city. This highlights the need for proper policy-making and implementation studies to address this issue. The review incorporates studies of key policies, including the Graded Response Action Plan (GRAP) for air quality management and the Odd–Even Scheme aimed at reducing vehicular emissions, etc. Additionally, some case studies of globally implemented actions to reduce pollution are included. Recent studies have consistently pointed out that urbanization plays a significant role in driving air pollution levels in Delhi. The multifaceted sources, including construction activities, industrial emissions, and waste burning, further underscore the complexity of the problem and necessitate diligent policy-making and effective implementation strategies.

Keywords Urban political ecology · Air pollution · Air pollution policies

B. Chaudhary (✉) · B. N. Tandel
Department of Civil Engineering, Sardar Vallabhbhai National Institute of Technology, Surat, India
e-mail: d22ce001@ced.svnit.ac.in

B. N. Tandel
e-mail: bnt@ced.svnit.ac.in

S. M. Shiva Nagendra et al. (eds.), *Select Proceedings of the 8th Indian International Conference on Air Quality Management (IICAQM 2023)*, Lecture Notes in Civil Engineering 582, https://doi.org/10.1007/978-981-96-2359-4_21

1 Introduction

Air pollution is a major environmental and health issue in many cities around the world. Delhi, the capital city of India, is one of the most polluted cities in the world, with high levels of particulate matter and other harmful pollutants in its air. The causes of air pollution in Delhi are multifaceted, including rapid urbanization, industrialization, transportation, and construction. At the same time, policies and regulations aimed at mitigating air pollution in the city have also played a critical role in shaping the urban environment [14].

This paper reviews the existing literature on the relationship between development, policy framing, and air pollution in Delhi's urban environment. The goal is to provide an overview of the current state of knowledge on this critical issue and identify gaps in our understanding that require further investigation. The rapid urbanization of Delhi has led to increased demand for housing, infrastructure, and services, which has put pressure on the city's resources and infrastructure. Rapid industrialization, driven by the city's growing economy, has also contributed to air pollution, with industries emitting a range of pollutants into the air.

Over the years, a range of policies and regulations have been introduced to address air pollution in Delhi, including measures such as banning diesel generators, implementing the odd–even rule for vehicles, and reducing emissions from industries [7]. However, the effectiveness of these policies and regulations in reducing air pollution in Delhi remains a subject of debate. By exploring the complex interplay between development, policy framing, and air pollution in Delhi's urban environment, this paper seeks to contribute to a better understanding of this critical issue. This, in turn, can inform the development of more effective policies and interventions to address air pollution and improve the well-being of people in the city. With a burgeoning population, rapid urbanization, and increasing industrial activities, air quality has emerged as a major public health and environmental concern in the region [6]. This overview provides a comprehensive understanding of the various aspects of air pollution in Delhi and its implications for the city's residents and ecosystem. Delhi's air quality is regularly monitored by various agencies, including the Central Pollution Control Board (CPCB) and the Delhi Pollution Control Committee (DPCC). Data from recent years consistently show alarming levels of particulate matter ($PM_{2.5}$ and PM_{10}), nitrogen dioxide (NO_2), sulfur dioxide (SO_2), and other harmful pollutants [6]. These pollutants often exceed the national and international air quality standards, posing significant health risks to the population [15].

The major contributors to air pollution in Delhi are vehicular emissions, industrial activities, construction dust, and biomass burning. The city's ever-increasing fleet of vehicles, particularly diesel-powered vehicles, is a significant source of particulate matter and nitrogen oxides [2]. Industrial zones, construction sites, and waste burning add to the pollution burden. The deteriorating air quality in Delhi has serious health consequences. High levels of air pollution have been linked to respiratory and cardiovascular diseases, lung cancer, and premature deaths. Children, the elderly, and those with pre-existing health conditions are particularly vulnerable. Air pollution also adversely affects the ecosystem, leading to damage to plants, water bodies, and biodiversity. The Indian government has implemented various policies and regulations to address air pollution in Delhi. The Graded Response Action Plan (GRAP) and the Odd–Even Scheme are some of the measures taken to control pollution during specific periods [6]. However, challenges in policy implementation, enforcement, and inter-agency coordination persist.

2 The Urban Development and Air Pollution Nexus

The relationship between urban development and air pollution is a complex and significant issue that has profound implications for the well-being of urban populations and the environment [20]. As cities grow and develop, they experience various changes in land use, transportation, and industrial activities, all of which can influence air quality and contribute to air pollution. One of the primary drivers of air pollution in urban areas is the rapid expansion of transportation systems, particularly the increase in the number of vehicles on the roads [10]. As urban populations grow, so does the demand for transportation, leading to a surge in vehicular emissions of pollutants such as nitrogen oxides (NOx), volatile organic compounds (VOCs), and particulate matter (PM). The use of fossil fuels in vehicles, especially in congested traffic conditions, exacerbates the emission levels, resulting in deteriorating air quality [13].

The research conducted by Hashmi et al. [11] delves into the dynamic relationship between urban agglomerations and environmental degradation in the top ten urban agglomerated countries. Employing innovative methods such as the quantile-on-quantile (QQ) approach and quantile Granger causality, the study reveals that urban agglomeration exerts a strong positive effect on CO_2 emissions in most of the sample countries, particularly at higher quantiles. However, certain countries display declining and negative trends, indicating the complexity of this association. The research underscores the necessity for individual-focused policies in urban planning and environmental sustainability. Kumar et al. [13] examine the interplay between green infrastructure, air quality, and human health in urban areas. While green infrastructure, such as trees and green roofs, offers various benefits, including air pollution mitigation, the study emphasizes the need for strategic deployment and design guidelines to avoid unintended consequences. Although urban vegetation can facilitate health benefits, further empirical evidence is needed to establish a direct link between these benefits and air pollution reduction by green infrastructure [12].

Effiong's study [8] investigates the environmental impact of urbanization in 49 African countries from 1990 to 2010. The findings demonstrate that urbanization leads to a reduction in environmental pollution, particularly in terms of particulate matter (PM10) emissions. However, the study does not support the inverted U-shaped curve relationship between urbanization and environmental pollution, as suggested by the Kuznets' hypothesis. The paper underscores the significance of strategic urban planning and infrastructure investment to promote a green environment. In a research, Zeng et al. [22] address the hazardous atmospheric pollution issues in

Beijing resulting from increased emissions due to population growth and industrialization. They propose a hybrid population–production–pollution nexus model (PPP) for air pollution management and air quality planning, considering factors such as employment structure, industrial layout, pollutant purification efficiency, and pollution mitigation schemes. The study facilitates the identification of optimized policies to alleviate population–production–emission conflicts and mitigate air pollution crises in the region.

3 Policy Framing and Air Pollution Management

Policy framing plays a critical role in the management of air pollution, as it provides the foundation for effective and targeted strategies to address this pressing environmental challenge [1]. With the ever-increasing levels of air pollution and its adverse impacts on human health and the environment, robust policy frameworks are essential to guide decision-making, set emission reduction targets, and promote sustainable practices. Effective policy framing begins with a comprehensive understanding of the sources and drivers of air pollution. This involves thorough research and data collection to identify the main contributors to pollution in a specific region or country. Once these sources are identified, policymakers can design tailored interventions to tackle the root causes of pollution.

Regulatory measures are a crucial component of air pollution management policies. Setting and enforcing stringent emission standards for industries, vehicles, and other pollution-emitting sources can significantly reduce pollution levels. Additionally, incentivizing the adoption of cleaner technologies and renewable energy sources can encourage businesses and individuals to transition away from fossil fuels and other polluting practices [4].

The paper by Burns et al. [5] discusses the use of "critical loads" as a policy tool to protect ecosystems from the effects of air pollutants. The critical-loads approach has been widely applied in Europe to shape air pollutant control policies, but its application in the USA has been limited. However, recent interest in applying this approach to manage sulfur and nitrogen air pollutants in the USA is growing, as evidenced by various conferences and initiatives. The authors describe the critical-loads concept, its limitations, and how it can better inform future air pollutant control policies in the USA. Tuinstra et al. [19] focus on the science–policy interaction in the context of the United Nations Economic Commission for Europe's Convention for Long Range Transboundary Air Pollution (CLRTAP). They analyze how participants in the assessment process divide and coordinate work between science and policy, which enhances credibility, legitimacy, and relevance with multiple audiences. The authors emphasize the importance of boundary work and co-production of science and policy in the CLRTAP process, which impacts credibility, legitimacy, and relevance. The combined framework used in this study provides insights into what happens at the science–policy interface.

Ravindra et al. [18] address air pollution in India and the need to bridge the gap between science and policy. They highlight the importance of statistical models and Big Data Analytics to estimate atmospheric concentrations, understand health impacts, and facilitate policy decisions. The study suggests the development of air quality networks and statistical models to forecast pollution levels, issue health advisories, and assess the cost and benefits of air pollution control measures. Statistical models and algorithms are seen as crucial tools to connect scientific knowledge with effective policy-making. Olofsson et al. [17] explore how the air pollution issue in Delhi is framed by news media and narrated by nonprofit organizations. The study uses both inductive and deductive approaches to analyze news coverage and nonprofit narratives. The findings confirm existing knowledge about the causes and effects of air pollution in Delhi, while perceptions vary regarding the government's ability to implement policy. The study provides insights into the priorities that influence public opinions, political decisions, and eventually public policies regarding air pollution in one of the world's largest cities.

4 Interplay Between Development and Policy

Following are some policies related to urbanization in Delhi aimed at controlling air pollution.

The Graded Response Action Plan (GRAP): The Graded Response Action Plan (GRAP) is a comprehensive policy response to control air pollution in Delhi and surrounding states. The GRAP is a set of measures that can be implemented by the government of Delhi and surrounding states depending on the severity of air pollution levels. The plan includes actions such as a ban on diesel generators, the closure of brick kilns, and restrictions on the use of private vehicles. The plan also includes measures such as the deployment of pollution control teams, the distribution of anti-pollution masks, and increased public awareness campaigns.

Green Budget: The Delhi government has also introduced a "Green Budget" that allocates funds for measures aimed at reducing air pollution. This includes investments in public transport, tree plantation drives, and subsidies for electric vehicles. The government has also banned the sale and use of firecrackers during the Diwali festival to reduce the level of air pollution caused by their combustion.

Ban on Firecrackers: The Delhi government has banned the sale and use of firecrackers during the Diwali festival to reduce the level of air pollution caused by their combustion.

Following are some policies related to transportation in Delhi aimed at controlling air pollution.

Delhi Electric Vehicle Policy: The Delhi government has launched an electric vehicle policy aimed at promoting the adoption of electric vehicles in the city. The policy offers incentives such as subsidies on the purchase of electric vehicles, exemption from road tax, and charging infrastructure.

Odd–Even Scheme: The Odd–Even Scheme is a traffic rationing policy that allows vehicles with odd-numbered license plates to operate on odd-numbered dates and even-numbered plates on even-numbered dates. This policy aims to reduce the number of vehicles on the road, thereby reducing air pollution.

Expansion of Public Transport: The Delhi government has taken several measures to improve and expand public transport in the city. This includes the introduction of more buses, the expansion of the metro rail network, and the development of dedicated bus lanes.

Phased Introduction of BS-VI Fuel: The Indian government has phased out the sale of BS-IV fuel and introduced BS-VI fuel, which has significantly lower levels of sulfur and other harmful pollutants. This has led to a reduction in the emission of harmful pollutants from vehicles, thereby contributing to improved air quality in the region.

Implementation of Bharat Stage Emission Standards: The Indian government has implemented Bharat Stage Emission Standards, which set limits on the emissions of harmful pollutants from vehicles. The implementation of these standards has led to the introduction of cleaner technologies and fuels in vehicles, thereby contributing to improved air quality in the region.

Following are some policies related to industrialization in Delhi aimed at controlling air pollution.

Ban on Polluting Industries: The Delhi government has banned the operation of several polluting industries in the city, such as brick kilns, furnace oil-based industries, and coal-based industries. This ban aims to reduce the emission of harmful pollutants from these industries.

Green Energy Fund: The Delhi government has set up a Green Energy Fund to promote the adoption of clean energy technologies in industries. The fund provides subsidies to industries to switch to cleaner technologies such as solar power, biogas, and biomass.

Inspection of Industries: The Delhi Pollution Control Committee (DPCC) regularly inspects industries in the city to ensure that they comply with pollution control norms. The DPCC also conducts surprise inspections of industries to ensure that they are not violating pollution control norms.

Pollution Control Boards: The Central Pollution Control Board (CPCB) and the State Pollution Control Boards (SPCBs) are responsible for regulating industrial pollution in the country. They set emission standards for industries and monitor their compliance with these standards.

Following are some policies related to fog in Delhi aimed at controlling air pollution.

Smog Towers: The Delhi government has recently announced plans to install smog towers in the city to combat air pollution during the winter season when the city experiences heavy fog. Smog towers are large structures that use advanced air filters to remove pollutants from the air. These towers are intended to improve air quality in specific areas of the city.

Artificial Rain: The Delhi government has experimented with cloud seeding to induce artificial rain to settle pollutants in the air. The technique involves spraying chemicals into the clouds to induce rain, which can help settle pollutants and improve air quality.

Crop Residue Management: The burning of crop residue in nearby states is a significant contributor to air pollution in Delhi during the winter months. The Delhi government has taken several measures to promote crop residue management, such as providing subsidies to farmers for purchasing equipment to manage crop residue and setting up biomass power plants to utilize crop residue as fuel.

The interplay between development and policy is a complex and crucial aspect of achieving sustainable growth and addressing various societal challenges. Case studies from different contexts can provide valuable insights into how development initiatives and policy decisions interact and influence each other. Here are some case studies that illustrate the interplay between development and policy.

Renewable Energy Policy in Germany: The German government introduced a series of policies and incentives to promote the adoption of renewable energy sources, such as solar and wind power, known as the "Energiewende" [3]. These policies included feed-in tariffs, tax incentives, and research funding. As a result, Germany became a global leader in renewable energy and reduced its dependence on fossil fuels. The Energiewende not only contributed to environmental sustainability but also stimulated economic growth in the renewable energy sector, creating jobs and driving technological advancements.

Land Use Policy in Singapore: Singapore's land use policy is an exemplary model of how effective urban planning can support economic development and address environmental challenges in a limited land area. The Singaporean government adopted a comprehensive land use plan that balanced industrial development, housing, green spaces, and infrastructure [21]. Strict zoning regulations, high-rise building construction, and efficient public transportation systems helped optimize land usage and promote sustainable urban growth. As a result, Singapore transformed from a developing nation with limited land resources into a highly developed and environmentally conscious city-state.

Green Building Policies in American Central Cities: Eugene Choi [9] conducted a study examining the effects of municipal policy on green building designations in America's central cities. The research focused on how local policies and regulations influenced the adoption and implementation of green building practices. Green building initiatives, including LEED (Leadership in Energy and Environmental Design) certifications, have gained traction as a means to promote sustainable and environmentally friendly construction and design. Choi's study analyzed the impact of municipal policies on encouraging the adoption of these green building practices, and the role of such policies in fostering sustainable urban development.

5 Challenges and Barriers

Challenges and barriers play a critical role in shaping the success and effectiveness of policies and initiatives aimed at addressing various issues. They can hinder progress, create obstacles, and necessitate strategic planning and innovative solutions. In the context of urban development and environmental policies, several challenges and barriers are commonly encountered:

1. One of the significant challenges is gaining alignment among various political entities and stakeholders. Differing priorities, interests, and perspectives can lead to conflicts, delays in decision-making, and difficulty in implementing cohesive policies.
2. The allocation of funds for sustainable urban development and environmental projects may be limited. High initial costs of eco-friendly infrastructure and technologies can deter investments, making it challenging to implement long-term solutions.
3. Inadequate institutional capacity and lack of coordination between relevant authorities and agencies may hinder the effective implementation of policies. Clear mandates and collaboration among institutions are essential for successful urban and environmental planning.
4. Engaging the public in policy development and fostering awareness about sustainable practices can be a barrier. Lack of public support or participation may impede the adoption of environmentally friendly policies.
5. The implementation of cutting-edge technologies and innovative approaches often requires specialized technical expertise. A shortage of skilled professionals may hinder the adoption of sustainable solutions.
6. Overcoming legal and regulatory hurdles can be challenging. Outdated or conflicting regulations, lack of enforcement, and cumbersome approval processes may slow down sustainable development initiatives.
7. Access to accurate and comprehensive data on environmental conditions, urban infrastructure, and demographic trends is crucial for evidence-based policy-making. Limited data availability can hinder informed decision-making.
8. Resistance to change from industries, businesses, and individuals accustomed to traditional practices can hinder the adoption of sustainable policies. Overcoming resistance and fostering buy-in are critical for successful implementation.

6 Future Directions and Recommendations

Moving toward sustainable solutions in urban development and environmental protection requires a comprehensive and integrated approach. Firstly, cities should prioritize climate resilience by incorporating adaptive measures into their urban planning and infrastructure development. Identifying vulnerable areas and implementing strategies to mitigate the impacts of climate change, such as extreme weather events and rising sea levels, is crucial. Additionally, expanding green infrastructure initiatives, including urban forests, green roofs, and permeable surfaces, can improve air quality, reduce the urban heat island effect, and enhance biodiversity. Embracing a circular economy model is another essential step toward sustainability. By promoting the reuse, recycling, and repurposing of resources, cities can minimize waste and reduce resource depletion. Encouraging businesses to adopt sustainable practices and reduce their environmental footprint is also vital for a more sustainable future. Sustainable transportation options, such as public transit, cycling lanes, and pedestrian-friendly streets, should be invested in to reduce emissions and alleviate congestion.

Implementing strict energy efficiency standards for buildings and incentivizing the use of renewable energy sources are key to reducing carbon emissions and ensuring a greener urban environment. Nature-based solutions, such as urban agriculture, wetland restoration, and natural water retention systems, should be embraced to enhance ecosystem services and improve overall urban well-being. Community engagement plays a significant role in successful sustainable development. Involving communities in the planning and decision-making process ensures that policies and projects align with their needs and preferences, fostering a sense of ownership and pride in sustainable initiatives. Green certifications and incentives can further encourage businesses and individuals to adopt eco-friendly practices.

The establishment of eco-industrial parks that promote resource efficiency and reduce pollution is critical for sustainable industrial development. Cities should also engage in international cooperation, collaborating with other countries and participating in global environmental agreements and initiatives to address worldwide challenges. Sharing best practices and experiences can foster collective action toward a sustainable future. Investing in environmental education and empowering the youth to be advocates for sustainable development are essential for long-term progress. Cross-sectoral policy integration and continuous research and innovation are key to addressing interconnected challenges and finding more efficient and eco-friendly solutions.

Ultimately, a collective effort from governments, businesses, communities, and individuals is required to build resilient, eco-friendly cities that provide a high quality of life for current and future generations. Such an approach will ensure the preservation and enhancement of our natural environment and pave the way for a more sustainable and thriving planet.

7 Conclusion

The interplay between development and policy in the context of air pollution and the urban environment is a critical and complex issue that demands urgent attention and concerted efforts. As evidenced by the case studies and literature, rapid urbanization, industrialization, and population growth have contributed to severe air pollution and environmental degradation in many cities, with detrimental impacts on public health and ecosystems. The challenges and barriers, such as inadequate infrastructure, policy implementation gaps, and resistance to change, underscore the need for innovative and holistic approaches to address the root causes of air pollution. The incorporation of green and sustainable practices, the adoption of clean technologies, and the integration of nature-based solutions can significantly improve air quality and promote environmental health. There is a need for stronger political will to ensure policy compliance. Policymakers and urban planners must work together to design and implement comprehensive strategies that prioritize climate resilience, circular economy principles, and nature conservation. Engaging with communities and fostering public awareness is vital in building a collective commitment to sustainable development. To achieve tangible progress, international cooperation and knowledge-sharing are essential, as air pollution and environmental challenges transcend national borders. By learning from successful policies and experiences in various countries, cities can benefit from a global pool of best practices. The future direction and recommendations point toward a sustainable path that considers environmental, social, and economic dimensions in urban development. It requires a paradigm shift toward greener, smarter, and more inclusive cities that prioritize the well-being of both people and the planet. This transformation demands bold leadership, innovative technologies, and active citizen participation.

References

1. Anderson A (2009) Media, politics and climate change: towards a new research agenda. Sociol Compass 3(2):166–182
2. Asif M, Haq RAU, Gulfreen E, Arshad S, Tasleem MW, Rajpoot S, Maqbool M (2022) Particulate matter emission sources and their control technologies. Pollut Res 41(2):696–706
3. Bechberger M, Reiche D (2004) Renewable energy policy in Germany: pioneering and exemplary regulations. Energy Sustain Dev 8(1):47–57
4. Brunel C, Levinson A (2016) Measuring the stringency of environmental regulations. Rev Environ Econ Policy 10(1):47–67
5. Burns DA, Blett T, Haeuber R, Pardo LH (2008) Critical loads as a policy tool for protecting ecosystems from the effects of air pollutants. Front Ecol Environ 6(3):156–159
6. Central Pollution Control Board (CPCB) (2020) Comprehensive environmental pollution index (CEPI) for Delhi. http://cpcb.nic.in/cepi-ranking/
7. Chatterji A (2021) Air pollution in Delhi: filling the policy gaps. Massach Undergr J Econ 17(1)
8. Effiong EL (2018) On the urbanization-pollution nexus in Africa: a semiparametric analysis. Qual Quant 52(1):445–456
9. Choi E (2010) Green on buildings: the effects of municipal policy on green building designations in America's central cities. J Sustain Real Estate 2(1):1–21

10. Gan L (2003) Globalization of the automobile industry in China: dynamics and barriers in greening of the road transportation. Energy Policy 31(6):537–551
11. Hashmi SH, Fan H, Fareed Z, Shahzad F (2021) Asymmetric nexus between urban agglomerations and environmental pollution in top ten urban agglomerated countries using quantile methods. Environ Sci Pollut Res 28(11):13404–13424
12. Hewitt CN, Ashworth K, MacKenzie AR (2020) Using green infrastructure to improve urban air quality (GI4AQ). Ambio 49:62–73
13. Kumar P, Druckman A, Gallagher J, Gatersleben B, Allison S, Eisenman TS, Hoang U, Hama S, Tiwari A, Sharma A, Abhijith KV, Morawska L (2019) The nexus between air pollution, green infrastructure and human health. Environ Int 133:105181
14. Mayer H (1999) Air pollution in cities. Atmos Environ 33(24–25):4029–4037
15. Ministry of Environment, Forest and Climate Change (MoEFCC) (2019) National clean air program (NCAP). https://www.india.gov.in/spotlight/national-clean-air-programme-ncap
16. NITI Aayog (2020) Transformative mobility solutions for Delhi. https://niti.gov.in/writeread data/files/document_publication/Transformative_Mobility_Solutions_Delhi.pdf
17. Olofsson KL, Weible CM, Heikkila T, Martel JC (2018) Using nonprofit narratives and news media framing to depict air pollution in Delhi, India. Environ Commun 12(7):956–972
18. Ravindra K, Sidhu MK, Mor S, John S, Pyne S (2016) Air pollution in India: bridging the gap between science and policy. J Hazard Toxic Radioact Waste 20(4):A4015003
19. Tuinstra W, Hordijk L, Kroeze C (2006) Moving boundaries in transboundary air pollution co-production of science and policy under the convention on long range transboundary air pollution. Glob Environ Chang 16(4):349–363
20. Van Kamp I, Leidelmeijer K, Marsman G, De Hollander A (2003) Urban environmental quality and human well-being: towards a conceptual framework and demarcation of concepts; a literature study. Landsc Urban Plan 65(1–2):5–18
21. Wang LH (1986) Land use policy of a city state: the Singapore case. Land Use Policy 3(3):180–192
22. Zeng XT, Tong YF, Cui L, Kong XM, Sheng YN, Chen L, Li YP (2017) Population-production-pollution nexus-based air pollution management model for alleviating the atmospheric crisis in Beijing, China. J Environ Manage 197:507–521

IICAQM 2022—Measurement, Modelling, Health Risk, and Public Policy

Investigation of Occupational Thermal Comfort in Urban Subway Stations

Amit Passi, S. M. Shiva Nagendra, and M. P. Maiya

Abstract Exposure to a thermally unstable environment causes work productivity loss, fatigue, stress, and respiratory complications. The current study investigated occupational thermal comfort in urban subway stations. The study was carried out in Chennai from March–June 2019. A monitoring campaign was conducted in seven urban subway stations using DeltaOhm HD32.3. The thermal parameters, temperature, relative humidity, and air velocity surpassed the regulatory standards given by ASHRAE-55 and ISO7730; the highest mean values approached 35 ºC, 80%, and < 0.1 m/s. Substantial low mean wet bulb temperature (< 26 ºC) shows that low room air temperature could be attained with evaporative cooling of air. Further, no significant impact of radiations and airflow was observed on comfort conditions. Thermal stress indicators, globe, mean radiant, and operative temperature values were similar to the dry bulb temperature. Furthermore, the estimation of comfort indices indicated subway occupational dissatisfaction of 36–99%. The Pearson's correlation analysis illustrated that station characteristics significantly influence comfort conditions.

Keywords Thermal comfort · Occupational exposure · Urban subway · Percentage of dissatisfaction

A. Passi · S. M. S. Nagendra (✉)
Department of Civil Engineering, Indian Institute of Technology Madras, Chennai, India
e-mail: snagendra@iitm.ac.in

A. Passi
e-mail: amitpassi@alumni.iitm.ac.in

M. P. Maiya (✉)
Department of Mechanical Engineering, Indian Institute of Technology Madras, Chennai, India
e-mail: mpmaiya@iitm.ac.in

S. M. Shiva Nagendra et al. (eds.), *Select Proceedings of the 8th Indian International Conference on Air Quality Management (IICAQM 2023)*, Lecture Notes in Civil Engineering 582, https://doi.org/10.1007/978-981-96-2359-4_22

1 Introduction

Thermal comfort is defined as the physical and psychological well-being of an individual in a particular space or environment. Poorly maintained thermal conditions impact our surroundings severely. It causes substandard indoor air quality conditions. Also, it affects human health sorely. Continual exposure to a thermally unstable environment results in increased sick-building syndrome, causing fatigue, mood disturbance, headache, nausea, migraine, irritation, upper respiratory complications, respiratory infections, cardiovascular diseases, and cognitive impairment [12]. Due to uncomfortable environmental settings, occupants spend time and effort trying to make themselves more comfortable rather than concentrating on their main activity [17]. Thus, improving environmental quality increases occupants'/workers productivity rate [4, 11].

Thermal comfort in an indoor space depends on various aspects, such as the surrounding environment, occupants' personal factors, and their behavioral control [5, 6]. The COVID-19 pandemic has made regulatory bodies more concerned about indoor thermal conditions, especially in public places. Subway stations are compact and airtight environments. The thermal environment is maintained with a central HVAC system. Poorly ventilated building and high energy cost results in inappropriate mechanical ventilation/air-conditioning [16]. Past studies indicated that ventilation is a prime influential parameter. More than 90% of discomfort rate was observed in the Budapest subway due to a lack of good ventilation [14]. Studies also reported that thermal comfort is not achievable in subway stations, especially on hot summer days [1, 2]. A study in Delhi subway stations also indicated moderate thermal stress due to high passenger density and unstable thermal conditions [18]. Further, Passi et al. [15] specified that inferior indoor thermal conditions cause air pollutants accumulation, resulting in high transmission of infectious diseases.

The adverse health impacts of exposure to such environments depend on the design of a building, air exchange rate, and exposure duration [19]. Exposure to sweltering heat gives rise to anxiety in occupants, dehydration, stroke, and lung infections. Lack of airflow causes difficulty breathing, asthma, and other respiratory diseases. High humidity causes mold growth and causes indoor air quality problems. Thus, the prevalent thermal conditions have a significant role in maintaining occupant comfort and environmental health [8]. Limited studies were conducted to understand the thermal environmental conditions in subway microenvironments.

This research study aims to understand occupational thermal comfort in subway stations. The objectives of this study are (i) to investigate the thermal environment in different subway stations, (ii) to study thermal stress indicators under prevalent thermal conditions, (iii) to evaluate thermal stress and comfort indices, and (iv) to investigate the influence of station characteristics on the thermal dissatisfaction rate of occupants.

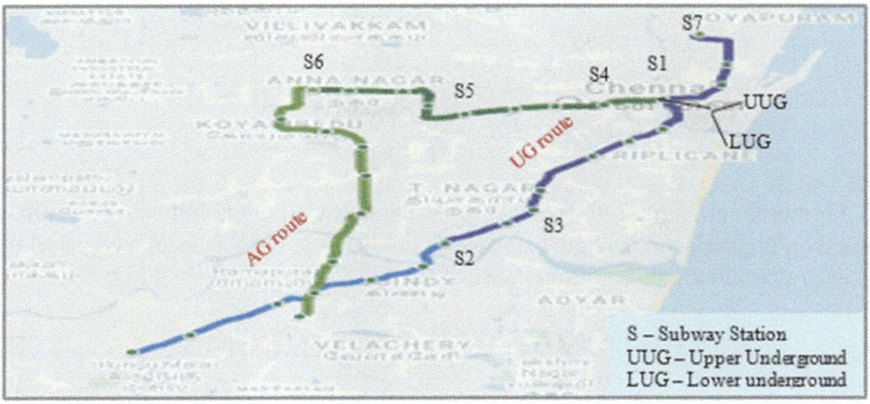

Fig. 1 Map of Chennai subway network

2 Research Approach

2.1 Study Area

The comfort monitoring study was conducted in the subway stations of Chennai, a cosmopolitan city in Southern India. The Chennai subway system started operating in 2015. It is the 4th largest metro system in India. In the city, two metro lines run, with a total network length of 54.15 km. The metro lines in Chennai are built half underground and partially aboveground. Figure 1 presents the map of the Chennai subway network. In the map, the portion of metro lines shown with dark color indicates an underground network. The segment of metro lines shown with light color presents an elevated network. The field measurements were conducted from March to June 2019. The occupational thermal stress monitoring was conducted in seven subway stations (Fig. 1). Data were collected for a week in each subway station.

2.2 Environmental Monitoring and Comfort Estimation

The thermal environment of subway stations was evaluated by measuring four fundamental parameters: room air temperature, relative humidity, room air velocity, and wet bulb temperature. HD 32.3 Delta Ohm instrument was used to conduct measurements using sensor probes such as pt100, capacitive, and hot wire sensors. Further, on the basis of elementary parameters, thermal stress indicators, black globe, mean radiant, and operative temperatures are calculated. A globe of 50 mm diameter consisting of a copper sphere coated with matt black paint, with an emissivity of 0.95, was used to measure globe temperature. The following equations were used for further calculations.

$$T_r = \left[(T_g + 273)^4 + \frac{1.1 \times 10^8 \times V_A^{0.6}}{0.95 \times D^{0.4}} (T_g - T_a) \right]^{\frac{1}{4}} - 273 \quad (1)$$

$$T_O = \frac{T_r + (T_a \times (10V_A)^{0.5})}{1 + (10V_A)^{0.5}} \quad (2)$$

Furthermore, thermal stress indices are calculated using the following equations based on fundamental parameters and stress indicators. PMV is estimated based on a seven-point thermal sensation scale (ASHRAE-55).

$$\text{WBGT} = 0.7T_w + 0.2T_g + 0.1T_a \quad (3)$$

$$\text{PPD} = 100 - 95 \times e^{(-0.03353 \times \text{PMV}^4 - 0.2179 \times \text{PMV}^2)} \quad (4)$$

2.3 Data Analysis

The data were analyzed using descriptive and inferential statistics. Interpreted data are compared against regulatory standards. Considering occupants' dissatisfaction rate as a prime parameter, the Pearson's correlation analysis is performed among selected subway stations to identify the influence of station characteristics on subway comfort conditions.

3 Results and Discussion

3.1 The Thermal Environment in Subway Stations

Thermal environment conditions in subway stations are presented by the four basic parameters, including room air temperature, relative humidity, room air velocity, and wet bulb temperature. These parameters have a significant impact on humans as they influence the thermal acceptability of an individual. The thermal acceptability of occupants varies with their personal characteristics, such as body area, activity rate, and level of clothing. Workers in Chennai subway stations have a full clothing level, i.e., pant, shirt, socks, and shoes. Their usual activity includes sitting and walking operation. Thus, the average body surface area of 1.7 m^2, the activity rate of 1.2 MET, and the clothing ratio of 0.8 CLO were adjusted in the data logger.

Subway room air temperature was noted to be high in all studied stations, with the highest average value of 35 °C in S5. Temperature values exceeded the ASHRAE-55 guidelines in all subway stations except S4. Further, relative humidity values were

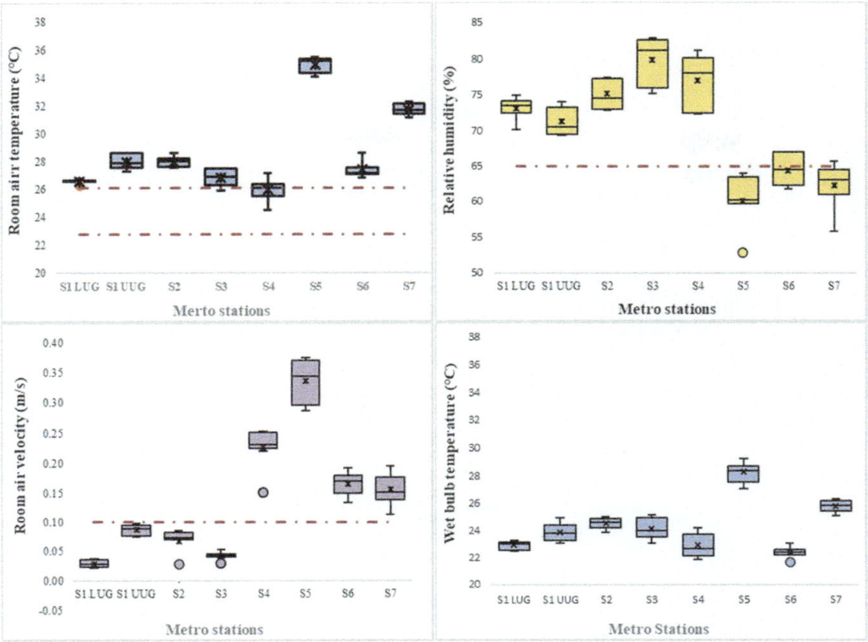

Fig. 2 Thermal environment conditions in subway stations **a** Room air temperature, **b** Relative humidity, **c** Room air velocity, **d** Wet bulb temperature

in the range of 60–80%. Subway stations along subway line 1 surpassed the upper limit of 65% (ASHRAE 62.1). The highest average value of 80% was observed in S3. Furthermore, observations of subway room air velocity indicated the least airflow among stations located along subway line 1 (air velocity < 0.1 m/s, ISO 7730). It suggests the lack of air exchange to maintain optimum airflow. Moreover, the substantial low range of wet bulb temperature (23–26 °C) compared to dry bulb temperature shows that low room air temperature could be attained with evaporative air cooling (Fig. 2).

3.2 *Thermal Stress Indicators*

Change in thermal conditions of an indoor environment produces thermal stresses on the occupants. Thermal stress is the state of mind that expresses dissatisfaction with the surrounding environment [2] (Fig. 3).

Globe temperature is one of the most familiar thermal variables representing heat stress [7]. It illustrates the compound effect of room temperature, air velocity, and radiation. It indicates the thermal conditions felt by the human body. In subway stations, globe temperature was noted in the range of 26–35 °C, with the highest

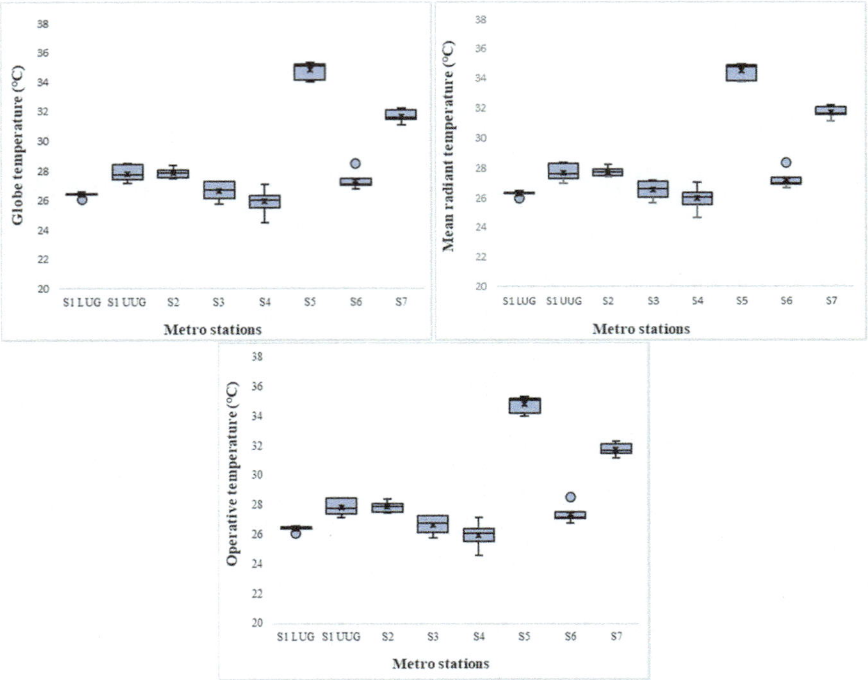

Fig. 3 Thermal stress indicators **a** Globe temperature, **b** Mean radiant temperature, **c** Operative temperature

value in S5. This range is similar to the dry bulb temperature range. Thus, there is no significant effect of radiation and airflow on comfort conditions.

Further, mean radiant temperature presents the area-weighted mean temperature of all the objects surrounding the body. It strongly influences thermophysiological comfort, which in turn impacts individuals' thermal sensations. It represents the combined effect of air temperature, globe temperature, and air velocity. It was observed in the range of 26–34.5 °C, with the highest value in S5. The range is similar to dry bulb and globe temperature.

Furthermore, operative temperature describes the average of radiant and air temperatures, weighted by their respective heat transfer coefficients. It is estimated with room air temperature, radiant temperature, and air velocity. In Chennai subway stations, it was detected in the range of 26–35 °C.

The dry, globe, radiant, and operative temperature ranges are similar. It indicates no substantial impact of radiant heat transfer indoors in a subway environment.

3.3 Thermal Comfort Indices

The Comfort Index is a numerical value representing how comfortable or uncomfortable our indoor environment is. There are number of indices to evaluate the thermal comfort of occupants. Here, we are dealing with three indices, wet bulb globe temperature, predicted mean vote, and predicted percentage of dissatisfaction.

Wet bulb globe temperature is a measure of heat stress on the human body with the influence of sunlight. It is a temperature equivalent perceived by humans and counts the impact of air temperature, humidity, air velocity, and visible and infrared radiations on humans. It is estimated as a compound effect of the wet bulb, dry bulb, and globe temperature. It was noted in the range of 23.8–30.2 °C, with the highest value in S5. As per ISO 7243, considering the combined sitting and walking operation of subway workers and assuming they are non-acclimatized to the heat, the value should be less than 26 °C. The values were within the limit except for stations S5 and S7, where index values approached 27.5–30.2 °C (Figs. 4 and 5).

The basis of human thermal comfort is an individual's skin temperature and sweat secretion. The environment feels comfortable if these two are balanced. The heat balance of a human being is impacted by clothing ratio, activity level, and thermal environment conditions. The predicted mean vote envisions the thermal sensation of occupants. Air temperature, air velocity, clothing insulation, metabolic rate, relative humidity, and mean radiant temperature provide the basis for calculation. As per ASHRAE-55, PMV is acceptable in range of − 0.85– + 0.85. In Chennai subway stations, these values were in the range of + 1.06– + 3.21.

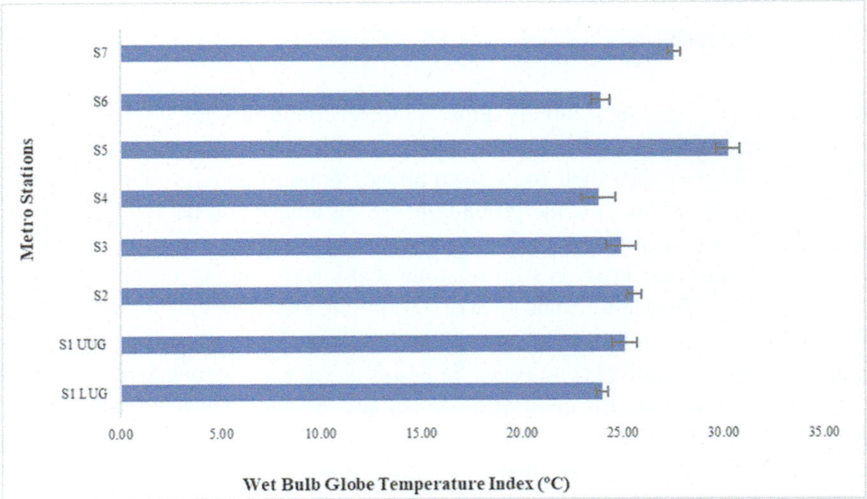

Fig. 4 Wet bulb globe temperature index in subway stations

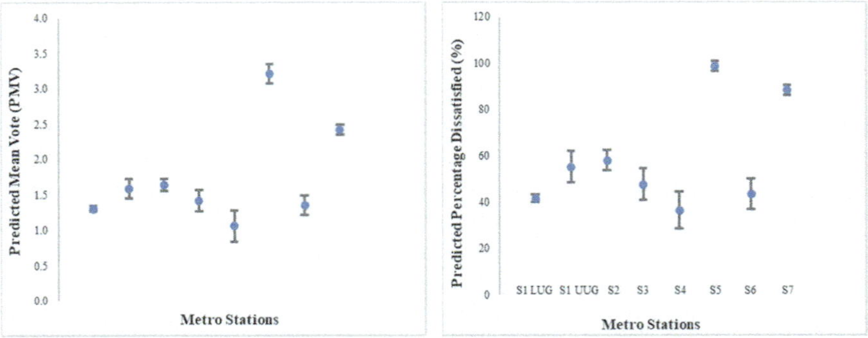

Fig. 5 Thermal comfort indices **a** Predicted mean vote, **b** Predicted percentage of dissatisfied

The prediction of thermal sensation does not illustrate the whole picture. We need to account for the level of satisfaction or dissatisfaction in a particular environment. The predicted percentage of dissatisfaction (PPD) demonstrates a quantitative projection of thermally dissatisfied occupants. As per ASHRAE-55, PPD should be less than 20%, which means at least 80% of occupants should be thermally satisfied. In Chennai subway stations, PPD values were observed in the range of 36–99%, with the highest dissatisfaction in S5. Thus, the rate of occupants' dissatisfaction has surpassed the ASHRAE-55 guidelines in all subway stations.

3.4 Influence of Station Characteristics on Comfort Conditions

Station characteristics substantially impact comfort conditions. Each selected station presents different features. S1 is a two underground level junction station. At S2 and S6, subway lines change their route from aboveground to underground. S3 and S5 are located in the middle of underground lines. S4 is adjacent to the junction station. And S7 is a terminal station.

Table 1 shows the Pearson's correlation coefficients among different selected subway stations. The significance level of 0.05 is considered for calculation. It can be observed from Table 1 that along subway line 1, S2 and S7 have a strong correlation, indicating similar comfort conditions. S2 and S3 have a strong negative correlation and show different comfort conditions. Along line 2, S4 and S5 have a moderate positive correlation and represent similar characteristics. S1 UUG and S4 have a strong negative correlation and show distinct characteristics due to different orientations. Further, across both subway lines, S3 and S5, S1 LUG and S1 UUG exhibit the same features, thus indicating similar variation in comfort conditions.

Table 1 Pearson's correlation coefficients among different subway stations

Subway lines	Subway stations	1				2			
		S2	S3	S1 LUG	S7	S1 UUG	S4	S5	S6
1	S2	1							
	S3	− 0.51*	1						
	S1 LUG	0.31*	0.06	1					
	S7	0.69*	− 0.20*	− 0.25*	1				
2	S1 UUG	− 0.17	0.64*	0.61*	− 0.23*	1			
	S4	0.43*	− 0.48	0.24	0.07*	− 0.52*	1		
	S5	− 0.07*	0.53*	− 0.07*	0.12*	− 0.06*	0.40*	1	
	S6	− 0.10*	− 0.01	− 0.10	0.29*	0.22*	− 0.14	− 0.08*	1

*Correlation is significant at a $p < 0.05$ (two-tailed)

4 Conclusions

In this study, occupational thermal comfort was investigated in Chennai subway stations. We studied the seven subway stations, which all consist of different station characteristics. The thermal parameters, room air temperature, relative humidity, and air velocity were compared against regulatory standards given by ASHRAE-55, 62.1, and ISO 7730. It was noted that thermal conditions in subway stations had surpassed the regulatory limits; the mean values approached 35 °C, 80%, and < 0.1 m/s. The substantially low range of wet bulb temperature (< 26 °C) compared to dry bulb temperature shows that low room air temperature could be attained with evaporative cooling of air. Thermal stress indicators, globe, mean radiant, and operative temperature variations were almost similar, in the range of 26–35 °C. Therefore, there was no significant impact on radiation and airflow in subway stations.

Further, the heat stress index was found to be < 26 °C except in S5, where it approached 30.2 °C. The comfort indices predicted mean vote was rated from a slightly hot to a very hot range of thermal sensation scale. Furthermore, the percentage dissatisfaction rate approached 99% in subway stations. Moreover, the Pearson's correlation analysis indicated that station characteristics impact comfort conditions significantly. Stations with the same characteristics were found to have similar comfort conditions.

Acknowledgements The authors would like to thank CMRL (Chennai Metro Rail Limited) for permitting to conduct of IAQ measurements in metro stations.

References

1. Abbaspour M, Jafari MJ, Mansouri N, Moattar, F, Nouri N, Allahyari M (2008) Thermal comfort evaluation in Tehran metro using relative warmth index. Int J Environ Sci Tech 5(3):297–304
2. ASHRAE 55 (2010) ASHRAE standard. Thermal environmental conditions for human occupancy. ANSI. ISSN:1041-2336. www.ashrae.org
3. ASHRAE 62.1 (2016) ASHRAE standard. Ventilation for acceptable indoor air quality. ANSI. ISSN:1041-2336. http://www.ashrae.org
4. Bueno AM, de Paula Xavier AA, Broday EE (2021) Evaluating the connection between thermal comfort and productivity in buildings: a systematic literature review. Buildings 11(6):244. https://doi.org/10.3390/buildings11060244
5. Dhaka S, Mathur J, Wagner A, das Agarwal G, Garg V (2013) Evaluation of thermal environmental conditions and thermal perception at naturally ventilated hostels of undergraduate students in composite climate. Build Environ 66:42–53. https://doi.org/10.1016/j.buildenv.2013.04.015
6. Dili AS, Naseer MA, Varghese TZ (2010) Thermal comfort study of Kerala traditional residential buildings based on questionnaire survey among occupants of traditional and modern buildings. Energy Build 42(11):2139–2150. https://doi.org/10.1016/j.enbuild.2010.07.004
7. Hajizadeh R, Dehghan SF, Golbabaei F, Jafari SM, Karajizadeh M (2017) Offering a Model for Estimating Black Globe Temperature According to Meteorological Measurements. Meteorol Appl 24(2):303–307. https://doi.org/10.1002/met.1631
8. He M, Li N, He Y, He D, Wang K (2017) Influences of temperature and humidity on perceived air quality with radiant panel workstation. Procedia Eng 205:765–772. https://doi.org/10.1016/j.proeng.2017.10.008
9. ISO 7243 (2017) Ergonomics of the thermal environment-assessment of heat stress using the WBGT (wet bulb globe temperature) index. https://files.isec.pt/DOCUMENTOS/SERVICOS/BIBLIO/Documentos%20de%20acesso%20remoto/ISO_7243_2017.pdf
10. ISO 7730 (2005) Ergonomics of the thermal environment-analytical determination and interpretation of thermal comfort using calculation of the PMV and PPD indices and local thermal comfort criteria. ISO 7730:2005(E) ISO 7730 Third edition. www.iso.org
11. Lamb S, Kwok KCS (2016) A longitudinal investigation of work environment stressors on the performance and wellbeing of office workers. Appl Ergon 52:104–111
12. Mendes A, Pereira C, Mendes D, Aguiar L, Neves P, Silva S, Batterman S, Teixeira JP (2013) Indoor air quality and thermal comfort—results of a pilot study in elderly care centers in Portugal. J Toxicol Environ Health A 76:333–344. https://doi.org/10.1080/15287394.2013.757213
13. Nakano J, Sakamoto K, Tanabe SI (2006) Thermal comfort conditions in train stations for transit and short-term occupancy. https://www.researchgate.net/publication/237612457
14. Ordódy (2000) Thermal comfort in the passenger areas of the budapest metro. Periodica Polytech Ser Mech Eng 44(2):309–317
15. Passi A, Shiva Nagendra SM, Maiya MP (2021) Assessment of exposure to airborne aerosol and bio-aerosol particles and their deposition in the respiratory tract of subway metro passengers and workers. Atmos Pollut Res 12(11):101218. https://doi.org/10.1016/j.apr.2021.101218
16. Passi A, Shiva Nagendra SM, Maiya MP (2022) Evaluation of comfort perception of passengers in urban underground metro stations. Energy Sustain Dev 68:273–288. https://doi.org/10.1016/j.esd.2022.04.003
17. Roskams M, Haynes B (2019) Predictive analytics in facilities management: a pilot study for exploring environmental comfort using wireless sensors. J Facil Manage 17:356–370
18. Sinha K, Rajasekar E (2020) Thermal comfort evaluation of an underground metro station in New Delhi using agent-based modelling. Build Environ 177. https://doi.org/10.1016/j.buildenv.2020.106924
19. Wen Y, Leng J, Shen X, Han G, Sun L, Yu F (2020) Environmental and health effects of ventilation in subway stations: a literature review. Int J Environ Res Public Health 17(3):1084. https://doi.org/10.3390/ijerph17031084

Measurement of Particulate Matter Levels in and Around the Mining Area Using Low-cost Sensors

Abhishek Penchala, Aditya Kumar Patra, Samrat Santra, Namrata Mishra, and Ravish Dubey

Abstract Mining activities produce large quantities of particulate matter (PM) of varying sizes, posing significant environmental and health risks. In this study, five OPC N3 PM sensors and one portable aerosol spectrometer were strategically deployed at different mining locations, such as mine periphery, belt conveyor, railway sidings, and unpaved roads within an iron ore mining complex to quantify PM mass concentration. The findings revealed that the mean concentration in these locations varied from 25–37 μg m^{-3} (PM1), 54–66 μg m^{-3} (PM2.5), and 122–165 μg m^{-3} (PM10). The highest coarse (PM_{10}) and fine ($PM_{2.5}$) concentrations were recorded at unpaved haul roads. At the same time, the railway sidings was found to be the predominant source of PM_1.

Keywords Low-cost PM sensors · Surface mine · Unpaved haul roads · Calibration

1 Introduction

The mining sector is critical to India's economic and social well-being, contributing around 2.5% to the Gross Domestic Product (GDP) [1]. However, the extraction of minerals through the surface mining method comes with various environmental hazards, including air pollution. Air quality in and around the mining area is greatly affected due to mining activities like drilling, blasting, excavating, crushing, grinding, and smelting. All activities produce large quantities of PM of varying sizes with potentially dangerous environmental and health impacts [2]. The impact is exacerbated in surface mining, where these activities are directly exposed to the ambient

A. Penchala · A. K. Patra (✉)
Department of Mining Engineering, Indian Institute of Technology Kharagpur, Kharagpur, India
e-mail: akpatra@mining.iitkgp.ac.in

A. K. Patra · S. Santra · N. Mishra · R. Dubey
School of Environmental Science and Engineering, Indian Institute of Technology Kharagpur, Kharagpur, India

S. M. Shiva Nagendra et al. (eds.), *Select Proceedings of the 8th Indian International Conference on Air Quality Management (IICAQM 2023)*, Lecture Notes in Civil Engineering 582, https://doi.org/10.1007/978-981-96-2359-4_23

atmosphere. These emitted particles interfere with the breathing pattern of air and are hazardous to the health and safety of mine workers. Long-term exposure to mining-generated PM has resulted in irreversible diseases like silicosis, coal workers' pneumoconiosis (CWP), asbestosis, and siderosis [3].

After the mineral is extracted from the mines, it is typically transported off-site via large trucks, belt conveyors, and railway wagons to reach the end user. The adjoining communities are exposed to different sizes of particulate matter generated by the transport network [4]. Therefore, monitoring PM is important to quantify the source strength and risk associated with these activities. Recently, low-cost PM sensors that use the light scattering principle have been extensively calibrated and used with standard regulatory instruments to monitor aerosol concentrations [5]. The sensors use the light scattering principle to detect particle size and provide real-time particle and mass concentrations [6]. However, monitoring mining-generated PM using these sensors remains a new technology in the mining industry. The standard practice is to calibrate low-cost PM sensors along with reference instruments in the desired field of deployment before installation [7]. Calibration is the process of adjusting the sensor's output to match a reference standard to ensure accurate data over time.

This study aims to assess the PM concentration near a surface iron ore mine, focusing on different mining activities. In addition, the study aims to evaluate the accuracy of sensors in the mining environment compared to research-grade reference instruments.

2 Methodology

2.1 Instruments

Particulate matter released from different mining activities is measured using low-cost OPC N3 PM sensors. The sensor is a small (75 mm $\times$ 60 mm $\times$ 65 mm) and light (105 g) optical particle counter that can identify particles with diameters ranging from 0.38 μm to 40 μm in 24 bin sizes. The particles are counted using an elliptical mirror and a dual-element photo-detector. The particle count is converted into mass concentrations by assuming a particle density of 1.65 g cm^{-3} and a refractive index of $1.5 + i0$ [8].

The manufacturer of the OPC N3 sensor provides software and necessary drivers to run the sensor from a computer or PC. The sensor is integrated with a Raspberry Pi microprocessor to create a compact and portable monitoring device suitable for monitoring in mines. An SPI to USB adapter connects OPC N3 to individual Raspberry Pi microprocessors. The Raspberry Pi is fueled by a 5 V, 20,000 mAh power bank, while the sensor is powered via the Raspberry Pi's USB serial connection. An external touchscreen monitor was attached to the Pi via an HDMI (High-Definition Multimedia Interface) connection to operate and display real-time sensor data (Fig. 1). The

Fig. 1 OPC N3 sensor and its assembly

Pi contains 2 GB of RAM (Random Access Memory) and built-in Wi-Fi and Bluetooth connectivity for real-time data transmission. The developed portable sensor module weighs less than 220 g and can be operated continuously with a battery.

A portable aerosol spectrometer (Grimm aerosol Technik, 11D) is a reference and standard instrument designed for real-time detection of airborne aerosol particles. It works on the light scattering at single particles with a diode for measuring particles ranging from 0.25 µm to 35.15 µm across 31 equidistant-size channels.

2.2 *Study Protocol*

The sensors were placed at different locations, such as mine periphery, unpaved haul roads, belt conveyors, and railway siding. The mine periphery location represents the overall PM emissions from the surface mine. The particle concentration varies with the activity involved, and the size fraction can show different signatures when the material or mineral is subjected to various mechanical processes. Five OPC N3 sensors (S1, S2, S3, S4, S5) and the reference instrument (G) were deployed close to each other and were placed near the activities to measure the PM mass concentration. The instrument setup was placed 1 m above the ground surface and away from the obstructions for a better sampling of airborne PM. The sensor generates a dataset for every 2 s interval, Grimm generates a dataset for every 6 s interval, and the sampling was done for a 4-h duration at each location. All the instruments were new, factory-calibrated, and pre-checked in the laboratory conditions before conducting the experiment. Data points recorded at the start of the sampling were eliminated considering the instrument warm-up, initialization, or settling time, and the rest of them were converted to a 1-min average to maintain a uniform scale for analysis.

The PM concentrations were assessed using the Grimm instrument (G) readings. The sensor-recorded concentrations were then compared with those from the Grimm instrument to evaluate accuracy based on the percentage deviation.

3 Results and Discussion

3.1 PM Concentration

The spatial distribution of PM concentration associated with different mining activities is shown in Fig. 2. A significant variation in mass concentration for different size fractions is observed depending on the nature of the mining activity. The concentration recorded at all the locations exceeded the NAAQS by several times. The PM mass concentrations were in the range of 25–37 μ g m^{-3} (PM_1), 54–66 μ g m^{-3} ($PM_{2.5}$), and 122–165 μ g m^{-3} (PM_{10}).

Among different activities, higher mean $PM_{2.5}$ (66 μ g m^{-3}) and PM_{10} (165 μ g m^{-3}) and lowest PM_1 (25 μ g m^{-3}) concentrations were recorded near the unpaved haul roads. The plying of mining machinery, especially high-capacity off-road dumpers, makes the underlying unconsolidated dry dust layer airborne and quickly forms a dust cloud. Previous emission inventory studies in Indian surface iron ore mines found that unpaved haul roads are the predominant PM source [9]. In another study, ore transportation by trucks using unpaved haul roads contributed to 78–97% of total emissions [10].

Railway sidings reported the highest very fine PM concentrations (PM_1–37 μ g m^{-3}) and lowest fine ($PM_{2.5}$–54 μ g m^{-3}) and coarse PM concentrations (PM_{10}–122 μ g m^{-3}). Material handled near the railway sidings is usually high, with continuous transit of mineral ore from one transport medium to another. The handling and movement of these materials can cause very fine particles to separate from the air stream due to air turbulence. These particles are easily carried away by the incoming wind into the atmosphere, causing air pollution in the mining area. The pollution intensity is directly proportional to the amount of material the railway wagons transfer (Fig. 3).

Further, $PM_1/PM_{2.5}$ and $PM_{2.5}/PM_{10}$ ratios are evaluated to investigate the dominant particle size fraction. Railway sidings reported a high $PM_1/PM_{2.5}$ ratio (~0.70),

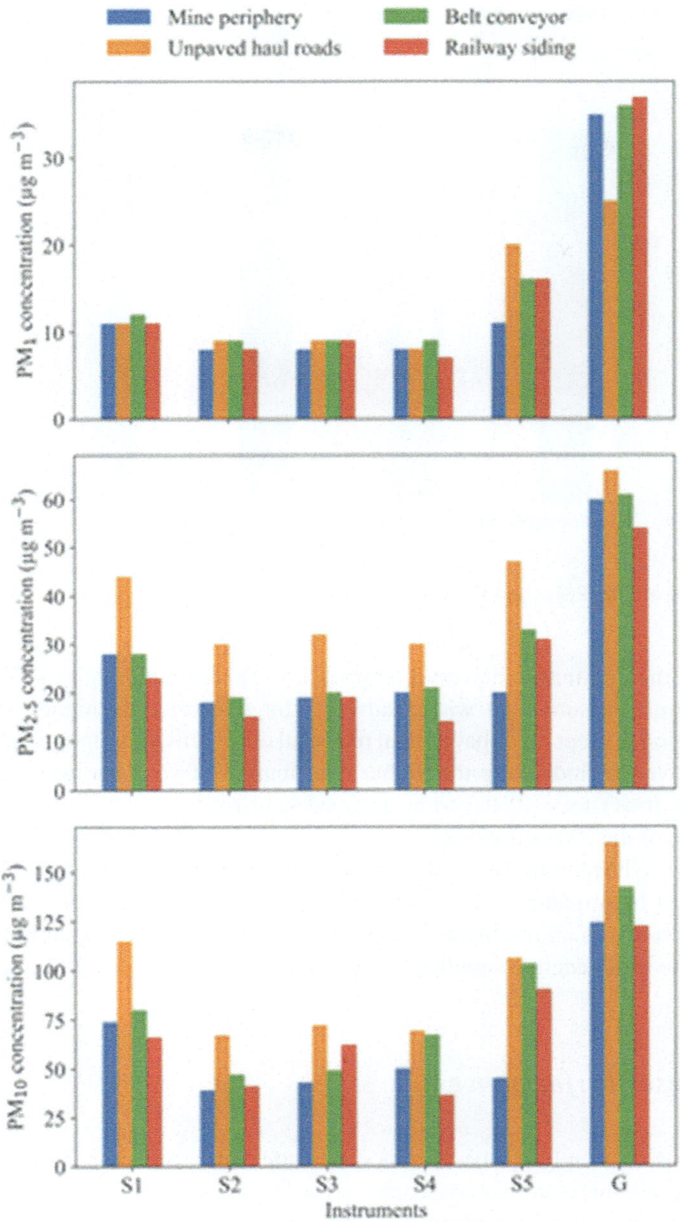

Fig. 2 PM concentration recorded by sensors and the Grimm at different mining activities

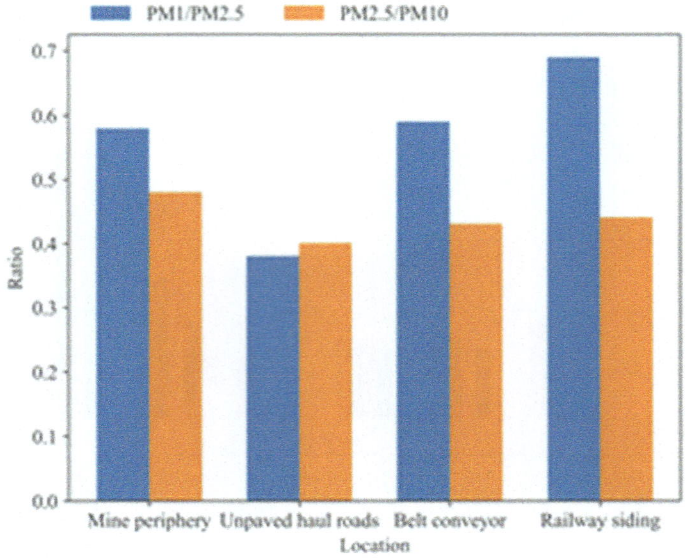

Fig. 3 Ratio of $PM_1/PM_{2.5}$ and $PM_{2.5}/PM_{10}$

indicating the dominance of very fine particles. This finding aligns with the mass concentration measurements where railway sidings reported the highest PM1 mass concentration. Except for unpaved haul roads, all other activities reported $PM_1/PM_{2.5}$ values above 0.5, indicating the higher contribution of very fine particle concentrations. A lower ratio (0.40) from unpaved haul roads signifies the proportional dominance of the $PM_{2.5}$ fraction.

The $PM_2.5/PM_{10}$ ratio for all the activities is less than 0.5, with the lowest being for unpaved haul roads (0.4). These findings indicate the dominance of a coarser size fraction across all mining activities. Several studies in the past have found that mining activities generate significant amounts of coarse particles (PM_{10}) [2].

3.2 *Sensor Performance*

The sensor consistently demonstrates a systematic underestimation of the mean mass concentration. This pattern is consistent across all the mining activities. Although the sensors are of the same manufacturer's model, they reported a significant difference in the recorded data. This difference is assessed in terms of the percentage of deviation between sensors and reference instruments (G). The percentage of deviation is high (81–20) for PM_1 compared to $PM_{2.5}$ (74–29) and PM_{10} (70–26). Out of five OPC N3 sensors (S1, S2, S3, S4, S5), three sensors (S2, S3, S4) deviation values are close in a small range PM_1 (81 -64), $PM_{2.5}$ (74–52), and PM_{10} (70–49) (Fig. 4).

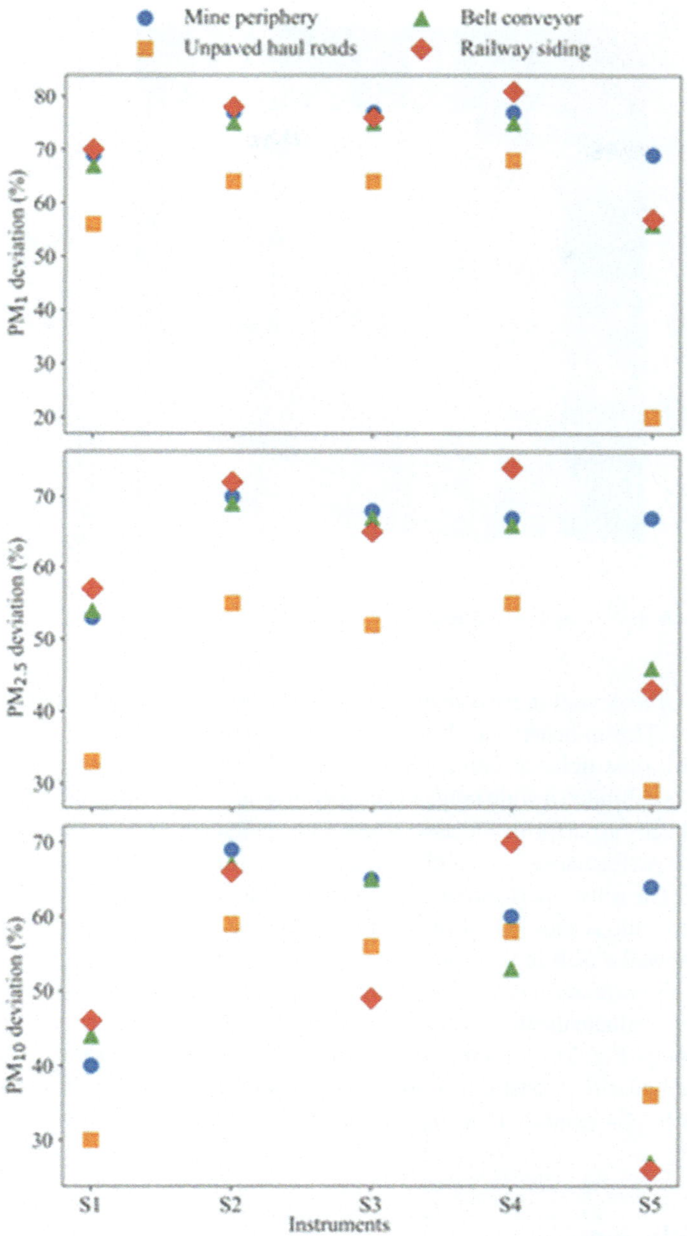

Fig. 4 Percentage deviation of sensors

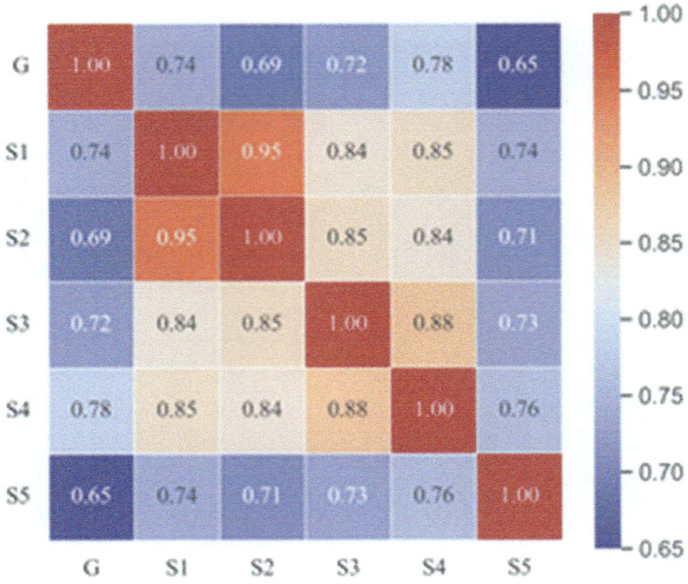

Fig. 5 Heat map showing $PM_{2.5}$ correlation

Overall, the deviation percentage was lowest near the unpaved haul roads for all the sensors. This indicates the ability of the sensors to capture and accurately report coarse (PM_{10}) particles as compared to fine ($PM_{2.5}$) and very fine (PM_1) particles. The sensors exhibited a wide range of sensor deviations, indicating the importance of individual sensor performance assessment rather than common calibration models. The sensor performance is also observed as a function of PM size and type of PM measured. The sensor performance when exposed to different aerosols also depends on the age of the sensor [11]. Long-term measurements of PM using these types of sensors showed a drift in their precision.

Pearson's correlation is applied for the entire $PM_{2.5}$ dataset across all four mining activities to evaluate the linear relation between sensors and the reference instrument. The heat map (Fig. 5) shows the correlation coefficient values for the entire dataset. A positive linear relationship is observed, with values reported to be close to 0.70, except for S5. S4 exhibited the highest linear correlation among the five sensors.

4 Conclusion

Assessment of air quality monitoring data by PM sensors in and around the mining area showed significant spatial and temporal variation during the study period. All sensors could withstand the harsh environment and recorded high and low PM levels prevailing in and around the mining area. They successfully measured continuous

PM (real-time) mass concentration levels throughout the sampling period without glitches. The activities performed in and around the mining area are significant sources of particulate matter for the mine workers and also local residents. Apart from mining operations inside a surface mine, other activities like transferring minerals through unpaved roads, belt conveyors, and railway wagons, which involve mineral handling, generate huge amounts of PM.

Unpaved haul roads and railway siding in the mining area were reported to be the major sources of coarse and fine PM mass concentration compared to other activities. In the case of PM levels near the mine periphery, PM_{10} emitted from mining activities is retained inside the mine, marking less concentration than concentration around the mine. Transfer of material near the railway sidings contributed to PM concentrations.

Sensors successfully captured the real-time PM concentration levels at different activities in and around the mining area. Moreover, the sensors recorded the variations in the concentration levels (highs and lows) during the sampling period. However, the concentrations recorded by the sensor underestimated the actual levels. Also, there is a variation in readings among the sensors. Therefore, the difference in the sensor reading and the reference instrument is evaluated in terms of the percentage of deviation. Among five sensors (S1, S2, S3, S4, S5), three sensors (S2, S3, S4) showed less inter sensor deviation. Moreover, the sensors are substantially less precise, indicating the need to treat each sensor independently when computing mass concentration. This is why sensors were calibrated with reference instruments before sampling or deploying at the site of interest. A correction factor is derived from the calibration results and adjusted to the sensor data. Many researchers have used this technique and advanced statistical tools like machine learning to understand the sensor data.

References

1. DGMS (2020) Director general of mines safety, ministry of labour and employment, Government of India. Accessed August, 2022. https://www.dgms.gov.in/UserView/index?mid=1287
2. Pandey B, Agrawal M, Singh S (2014) Assessment of air pollution around coal mining area: emphasizing on spatial distributions, seasonal variations, and heavy metals, using cluster and principal component analysis. Atmos Pollut Res 5(1):79–86
3. Patra AK, Gautam S, Kumar P (2016) Emissions and human health impact of particulate matter from surface mining operation—A review. Environ Technol Innov 5:233–249
4. Tripathy DP, Dash TR (2018) Assessment of particulate and trace element pollution in airborne dust around a highly mechanized opencast coal mine in Talcher, Odisha. J Min Sci 54(4):697–708
5. Crilley LR, Shaw M, Pound R, Kramer LJ, Price R, Young S, Lewis AC, Pope FD (2018) Evaluation of a low-cost optical particle counter (Alphasense OPC-N2) for ambient air monitoring. Atmos Meas Tech 11(2):709–720
6. Alfano B, Barretta L, Del Giudice A, De Vito S, Di Francia G, Esposito E, Formisano F, Massera E, Miglietta ML, Polichetti T (2020) A review of low-cost particulate matter sensors from the developers' perspectives. Sensors 20(23):6819
7. Badura M, Batog P, Drzeniecka-Osiadacz A, Modzel P (2018) Evaluation of low-cost sensors for ambient PM2. 5 monitoring. J Sens

8. Alphasense (2018) Alphasense OPC N3. https://www.alphasense.com/products/optical-particle-counter/. (Accessed August 2022)
9. Chaulya SK, Trivedi R, Kumar A, Tiwary RK, Singh RS, Pandey PK, Kumar R (2019) Air quality modelling for prediction of dust concentrations in iron ore mines of Saranda region, Jharkhand India. Atmos Pollut Res 10(3):675–688
10. Thompson RJ, Visser A (2007) Selection, performance and economic valuation of dust palliatives on surface mine haul roads. J Southern African Inst Mining Metallurgy 107(7):435–450
11. Tryner J, Mehaffy J, Miller-Lionberg D, Volckens J (2020) Effects of aerosol type and simulated aging on performance of low-cost PM sensors. J Aerosol Sci 150:105654

Measurements of Size-Resolved Nanoparticle Concentration and Its Distribution in Polluted Urban Environment During Induced Firework Event

Kanagaraj Rajagopal, S. Ramachandran, and Rajeev Kumar Mishra

Abstract Particle number concentration measurements of nanoparticles size ranging from 10 to 1090 nm was performed during the major firework event, Diwali, in the urban city Delhi. The fireworks during the festival emitted particles concentration ~9×10^4 cm^{-3} throughout the day. During the peak fireworks period, i.e., 18–20 hrs concentration was observed to be around 2 to 3×10^5 cm^{-3}. During the pre- and post-Diwali period, the concentration of the particles was observed in the range of 6–7×10^4 cm^{-3}. Accumulation mode particles contributed around 57% to the total concentration during Diwali day, and during peak hour of crackers bursting, the contribution reached to a maximum of about 60–70%. During normal period, contribution of accumulation mode to total particle concentration was around 31–40%. The particle size distribution showed peak concentration of particles around 150 nm on Diwali day, and on normal days, the peak found was around 100 nm. Post-Diwali period showed a 19% increase in total concentration of particles compared to the pre-Diwali period. NOx, NO_2, and NO showed moderate correlation with accumulation mode ($r^2 = 0.4$ to 0.6), and highest correlation ($r^2 = 0.7$) was seen during the post-Diwali phase with large Aitken number concentration. Accumulation mode number concentration showed good correlation with CO ($r^2 = 0.6$ to 0.7) during the pre-Diwali phase. The outcomes of the work will be helpful in distinguishing the different sources of nanoparticle emissions in the atmosphere. The findings of the study can be utilized to analyze the potential health effect due to inhalation of these particles and can help policymakers and authorities to formulate policies related to emissions during such firework events.

K. Rajagopal · R. K. Mishra (✉)
Department of Environmental Engineering, Delhi Technological University, New Delhi, India
e-mail: rajeevkumarmishra@dtu.ac.in

S. Ramachandran
Physical Research Laboratory, Space and Atmospheric Sciences Division, Ahmedabad, Gujarat, India

S. M. Shiva Nagendra et al. (eds.), *Select Proceedings of the 8th Indian International Conference on Air Quality Management (IICAQM 2023)*, Lecture Notes in Civil Engineering 582, https://doi.org/10.1007/978-981-96-2359-4_24

Keywords Fire work emissions · Nano particles · Number concentration · Polluted urban city

1 Introduction

Diwali is one of the major festivals celebrated every year in India in almost all parts of the country [12]. It is celebrated by bursting of crackers and lighting of lamps [2, 20, 21]. The crackers used for firework contains a large number of chemicals such as aluminum, sulfur, potassium nitrate, barium nitrate, charcoal, manganese, strontium nitrate, potassium, and iron dust powder as a composition material for the manufacturing of firecrackers [9, 11, 16]. The lighting of fireworks emits different types of emissions into the atmosphere. It releases particulate matter, gaseous pollutants, and toxic metals of significant quality [7]. These particles stay in the atmosphere for a few days causing the formation of toxic smog. The pollutants are hazardous in nature and cause serious health effects as well [6]. The study location is the national capital of India, New Delhi. The city is facing severe pollution events throughout the year, and majority of the time, the air quality remains in the poor category [17, 19] and it exceeds national ambient air quality standards [5]. The event Diwali usually celebrated in the post-monsoon season (beginning of the winter season) from October to November every year [4]. During winter, air quality becomes poor due to the prevailing meteorological conditions such as relative humidity, boundary layer height [13], and ventilation [15]. Further increased emissions on Diwali also add more pollutants to the atmosphere [14]. Recent studies found that air quality worsens during Diwali [1, 10, 20, 21]. Apart from Diwali, Delhi is facing emission issues from different sources such as vehicles, construction sites, road dust emission, industry sector, waste burning [18] and also due to long-range transportation of pollutants of different geographical origins [8]. Several studies in Delhi analyzed the impacts of crackers on the local air quality, and only very few studies reported the status of particle number concentration [3, 10, 16]. This study analyzed the concentration and temporal variation of nano-size particles ranging from 10 to 1090 nm. The wide-range nanoparticles are classified into four different sub-categories such as nucleation (10–30 nm), small Aitken (30–50 nm), large Aitken (50–100 nm), and accumulation mode (100–1000 nm). The study focuses on the size distribution of different size range particles during the event of Diwali and the period adjacent to it. The study helps in understanding the pattern of emission of nanoparticles during the firework event. Although the use of firecrackers is restricted in the city but few firework events still occurs on the evening of Diwali as a part of the tradition.

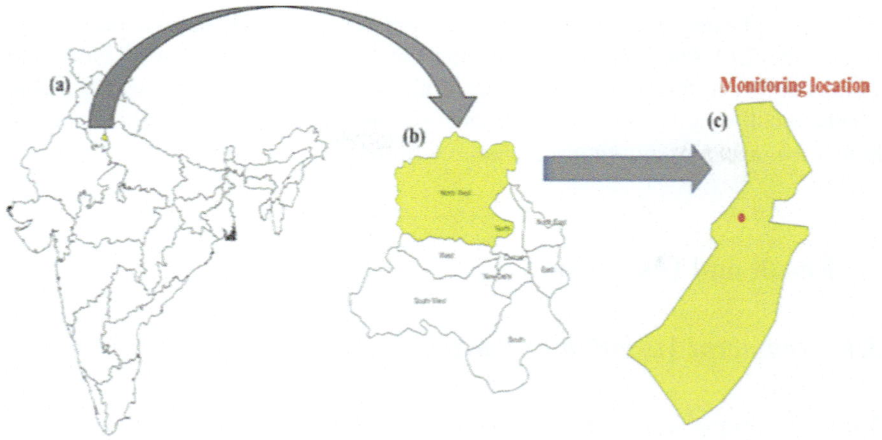

Fig. 1 Map showing the study location **(a)** map of India showing all states, **(b)** map showing all the districts of Delhi, and **(c)** map showing study location in Rohini District

2 Materials and Methods

2.1 Description of Monitoring Location

The nanoparticle measurement was carried out in the university premises (Delhi Technological University) (Fig. 1). The university is located in the Northwest district, which comes under the Rohini region. The latitude and longitude of the monitoring location are 28° 36′ N and 77° 33′ E, and it is located 216 m above the mean sea level. The monitoring location is surrounded by residential areas, educational institutions and road networks. The city comes under the territory of the Indo-Gangetic Plain. The sampling was conducted 2 m above the ground level and represents the particles present at the inhalable level.

2.2 Instrumentation, Data Collection, and Analysis

The scanning mobility particle sizer SMPS (GRIMM make) was used to measure nanoparticles from 10 to 1090 nm. The diffusion dryer was used along with the SMPS for reducing the relative humidity in the air sample. Gaseous pollutants are measured using Horiba gas analyzer (CO and SO_2), and 2B technologies NO, NO_X, NO_2, and O_3 analyzer. The study was carried out from October 30, 2022, to November 10, 2022. The period from October 30 to November 3, a total of five days, was considered as pre-Diwali period, November 4 was considered as Diwali period, and November 5 to 10 (5 days) was considered as post-Diwali period. The total monitoring period was 11 days. The data was collected for 24 h in all 11 days and analyzed using software such as MATLAB version 2016b and R statistical tool version (5.12.8). The data was unavailable from 31st October 18 hrs to 1st November 10 hrs due to instrumentation issues, and on all remaining days, the data was available. The data collected from the gaseous analyzer and SMPS are evenly converted into hourly average concentration and analyzed further.

3 Result and Discussion

3.1 *Temporal Variation of Particle Number Concentration*

The hourly average concentration of the total particle number concentration (PNC) of the nanoparticles was found high during Diwali, especially during the evening hours when the fireworks event occured significantly. During normal days the transportation sector played a significant role and it resulted in increased particle number concentration during the morning and evening peak hours. The concentration on Diwali day reached a maximum of about 9×10^4 cm^{-3} (Fig. 2), whereas the maximum concentration attained during the pre- and post-Diwali was found in the range of 6 to 7×10^4 cm^{-3}(Fig. 2). During Diwali, the concentration of PNC increased about 30% compared to pre-Diwali period and 19% compared to post-Diwali period. The change in the percentage of concentration during the pre- and post-Diwali periods was due to the existence of particles from the Diwali emissions especially in the post-Diwali period. The geometric mean diameter of the particle also showed an increase in their mean diameter during the Diwali period, and the maximum concentration of the GMD was obtained during the post-Diwali phase compared to other phases due to the coagulation of the particles after the Diwali day. The different size particles such as N_{nuc} and N_{satk} were also observed in less concentration during the Diwali period since they are not directly associated with Diwali emissions, N_{latk} (50–100 nm) is slightly associated with Diwali emissions, and N_{acc} was observed in higher concentration that showed that majority of the particles found above 100 nm were from the Diwali emissions.

3.2 *Size-Resolved Particle Distribution*

The size distribution of the different size-resolved particles was analyzed for three periods, namely the pre, post, and Diwali periods. The size distribution of the particles showed a clear single peak of higher concentration during the Diwali day in the size range of particles around 150 nm in the accumulation mode particles. The pre- and post-Diwali concentrations showed a gradual rise in the concentration from smaller to larger size, attaining a peak in the accumulation mode. Whereas on Diwali day, the N_{nuc} concentration decreased to a certain extent, and then it starts to increase

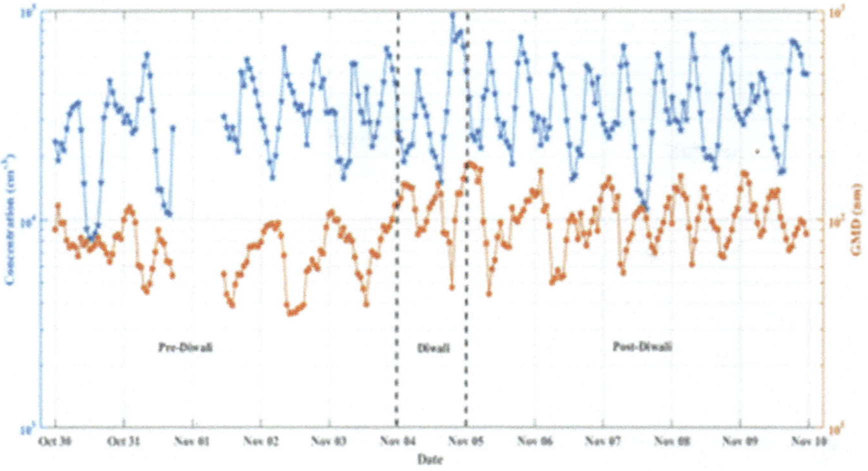

Fig. 2 Daily variation of hourly averaged particle number concentration and GMD from October 30 to November 10, 2022

gradually from N_{satk} to N_{acc} (Fig. 3). This showed that the presence of N_{nuc} particles was less during the Diwali period compared to pre- and post-Diwali period. The trend in concentration of the particle was totally different from N_{nuc} to $N_{acc.}$ During the pre-Diwali period, the lower size particle till N_{latk} mode concentration was found to be high, followed by the post-Diwali period and then by the Diwali period (Fig. 3). In N_{acc} the pattern was totally inverse, i.e., higher concentration on Diwali period followed by a post-Diwali period and then by Diwali period. This indicates a clear representation of the particles and their sources. The smaller size particle till 100 nm is directly associated with vehicular emissions found in the pre- and post-Diwali period. During Diwali period, particles above 100 nm are found in higher concentrations and are associated with Diwali emissions.

3.3 Particle Composition in Total Number Concentration

The composition of different size particles throughout the monitoring period was done to analyze the role of different size particles in the total number concentration. This showed that the percentage contribution of different size particles were different in all the periods. During the pre-Diwali phase, the smaller size range N_{nuc} contributed to a higher extent in total concentration, whereas on Diwali period the trend was opposite, i.e., accumulation mode contributed to larger extent of around 57% to the total concentration (Fig. 4). During the post-Diwali phase, the contribution of accumulation mode to total concentration found in the range of 48–55%. The contribution of the fire emissions during the Diwali and post-Diwali was seen clearly in the figure. The total PNC concentration was based on the contribution

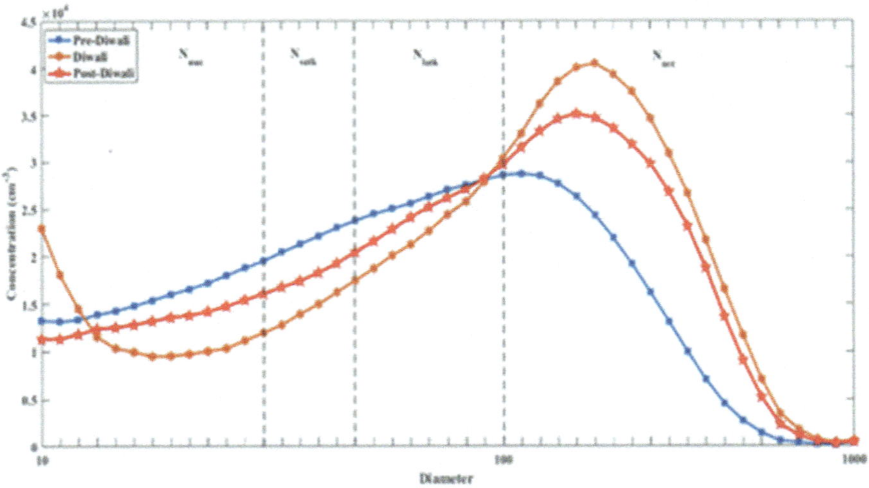

Fig. 3 Particle number size distribution of N_{nuc}, N_{satk}, N_{latk}, and N_{acc} during October 30 to November 10, 2022

from different size fractions emitted from different sources such as engine exhaust, biomass burning, and fire emissions in the urban regions. The hourly contribution of the different sizes to the total PNC was also analyzed to clearly understand the emissions. The hourly emission contribution of the different size particles to the total PNC showed on Diwali day, the accumulation mode dominated around 60 to 70% of the total PNC especially during the Diwali evening and the next morning (Fig. 5b).

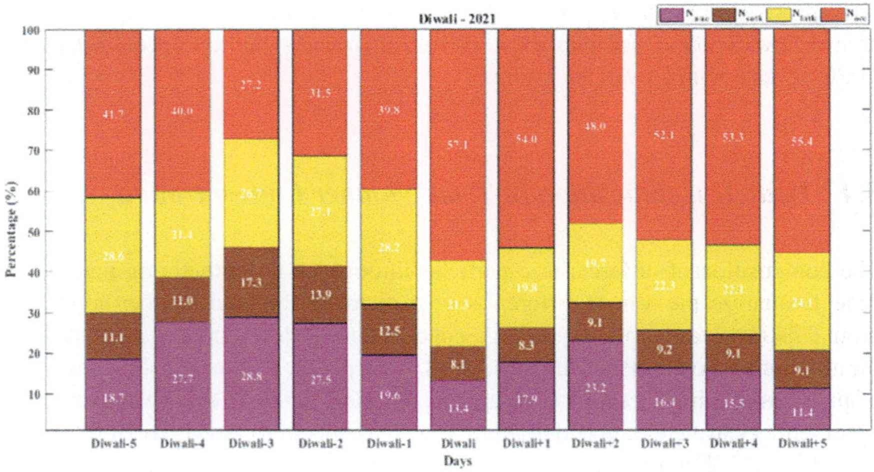

Fig. 4 Percentage contribution of N_{nuc}, N_{satk}, N_{latk}, and N_{acc} in total PNC (October 30 to November 10, 2022)

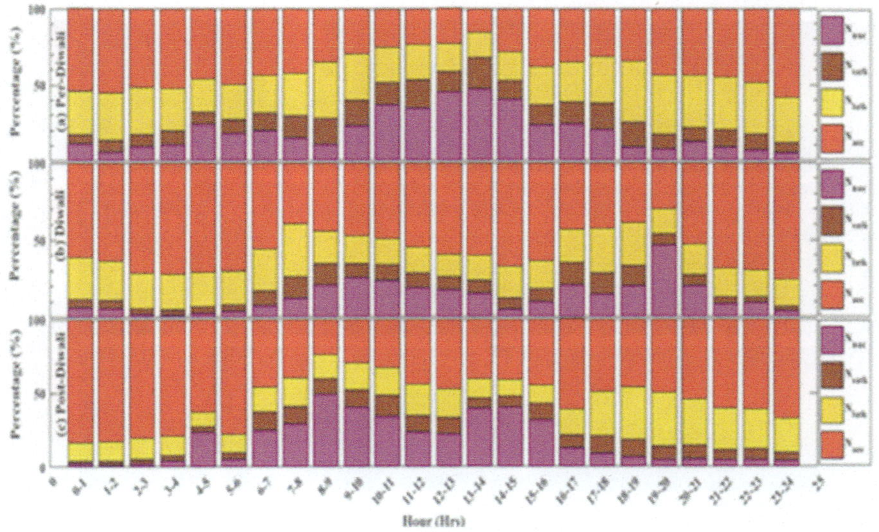

Fig. 5 Hourly classified percentage contribution of N_{nuc}, N_{satk}, N_{latk}, and N_{acc} in total PNC during **(a)** pre-Diwali, **(b)** Diwali and **(c)** post-Diwali from November 3 to November 5, 2022

The pre-Diwali period showed an increase in the concentration of the nucleation mode particles during the afternoon hour due to the effect of sun rays which causes nucleation burst events [7]. Since the monitoring period falls under winter seasons, the higher concentration of solar radiation and ventilation was observed during the afternoon in Delhi (Fig. 5a). This can be clearly visualized during the post-Diwali day when the solar radiation started to increase around 6 am, the concentration of the nucleation mode particles also increased (Fig. 5c). Whereas during the off-sunshine hours, the nucleation mode particles again started to decrease, and the concentration of PNC was dominated by the accumulation mode particles due to coagulation.

3.4 Heat Map Analysis

The heat map analysis was used to evaluate different size particles with respect to time. During the pre-Diwali period, in the early morning and late evening, particles around 100 nm was found (Fig. 6a) in the concentration range of 1×10^5 cm^{-3} due to the accumulation of particles under low boundary layer conditions and less dispersion of particles. During the Diwali period in the evening around 18–20 hrs, there was observed a clear visibility of particle emissions from fireworks in the range of 3 to 2×10^5 cm^{-3} (Fig. 6b), and the concentration of particles on pre-Diwali was seen in a similar pattern in the late evening and early morning on Diwali day with higher concentration due to the pollution load increased by Diwali. During the post-Diwali phase, during the morning hours, another emission source of around 1×10^5 cm^{-3}

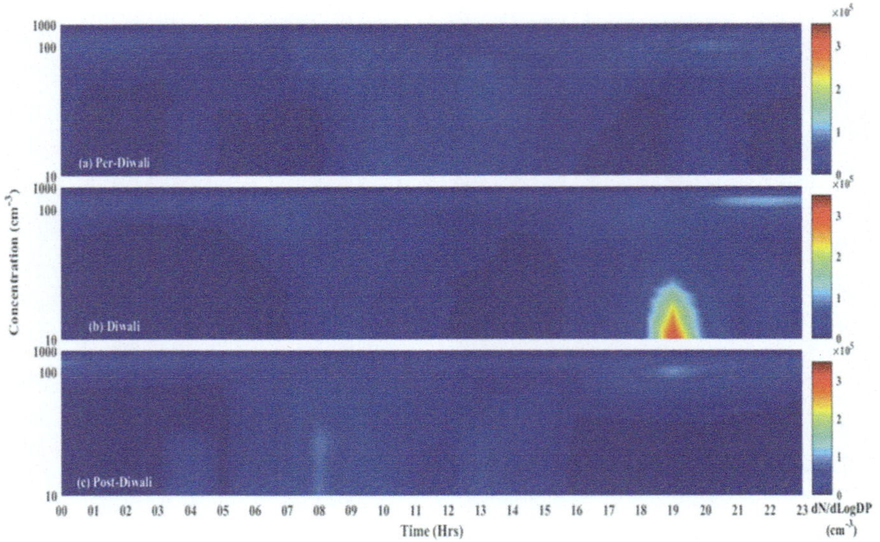

Fig. 6 Heat map analysis of size-resolved particle distribution and GMD during **(a)** pre-Diwali, **(b)** Diwali and **(c)** post-Diwali from November 3 to November 5, 2022

was found due to the vehicular emissions, which added more particles apart from preexisting Diwali emission particles.

3.5 Correlation Analysis

The correlation analysis of particles with gaseous pollutants showed that the N_{acc} and N_{latk} correlated moderate to good correlation with particles associated with Diwali emissions, and the correlation values were also found to be higher during the post-Diwali phase. Certain pollutant, such as O_3, showed very less correlation ($r^2 = 0.1$ to 0.2) (Fig. 7c) throughout the Diwali period, since O_3 is not directly associated with Diwali emissions. The nitrogen components such as NOx, NO_2, and NO showed the highest correlation values. The r^2 values ranged from 0.4 to 0.6 (Fig. 7d–f) for the N_{acc} mode particles, and for particles in N_{latk} mode, the correlation values found to be high in the post-Diwali phase ($r^2 = 0.7$) compared to the pre-Diwali ($r^2 = 0.2$ to 0.3) and Diwali phase ($r^2 = 0.01$ to 0.02) (Fig. 8d–f). The r^2 of CO ranged from 0.6 to 0.7 (Fig. 7a) for the N_{acc} and for N_{latk} mode ($r^2 = 0.2$–0.4) (Fig. 8a). The carbon emissions were high before Diwali, on Diwali, the concentration was found less, and then, during the post-Diwali phase, the values started to resume back to its original concentration. The correlation values are insignificant for the smaller sizes, such as N_{nuc} and N_{satk} mode particles.

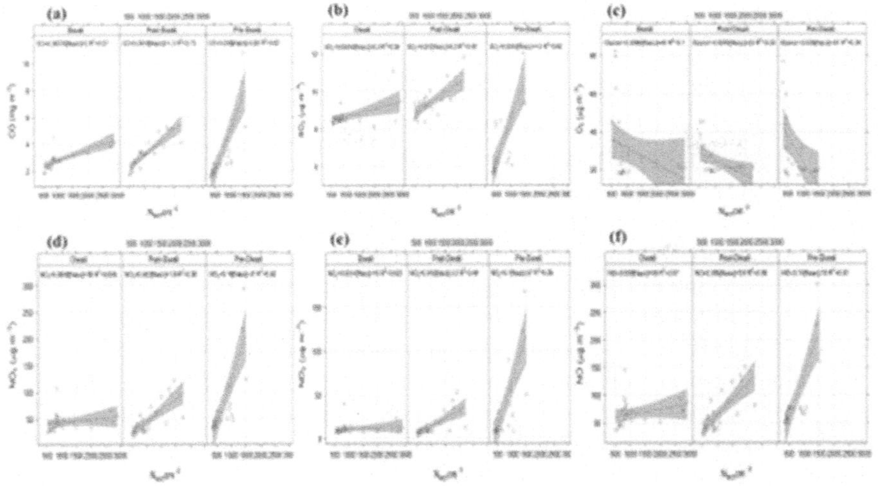

Fig. 7 Correlation analysis of N_{acc} with **(a)** CO, **(b)** SO_2, **(c)** O_3, **(d)** NO_x, **(e)** NO_2, **(f)** NO during pre, post, and Diwali period from October 30 to November 10, 2022

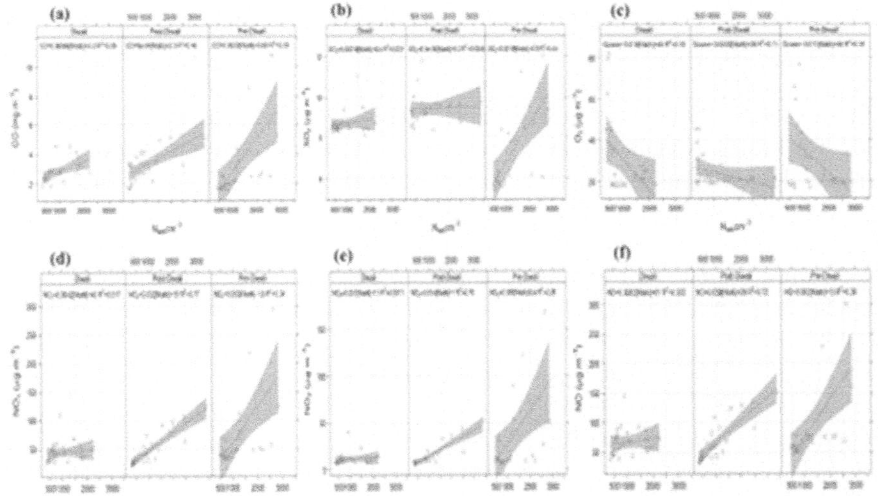

Fig. 8 Correlation analysis of N_{latk} with **(a)** CO, **(b)** SO_2, **(c)** O_3, **(d)** NO_x, **(e)** NO_2 and **(f)** NO during pre, post, and Diwali period from October 30 to November 10, 2022

4 Summary and Conclusion

The particle number concentration of the nanoparticles ranging from 10 to 1090 nm was measured from October 30 to November 10, 2022. This period covers the major festival of the country Diwali, which is usually celebrated by the bursting of fire-crackers. Diwali day received the maximum particle number concentration of about 9×10^4 cm^{-3}. In the pre and post Diwali period concentrations ranged from 6 to 7×10^4 cm^{-3}. On Diwali day, a 30% increase in the concentration of the particles was observed compared to the normal day emissions. The increase in concentration was due to the fireworks emissions that occurred. In the post-Diwali period, the concentration was increased by 19% compared to the pre-Diwali period due to the addition of Diwali emissions along with normal emissions. The particles in size range > 100 nm, i.e., N_{acc} mode particles were seen in higher concentration. The size distribution of the particles showed a peak concentration of particles around 150 nm on Diwali day. Before the Diwali event, the smaller size particles in the range of N_{atk}, i.e., till 100 nm size particles were found in peak concentration. These size ranges are directly associated with exhaust emissions in urban regions. N_{acc} mode particles were observed in higher concentrations during Diwali day, and the concentration was about 57% of the total concentrations, whereas before Diwali, the contribution was around 31–40%. During the evening hours, when the firework events occured, the percentage contribution of N_{acc} mode particles to the total concentration reached around 60–70%. During the pre-Diwali period, N_{nuc} particles in the mid-day received a higher concentration of about 25–28% due to the burst of particles. The heat map analysis showed a clear evidence of particle emissions during the morning and evening peak hours resulting from vehicular emissions. During Diwali day, especially during 18–20 hrs, the particles' concentration was found around 2–3 $\times 10^5$ cm^{-3}. The correlation of gaseous pollutants showed a good correlation with N_{acc} and N_{atk} mode particles, and for other smaller sizes, the correlation was not significant. NOx, NO_2, and NO showed a good correlation with N_{acc} ($r^2 = 0.4$ to 0.6). The highest correlation of NOx, NO_2, and NO found throughout the monitoring period was ($r^2 = 0.7$) with N_{latk} mode particles during the post-Diwali phase. The correlation of N_{acc} mode with CO was observed to be high during the pre-Diwali period ($r^2 = 0.6$ to 0.7) and less during the Diwali period. The study revealed the role of Diwali festival emissions on particle number concentration of nanopollutants in the atmosphere.

Acknowledgements The authors are thankful to Advance Air and Acoustics Research Laboratory, Department of Environmental Engineering, Delhi Technological University, for providing an instrumentation facility to conduct this study.

References

1. Chatterjee A, Sarkar C, Adak A, Mukherjee U, Ghosh SK, Raha S (2013) Ambient air quality during Diwali festival over Kolkata—A mega-city in India. Aerosol Air Quality Res 13(3):1133–1144. https://doi.org/10.4209/aaqr.2012.03.0062
2. Ganguly ND (2015) Short term change in relative humidity during the festival of Diwali in India. J Atmos Solar Terr Phys 129:49–54. https://doi.org/10.1016/j.jastp.2015.04.007
3. Garg A, Gupta NC (2018) Episodic Levels of PM 10 , PM 2 . 5 and PM 1 during Diwali : A study in urban area of Delhi , India. Int Res J Environ Sci 7(3):25–30
4. Ghei D, Sane R (2018) Estimates of air pollution in Delhi from the burning of firecrackers during the festival of Diwali. PLoS ONE 13(8):1–11. https://doi.org/10.1371/journal.pone.0200371
5. Goyal P, Gulia S, Goyal SK (2021) Identification of air pollution hotspots in urban areas—an innovative approach using monitored concentrations data. Sci Total Environ 798. https://doi.org/10.1016/j.scitotenv.2021.149143
6. Kanawade VP, Shika S, Pöhlker C, Rose D, Suman MNS, Gadhavi H, Kumar A, Nagendra SMS, Ravikrishna R, Yu H, Sahu LK, Jayaraman A, Andreae MO, Pöschl U, Gunthe SS (2014) Infrequent occurrence of new particle formation at a semi-rural location, Gadanki, in tropical Southern India. Atmos Environ 94:264–273. https://doi.org/10.1016/j.atmosenv.2014.05.046
7. Kompalli SK, Babu SS, Udayasoorian C, Jayabalakrishnan RM (2018) Role of anthropogenic emissions and meteorology on ultrafine particle bursts over a high altitude site in Western Ghats during pre-monsoon. J Atmos Solar Terr Phys 179:378–388. https://doi.org/10.1016/j.jastp.2018.09.001
8. Kumar P, Khare M, Harrison RM, Bloss WJ, Lewis AC, Coe H, Morawska L (2015) New directions: air pollution challenges for developing megacities like Delhi. In: Atmospheric Environment, vol 122. Elsevier Ltd. , pp 657–661. https://doi.org/10.1016/j.atmosenv.2015.10.032
9. Nishanth T, Praseed KM, Rathnakaran K, Satheesh Kumar MK, Ravi Krishna R, Valsaraj KT (2012) Atmospheric pollution in a semi-urban, coastal region in India following festival seasons. Atmos Environ 47:295–306. https://doi.org/10.1016/j.atmosenv.2011.10.062
10. Parkhi N, Chate D, Ghude SD, Peshin S, Mahajan A, Srinivas R, Surendran D, Ali K, Singh S, Trimbake H, Beig G (2016) Large inter annual variation in air quality during the annual festival "Diwali" in an Indian megacity. J Environ Sci (China) 43:265–272. https://doi.org/10.1016/j.jes.2015.08.015
11. Perrino C, Tiwari S, Catrambone M, Torre SD, Rantica E, Canepari S (2011) Chemical characterization of atmospheric PM in Delhi, India, during different periods of the year including Diwali festival. Atmos Pollut Res 2(4):418–427. https://doi.org/10.5094/APR.2011.048
12. Rajagopal K, Mohan V, Mishra RK (2024) Are Delhi residents exposed to lesser particle number concentration due to the firework ban in the city? Air Qual Atmos Health 17:1617–16271. https://doi.org/10.1007/s11869-024-01532-3
13. Rajagopal K, Ramachandran S, Mishra RK (2024) Influence of local meteorology and gaseous emission on atmospheric nanoparticle concentrations in the pedestrian way in the urban region. Atmos Pollut Res 102358. https://doi.org/10.1016/j.apr.2024.102358
14. Rajagopal K, Ramachandran S, Mishra RK (2024) Size resolved particle contribution to vehicle induced ultrafine particle number concentration in a metropolitan curbside region. Atmos Environ 337:120773. https://doi.org/10.1016/j.atmosenv.2024.120773
15. Ramachandran S, Rupakheti M (2022) Trends in the types and absorption characteristics of ambient aerosols over the Indo-Gangetic Plain and North China Plain in last two decades. Sci Total Environ 831:154867. https://doi.org/10.1016/j.scitotenv.2022.154867
16. Sateesh M, Soni VK, Raju PVS (2018) Effect of Diwali Firecrackers on Air Quality and Aerosol Optical Properties over Mega City (Delhi) in India. Earth Systems and Environment 2(2):293–304. https://doi.org/10.1007/s41748-018-0054-x

17. Sharma K, Goyal M, Mishra RK, Vijayakumar T, Kumar P, Rajagopal K (2025) Air quality status and estimated health exposure in major metropolitan Indian cities. Aerosol Sci Eng https://doi.org/10.1007/s41810-024-00278-w
18. Sharma M, Khare M, Mishra RK (2023) Air quality changes in Delhi due to open waste burning: an accidental fire in Bhalswa landfill. Int J Environ Sci Tech 21:655–664. https://doi.org/10.1007/s13762-023-04921-w
19. Sharma M, Khare M, Mishra RK (2025) Indoor particle number concentration measurements across different income group households in urban region. Urban Clim 102329. https://doi.org/10.1016/j.uclim.2025.102329
20. Yadav SK, Kumar M, Sharma Y, Shukla P, Singh RS, Banerjee T (2019) Temporal evolution of submicron particles during extreme fireworks. Environ Monit Assess 191(9). https://doi.org/10.1007/s10661-019-7735-2
21. Yadav SK, Mishra RK, Gurjar BR (2022) Fireworks induced quasi-ultrafine particle number concentration and size-resolved elemental distribution in megacity Delhi. Arab J Geosci 15(1):1–13. https://doi.org/10.1007/s12517-021-09385-1

Emissions Assessment and Health Impacts of PM2.5 Due to Burning Sugarcane Residue in Mandya District, Karnataka

Sarup Das and S. M. Shiva Nagendra

Abstract One of the major sources of particulate matter (PM) emissions in India is open biomass burning (BB). This study focuses on the amount of PM released into the atmosphere due to the open biomass burning in the dry season. The long-term health exposure evaluation for adults (25^+ age) was also studied with the help of AirQ+ . The field monitoring showed an overall increment of 66.26, 83.06, and 107.50 μg/m^3 for PM_1, $PM_{2.5}$, and PM_{10} in the ambient atmosphere. Hence, the mortality rates of $PM_{2.5}$ due to COPD, LC, lung cancer, IHD, stroke, and all other natural causes were found for adults (25^+ age groups) as 29%, 27.12%, 46.72%, 28.35%, 26.69%, and 35.56%, respectively. Results showed a severe mortality rate due to the exposure of biomass burning for adults (for 25+ age) in the locality.

Keywords Particulate matter · AirQ+ · Mortality · COPD · LC · Lung cancer · IHD · Stroke

1 Introduction

Particulate matter concentration has a long-term effect on human health as well as on the environment. Smaller particulate matter size has a higher and long-term effect on the human body [5]. That is why the measurement of PM is now considered throughout the country. India is having a total of 804 monitoring stations in 344 cities in 28 districts. Most of the monitoring stations mainly measure PM_{10} & $PM_{2.5}$ both, or at least one among them [3]. As per the global scenario, $PM_{2.5}$ & PM_{10} concentrations were found 30.85 μg/m^3 (minimum–maximum: 1–217) μg/m^3 and (minimum–maximum: 1–540) μg/m^3 (WHO 2018). Particulate matter is a serious health issue not only for its size, also because it may be of different chemical compositions or different trace metals as well.

S. Das · S. M. S. Nagendra (✉)
Department of Civil Engineering, Indian Institute of Technology Madras, Chennai, India
e-mail: snagendra@iitm.ac.in

S. M. Shiva Nagendra et al. (eds.), *Select Proceedings of the 8th Indian International Conference on Air Quality Management (IICAQM 2023)*, Lecture Notes in Civil Engineering 582, https://doi.org/10.1007/978-981-96-2359-4_25

An aerodynamic diameter of less or equal to 10 μ can enter the human respiratory system through the oral or nasal pathway during breathing [5]. The inhaled particulate matter can be deposited in different layers of the lungs or the other parts of the respiratory tract. The deposited very small particulate matter is causing many short-term as well as long-term health effects [3].

Several studies have been conducted on the health impact on human health for the particulate concentration in ambient as well as in the indoor environment. One of the studies has found that around 7 million deaths across the world are directly or indirectly related to particulate pollution. India is among the countries having an urbanization rate of approx. 31% and the expected urban population to be added by 416 million by the end of 2050 [3].

Among the different sources of PM emission, biomass burning is considered one of the highly contributing sources. Since India is one of the highly populated countries, for meeting the food criteria and for the decreasing rate of cultivation lands, farmers are trying to increase crop production. Again, Indian farmers are getting two seasons to produce their crops. But the difference between these two seasons is very less (around 15 days) for which farmers have to prepare their field just after one crop in a very short period [4]. Hence, they do burn the residue in the field itself which generates a huge amount of finer particulate matter in the ambient air. Again, the burning activity is conducted during the dry season when the particle though takes a longer duration for the deposition. Hence, it travels longer distances as compared to particle deposition during the wet season.

A very recent study has found that 1 million annual deaths in India are caused due to PM pollution [3]. Very few studies are there for the health risk assessment of village residents due to the open residue burning. This study focuses on the contribution of PM in the ambient atmosphere due to open biomass burning. Also, to examine the mortality rate associated with biomass burning due to chronic obstructive pulmonary disease (COPD), lung cancer (for adults of age 25–39 years and also for 40+), ischemic heart disease (IHD), stroke, and all other natural diseases with AirQ+ software.

2 Materials and Methodology

2.1 Study Area

The Manikyanahalli is a small village in the Mandya district with a total area of 519.04 hectares, located toward the southmost part of Karnataka state in India. It has geographical coordinates (12.6141 °N, 76.6615 °E) and a population of about 2906 where the number of males is 1489 and the number of females is 1417. People living there have an agriculture-based economy mainly cultivating maize, sugarcane, chamanthi flower, and many other vegetables. During the post-harvesting, the farmer keeps the residue stack-by-stack to get it dry and hence does a manual burning. Burnt residues and the residual ash do fly at higher altitudes due to the surrounding air getting lighter weight during burning.

2.2 Locating the Sampling Site

HYSPLIT forward trajectory model (web version) was run to find out the overall downwind direction for the burning period. GDAS (1 degree, global, 2006-present) meteorological data was used for the simulation of forward trajectory for an average of 6 h. Instruments were set up before burning, during burning, and after burning toward the downwind direction. Background samples were collected from the upwind directions during burning.

2.3 Instrumentation

During the open burning, we run the GRIMM 1.108 for the real-time monitoring of 16 different size PM (unit: $\mu g/m^3$). From the dataset, we further calculate the PM_1, $PM_{2.5}$, and PM_{10} concentrations. The GRIMM 1.108 runs with a flow rate of 1.2 lt/min which was considered while calculating the particle count and mass of particles monitored.

We have used the thermal comfort meter (TCM) for collecting the meteorological data (WS, RH, and ambient temperature) half an hour early to the start of burning to collect the background data. The GRIMM 1.108 was set to collect data at every 6-s interval, starts after the burning started. Whole setup of the instrument was kept at 1.5 m high from the ground, at a distance of 10 m from the burning place. 1 h of average PM concentration data was collected, 20 min during the burning and 30 min before and after the burning. The thermal comfort meter I have kept on a plane surface in a vertically upward direction. It has collected meteorological data such as wind speed, relative humidity, and the different atmospheric temperatures.

The AirQ+ (v. 2.1.1) was used to study the long-term health impacts of biomass burning on the local people who were highly exposed to the huge air pollution due to open biomass burning.

3 Result and Discussion

Thermal Comfort Meter: The meteorological data (wind speed, RH, and temperature) was collected from the biomass burning field with thermal comfort meter. Before burning (30 min), during burning (20 min), and after burning (30 min) data are shown below in Table 1.

Table 1 Showing meteorological data (WS, RH, and temp) in different burning conditions

Burning conditions	Wind speed (m/s)	RH (%)	Ambient temp. (°C)
Before burning	0.3795	51.686	32.124
During burning	0.1467	63.13	30.53
After burning	0.0773	94.193	28.63

GRIMM 1.108: The inhalable particulate matter was studied during the biomass burning. The average concentration found for PM_1, $PM_{2.5}$, and PM_{10} was 75.65 $\mu g/m^3$, 94.12 $\mu g/m^3$, 137.77 $\mu g/m^3$, respectively. The observational data is shown below in Fig. 3 for the day of biomass burning.

HYSplit Forward Trajectory: The predominant wind direction (6 h average) simulated by the hybrid single-particle Lagrangian integrated trajectory (HYSPLIT) model (web version) was from west to east, toward the coastal area (Fig. 4). Hence, the instruments were set up toward the predominant downwind direction to get the maximum concentration of PM generated.

Health risk analysis: The mortality rate for different age groups for the different diseases (disease burden profile) for Karnataka state is shown below in Fig. 5. The data was further used for the health risk simulation with the AirQ+ software.

Morality due to COPD for adults: The method used for calculating the mortality rate due to COPD is GBD 2015/2016 (integrated function 2016 vs WHO AQG value). The relative risks associated with COPD with the increase in PM concentration were simulated (in Fig. 6). Four thousand nine hundred adults were affected due to COPD out of 100,000 in Karnataka state. Around 29% of people are having long-term effects due to COPD (shown in Table 2).

Morality due to LC for adults: The method used here for the analysis is GBD 2015/2016. The number of adults affected due to lung cancer for 25+ age in Karnataka state per 100,000 people was taken as 13,900. Around 27.12% of people are having long-term effects due to lung cancer (Table 3).

Morality due to lung cancer: The method used here for the analysis is log-linear for the age group more than 30 years. The number of adults affected due to lung cancer for 30 + age in Karnataka state per 100,000 people has been taken as 10,600. Around 46.72% of people are having long-term effects due to lung cancer (Table 4).

Morality due to IHD for adults: The method used here for the analysis is GBD 2015/2016 (integrated function 2016 vs WHO AQG value) for the age group of 25+ adults. The number of adults affected due to IHD for the 25+ age groups in the Karnataka state per 100,000 people has been taken as 11,000. Around 28.35% of people are having long-term effects due to IHD (Table 5).

Morality due to stroke for adults: The method used here for the analysis is GBD 2015/2016 (integrated function 2016 vs WHO AQG value) for the age group of 25+ adults. The adults affected due to stroke for the 25+ age groups in the Karnataka state per 100,000 people have been taken as 3800. Around 26.69% people are having long-term effect due to stroke (Table 6).

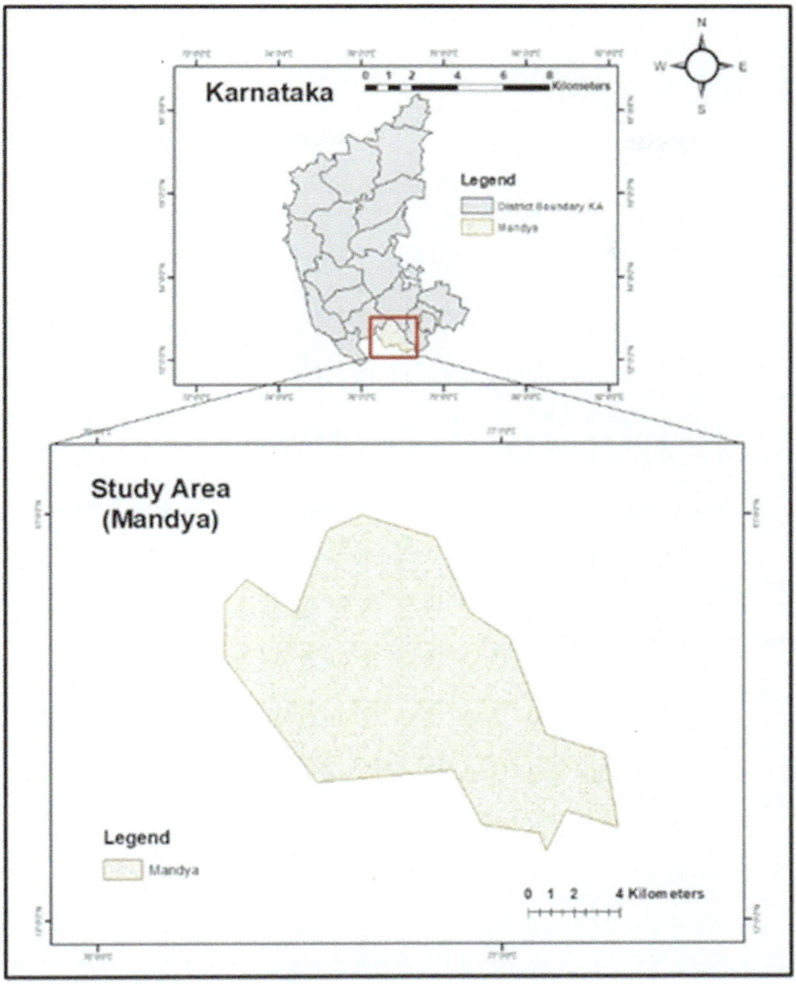

Fig. 1 Study area—Manikkanahalli village, Mandya

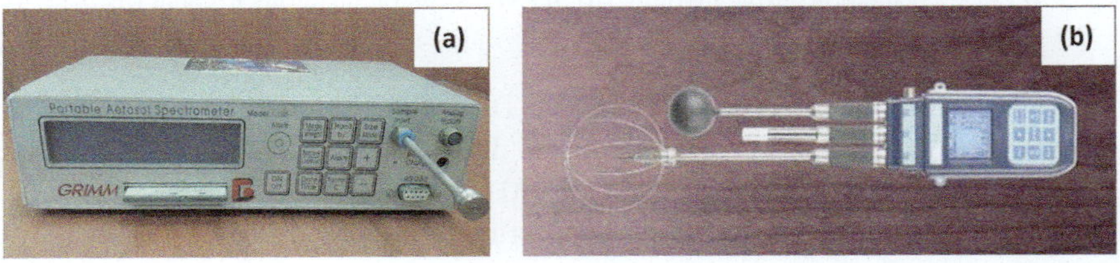

Fig. 2 **a** GRIMM 1.108, **b** Thermal comfort meter

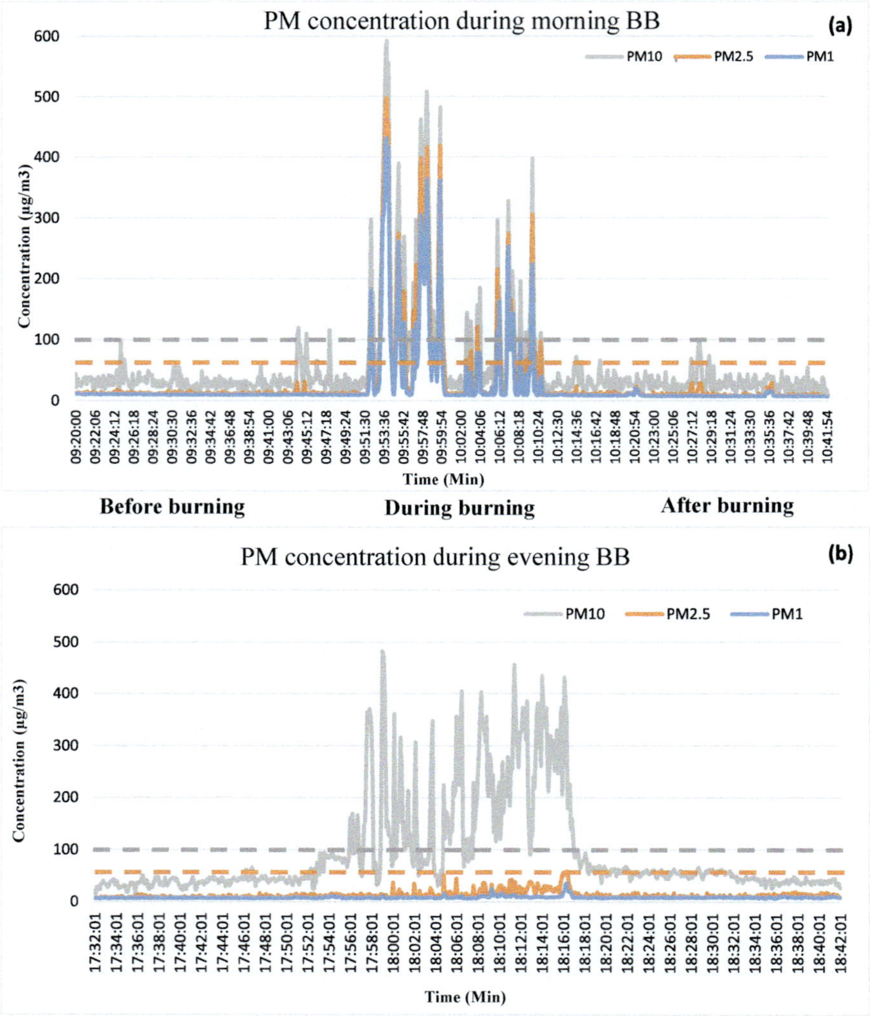

Fig. 3 **a** PM concentration during morning BB, **b** PM concentration during evening BB

Morality due to other natural causes: The method used is log-linear for the age group of 30+ adults. The adults affected due to all other natural causes for the 30+ age groups in the Karnataka state per 100,000 people have been taken as 5000. Around 35.56% of people are having long-term effects due to stroke (Table 7).

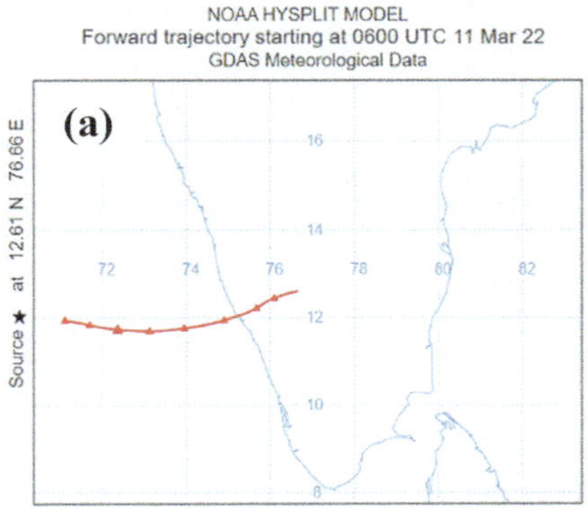

Fig. 4 HYSPLIT forward trajectory indicating predominant downwind direction at **a** 06:00 a.m. to 12:00 p.m. (UTC), and **b** 12:00 p.m. to 06:00 p.m. (UTC) on March 11, 2022

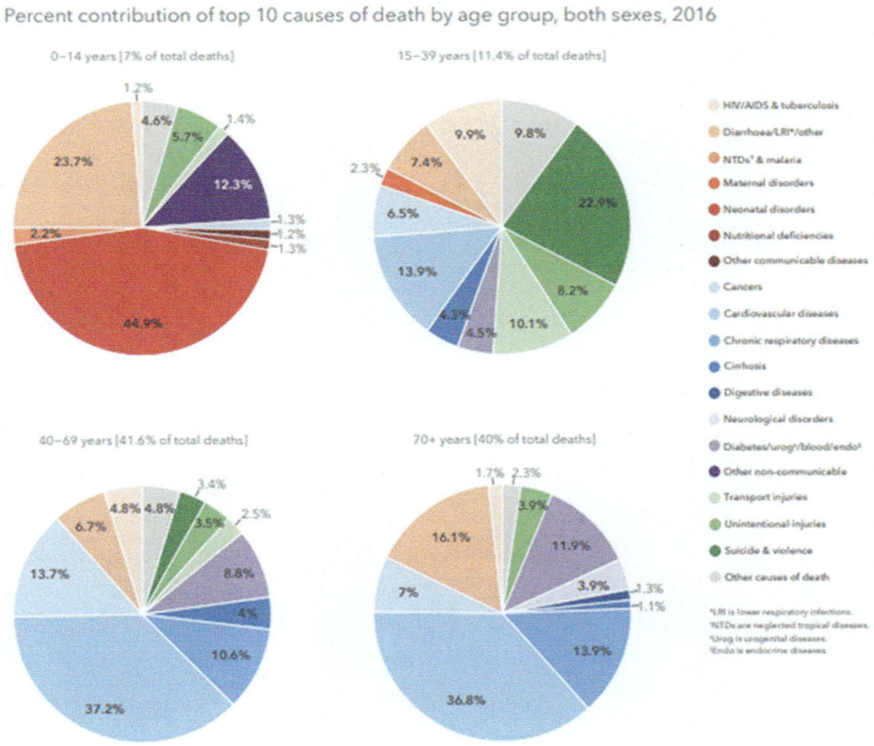

Fig. 5 Global burden disease profile for different age group people (in 2016) of Karnataka (*Source* ICMR, PHFI 2017)

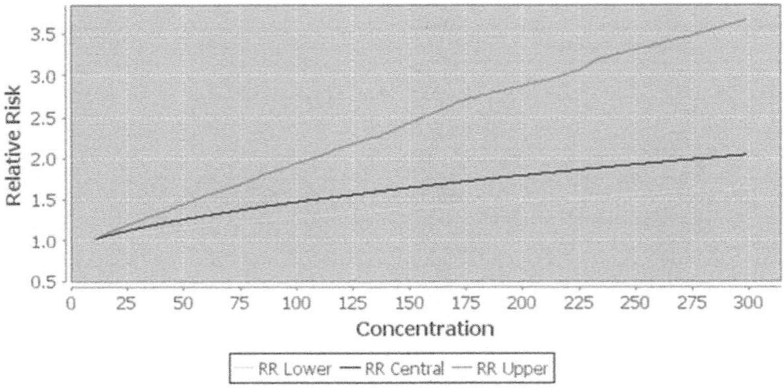

Fig. 6 Relative risks due to COPD (for age 25+) associated with the avg. PM concentration

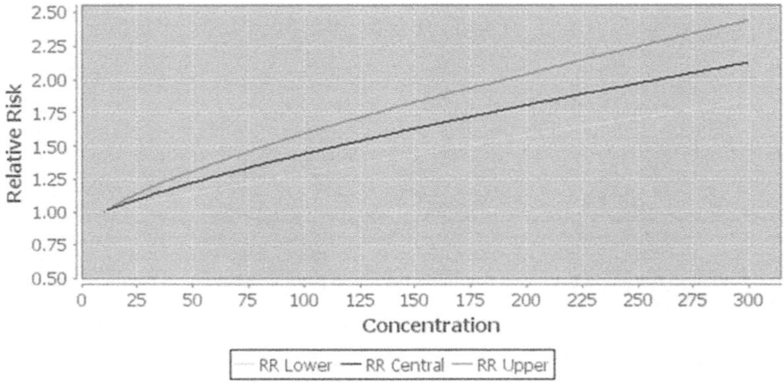

Fig. 7 Relative risks due to LC (for age 25+) associated with the avg. PM concentration

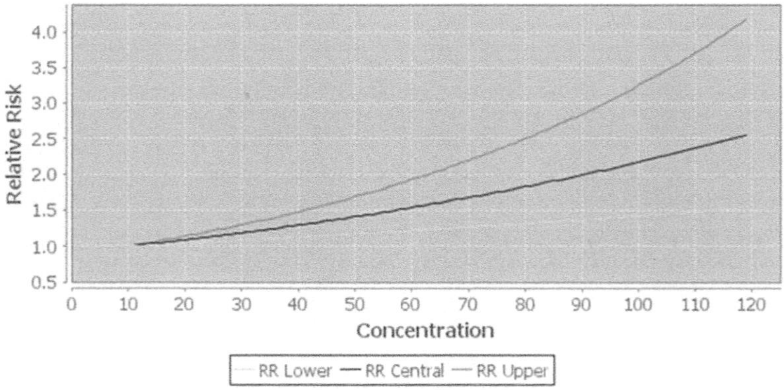

Fig. 8 Relative risks due to lung cancer (for age 30+) associated with the avg. PM concentration

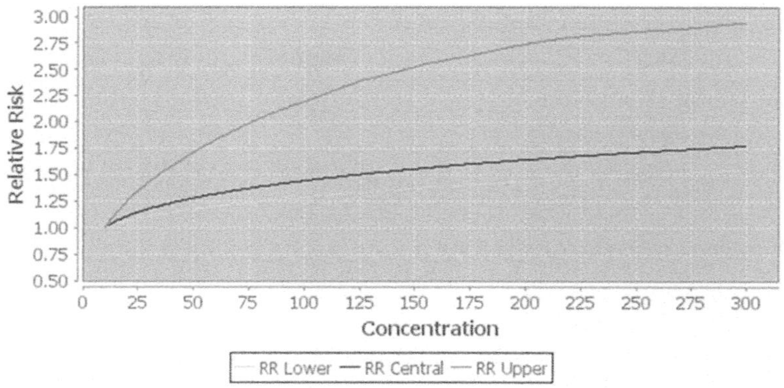

Fig. 9 Relative risks due to IHD (for age 25+) associated with the avg. PM concentration

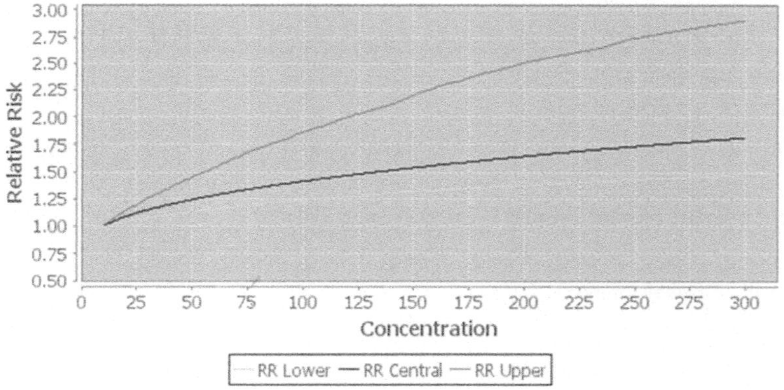

Fig. 10 Relative risks due to stroke (for age 25+) associated with the avg. PM concentration

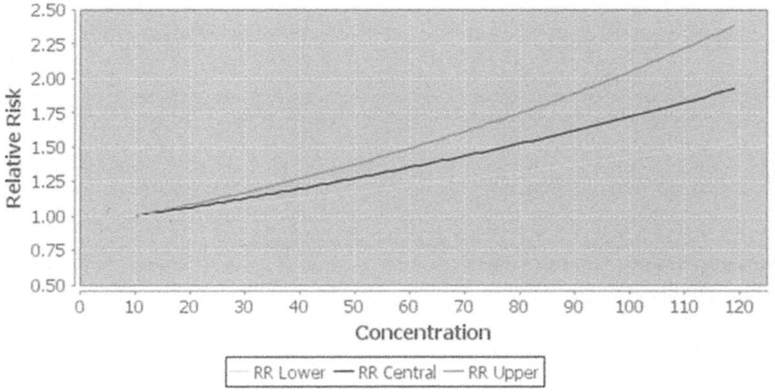

Fig. 11 Relative risks due to other natural diseases (for age 30+) associated with the avg. PM concentration

Table 2 Estimated mortality in adults due to COPD

Details	Central	Lower	Upper
Estimated attributable proportions (%)	29	19.4	43.73
Estimated no of attributable cases	41	28	62
Estimated no of attributable Cases per 100,000 population at risk	1420.92	950.69	2142.97

Table 3 Estimated mortality in adults due to LC

Details	Central	Lower	Upper
Estimated attributable proportions (%)	27.12	19.51	33.55
Estimated no of attributable cases	110	79	136
Estimated no of attributable Cases per 100,000 population at risk	3769.02	2711.48	4662.80

Table 4 Estimated mortality in adults (30+ age) due to lung cancer

Details	Central	Lower	Upper
Estimated attributable proportions (%)	46.72	24.91	61.61
Estimated no of attributable cases	144	77	190
Estimated no of attributable Cases per 100,000 population at risk	4952.35	2640.97	6530.33

Table 5 Estimated mortality in adults (25+ age) due to IHD

Details	Central	Lower	Upper
Estimated attributable proportions (%)	28.35	20.07	51.59
Estimated no of attributable cases	91	64	165
Estimated no of attributable Cases per 100,000 population at risk	3118.95	2207.28	5675.00

Table 6 Estimated mortality in adults (25+ age) due to stroke

Details	Central	Lower	Upper
Estimated attributable proportions (%)	26.69	16.96	42.54
Estimated no of attributable cases	29	19	47
Estimated no of attributable cases per 100,000 population at risk	1014.16	644.50	1616.70

Table 7 Estimated mortality in adults (30+ age) due to other natural diseases

Details	Central	Lower	Upper
Estimated attributable proportions (%)	35.56	24.91	44.15
Estimated no of attributable cases	52	36	64
Estimated no of attributable cases per 100,000 population at risk	1778.16	1245.74	2207.62

4 Conclusion

Study shows that the different sizes of PM exceeded the NAAQs permissible limit (PM_{10}: 100 $\mu g/m^3$, $PM_{2.5}$: 60 $\mu g/m^3$) by several times due to burning of sugarcane residue. It is found that the smaller size particles will affect human health more which may go deeper and deposit to the alveolar region [1]. Again, the health assessment study done for adults (25+ age) using the AirQ+ software exhibits higher mortality rates due to COPD, LC, lung cancer, IHD, stroke, and other natural diseases. The relative risks associated with the increase in the PM concentration were also examined. Hence, it is clear that the mortality rate is increasing with the increase in PM concentration. The age of the person is also an important factor in simulating the mortality rates. The duration of exposure in air pollution hot spots is another factor for which the mortality rate increases for higher-aged people.

References

1. Deng Q, Deng L, Miao Y, Guo X, Li Y (2019) Particle deposition in the human lung: health implications of particulate matter from different sources. Environ Res 169:237–45. https://doi.org/10.1016/j.envres.2018.11.014
2. ICMR, PHFI, and IHME (2017) "Karnataka: Disease Burden Profile, 1990 to 2016." India: Health of the Nation's States—The India State-Level Disease Burden Initiative. New Delhi: 5–8 (Accessed on: 09.09.2022)
3. Manojkumar N, Srimuruganandam B (2021) Health effects of particulate matter in major Indian cities. Int J Environ Health Res 31(3):258–70. https://doi.org/10.1080/09603123.2019.1651257
4. Manojkumar N, Srimuruganandam B (2022) Spatio-temporal health benefits attributable to PM2.5 reduction in an Indian City. Int J Environ Health Res. https://doi.org/10.1080/09603123.2022.2040959/SUPPL_FILE/CIJE_A_2040959_SM7532.DOCX
5. Prabhu V, Singh P (2022) Characteristics and health risk assessment of fine particulate matter and surface ozone: results from Bengaluru, India. Environ Monit Assess. https://doi.org/10.1007/s10661-022-09852-6

Chemical Characteristics, Source Apportionment and Health Risk Analysis of Atmospheric Particulates at the Urban Traffic Site in Vijayawada City, Andhra Pradesh India

Manuj Sharma, N. Manesha Manoon Mani, P. Hema, and Suresh Jain

Abstract The present study collected 24 hourly PM_{10} and $PM_{2.5}$ samples in Vijayawada city of Andhra Pradesh, India. The average concentration of PM_{10} and $PM_{2.5}$ at the monitoring site were 122 ± 29 and 70 ± 14 $\mu g/m^3$ during the winter season, which exceeds the 24 h average National Ambient Air Quality Standards (NAAQS). PM concentration during summer was well below the NAAQS. The CMB model deciphered the dataset into vehicular, industries, construction, road dust, secondary inorganic aerosols, coal, and biomass burning as the major sources of both the size fraction with seasonally varied contributions over Vijayawada city. Health risk assessment of particulate bound elements was also estimated. For non-carcinogenic risk (Hq), all the metals hold a value less than 1 (Hq < 1), indicating acceptable risk among adults and children, except Mn (Hq > 1). Carcinogenicity exposure to Cr was in the range of 10^{-5} and 10^{-4} in adults and children with higher cancer risk compared to USEPA standards, i.e. 1×10^{-6}.

Keywords Ambient air quality monitoring · Chemical characteristics · Source apportionment · Chemical mass balance model · Non-carcinogenic and carcinogenic risk

1 Introduction

Airborne particulates are ubiquitously present in the urban atmosphere of developing countries. Particulate matter (PM) has variable composition and size fraction, which originates from the primary (natural or anthropogenic) and secondary (aerosol

---

M. Sharma (✉) · N. M. M. Mani · P. Hema · S. Jain
Department of Civil and Environmental Engineering, Indian Institute of Technology Tirupati, Tirupati, Andhra Pradesh, India
e-mail: manuj.sharma73@gmail.com

S. M. Shiva Nagendra et al. (eds.), *Select Proceedings of the 8th Indian International Conference on Air Quality Management (IICAQM 2023)*, Lecture Notes in Civil Engineering 582, https://doi.org/10.1007/978-981-96-2359-4_26

formation from precursor gases reaction in the atmosphere) sources. Rapid urbanization and surge in industrial, vehicular emission, construction and demolition, biomass burning, soil and road dust, open waste burning, etc., increase the fine and coarse PM emissions in the ambient air. These particulates remain in the atmosphere for longer and severely impact the global climate, vegetation, economy, and human health [2, 14]. PM composition mainly holds ionic, elemental, and organic matter, including total carbon (TC) and volatile compounds like polycyclic aromatic hydrocarbons (PAHs) [33]. Prolonged exposure of PM and toxic elements like Pb, Co, Zn, Cu, Cd, Mn, V, As, Fe, Ni, and Cr exaggerates the toxicity and carcinogenic risk of the urban population [3]. A latest study on Global Burden of Diseases (GBD) reported 6.7 million premature fatalities worldwide due to air pollution exposure [9]. Numerous studies show that PM-bound trace elements are associated with chronic respiratory diseases, oxidative stress, genetic mutations, cardiovascular diseases, dermatitis, lung inflammation, and cytotoxicity [8]. Along with the health risk studies, the source apportionment (SA) technique is often employed to assess PM sources' contribution to the urban atmosphere [23]. SA study permits better delineation of air pollutants characteristics by identifying and quantifying the source's contribution, which is essential in developing air pollution control strategies. In India, SA technique was performed by employing different receptor models like chemical mass balance (CMB) [32], positive matrix factorization (PMF) [15], and principal component analysis (PCA) [20] depending upon the available data. Numerous studies were conducted in the northern urban region like New Delhi [15], Amritsar [19], Dharamshala [20], and Indo-Gangetic Plan (IGP) [23], whereas few research were focussed in the south urban region like Mangalore [18], Chennai [10], and Vishakhapatnam [12]. Therefore, the present study performed the SA study to characterize the atmospheric particulate emission sources by employing the CMB receptor model and also evaluate the human health risk ascribed to particulates-bound elements at the traffic location in Vijayawada city, Andhra Pradesh. Along with the CMB model, HYSPLIT backward trajectory model was performed to identify and trace the PM_{10} and $PM_{2.5}$ trans-boundary movement and their potential sources of origin at the receptor location. This study can form the preliminary knowledge base for the researchers and scientists and also facilitate the policymakers in pollution abatement strategies.

2 Methodology

2.1 Site Description and Monitoring Procedure

Ambient PM_{10} and $PM_{2.5}$ monitoring was performed in Vijayawada city (16° 31′ 9.48″ N, 80° 37′ 49.8″ E), Andhra Pradesh, India. The city lies 11 m asl at the Krishna riverbank with a 1.03 million populace spread over 138 km^2. The city exhibits a tropical climate with the annual average temperature ranging between 23 and 34 °C, ~ 978 mm annual average rainfall, and ~ 73% relative humidity [35].

The city's average air mass trajectory seasonally varied from the northwest (NW) and northeast (NE) to southeast (SE) direction. Daily 24 h ambient air samples were collected for measuring the concentration of $PM_{10,}$ and $PM_{2.5}$ for 15 days in the January and April months, respectively, near Siddhartha Medical College (SMC) (16° 31′ 0.23″ N, 80° 40′ 16.34″ E), as a traffic site in Vijayawada city (Fig. 1). The site was adjacent to the National Highway (NH-16) with heavy traffic flow, hospitals, and educational institutes with growing residential and commercial activities around this highway.

Ambient air quality monitoring was performed with the low volume samplers (flow rate: 16.7 lpm) for collecting particulates samples on the polytetrafluoroethylene (PTFE) and pre-condition quartz (QMA) filters of 47 mm diameter. Pre- and post-conditioning were performed per the standard protocols [28]. After the post-conditioning, particulate mass concentration was estimated by the employing gravimetric analysis technique.

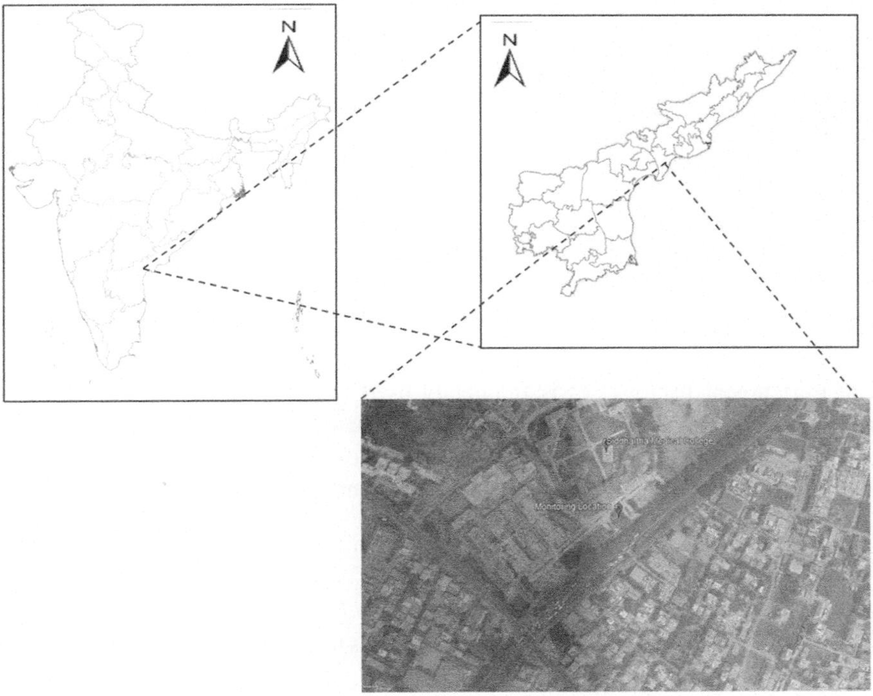

Fig. 1 Monitoring location in Vijayawada city

2.2 Chemical Characterization

The PM concentration was deciphered into ionic, elemental, and carbon (OC/EC) contents. PM samples accumulated on the PTFE filters were cut into two halves for ionic and elemental composition analysis. The first segment used for water-soluble ionic species was placed in 20 ml deionized water (Type 1, MilliQ, 18.2 MΩ-cm) and ultrasonicated for 60 min. The samples were further filtered through 0.45 μm PVDF filter discs, and the filtrate was analysed for water-soluble ionic species (WSISs) (F^-, Cl^-, NO_2^-, NO_3^-, SO_4^{2-}, Na^+, NH_4^+, K^+, Ca^{2+}, and Mg^{2+}) measurement using Ion Chromatography (IC) (Metrohm USA Inc.). The second half of the filter was extracted with 4 ml HNO_3 and 4 ml DI water and subjected to a closed digestion process (microwave digestor; 180 °C for 50 min). Sample was filtered through nylon filter discs (pore size ~ 0.2 μm) and subjected to Inductively Coupled Plasma Mass Spectrometry (ICPMS) (NexION 2000) for the elemental analysis (Ag, Al, As, Ba, Be, Bi, Ca, Cd, Co, Cr, Cs, Cu, Fe, Ga, In, K, Li, Mg, Mn, Na, Ni, Pb, Rb, Se, Sr, Tl, U, V, Zn, Hg, S, Si, Ge, Mo, and Ti) [18]. DRI-Model 2001 TOR/TOT Carbon Analyzer was used for the OC/EC analysis in the PM samples. In brief, 0.53 cm^2 of the QM/A filter papers were exposed at various temperatures level (140, 280, 480, and 580 °C) in pure helium for OC estimation and (580, 740, and 840 °C) in 98% helium and 2% oxygen for EC estimation [30].

2.3 Source Identification and Apportionment

2.3.1 Enrichment Factor

The Enrichment Factor (EF) is frequently used technique to assess the pr intensity of polluting sources by their degree of associated anthropogenic activities. The particulate sample's chemical species to reference element ratio is compared to the equivalent percentage in the reference system using EF. It is given as:

$$\mathrm{EF} = \frac{(C_n/C_{\mathrm{Ref}})}{(B_n/B_{\mathrm{Ref}})} \tag{1}$$

C_n is the concentration of nth species and C_{Ref} is the concentration of the reference element in the airborne particulate sample, and B_n and B_{Ref} are the concentration of nth species and reference element in the earth's upper crust. This study considered Al as the reference element due to its chemical properties and low spatial variability [24, 29]. EF differentiates the elements between crustal and anthropogenic activities. Elements of EF less than 5 are considered crustal and natural origin sources, whereas elements with $5 < \mathrm{EF} < 10$ represent natural and anthropogenic origin sources. $\mathrm{EF} > 10$ represents the elements of anthropogenic characteristics [15].

2.3.2 Chemical Mass Balance Receptor Model

The observed size segregated PM concentration, and respective chemical composition was apportioned to the emission sources using the EPA CMB8.2 model [7]. The CMB model is a widely used tool for emission contributions assessment of the particulates and VOCs concentration at the receptor locations [31]. The model used the effective variance least-squares fitting approach to solve the mass balance equation presented as (27.22) [37]. It assumes that the signatory species do not undergo any phase partitioning during the sampling period, all the potential contributing sources have been identified, and the sources should not transcend the identified chemical species in the ambient samples [4, 32].

$$x_{ij} = \sum_{p=1}^{P} g_{ik} f_{kj} + e_{ij}, \tag{2}$$

where, x_{ij} is the jth species concentration measured in the sample i of the source k, g_{ik} represent the individual source contribution in the sample i. f_{kj} is the selected source profile, and e_{ij} is the residual for each sample/species.

2.4 *HYSPLIT Model*

Backward trajectory analysis was performed by employing NOAA Air resources laboratory's Hybrid Single Particle Lagrangian Integrated Trajectory (HYSPLIT) model to determine the potential source regions and pollutants pathways [39]. The HYSPLIT model runs twice for seven days at 500 m above the ground level for backwards trajectory analysis to cover the monitoring period during winter and summer at the receptor location.

3 Health Risk Analysis

The study performed the health risk analysis of airborne particulate-bound heavy metals by following the USEPA recommended model USEPA [34]. Health risk analysis through population exposure pathways of particulate bound elements can be determined by three routes, (a) ingestion, (b) inhalation through mouth and nose, (c) direct contact with skin (dermal) [13, 16]. The present study explored the respiratory pathway related to the carcinogenic and non-carcinogenic risk of selected elements (Mn, Cr, Cu, Zn, Ni, Pb, Cd, and Co) [22]. Zn, Mn, Pb, Cu, and Co show the non-carcinogenic risk, whereas Cd, Ni, and Cr showed the carcinogenic risk in the human population [27]. Based on exposure factor, lifetime average daily dose (LADD) was calculated for adult and child populace by using:

$$\text{LADD} = \frac{C \times \text{IR} \times \text{EF} \times \text{ED}}{\text{AT} \times \text{BW}}, \tag{3}$$

where C = elemental concentration (mg m^{-3}), IR = inhalation rate (m^3d^{-1}), ED = exposure duration (in years), EF = exposure frequency (days/year), AT = average time (day), BW = body weight (kg).

Carcinogenic and non-carcinogenic risks associated with each element were also estimated using:

$$\text{HQ}_n = \text{LADD}_n/\text{RfD}_n \tag{4}$$

$$\text{Carcinogenic risk} = \text{LADD} \times \text{SF}, \tag{5}$$

where HQ (non-carcinogenic risk) is the hazardous quotient of the nth element, and Rfd_n and SF represents the reference dose (mg kg^{-1} day^{-1}) and cancer slope factor (mg kg^{-1} day^{-1}) According to USEPA [34], HQ < 1 indicates that the human populace is not susceptible to adverse health impacts; on the contrary, HQ > 1 indicates greater susceptibility to health hazards. The acceptable carcinogenic risk is less than 10^{-6} [40].

4 Results and Discussions

4.1 Ambient Air Quality Monitoring

Figure 2 illustrates the ambient PM_{10} and $PM_{2.5}$ concentrations at the SMC traffic site during the sampling period. In the winter and summer seasons, the average 24 h PM_{10} and $PM_{2.5}$ concentrations at the monitoring site were 122 ± 29 $\mu g/m^3$ and 70 ± 14 $\mu g/m^3$ and 76 ± 21 $\mu g/m^3$ and 38 ± 09 $\mu g/m^3$, respectively. Figure 2a indicates that PM_{10} and $PM_{2.5}$ concentration during the winter season exceeds the 24 h (NAAQS) ($PM_{10} = 100$ $\mu g/m^3$ and $PM_{2.5} = 60$ $\mu g/m^3$) during the entire monitoring period. PM_{10} concentration exceeds the standards by 1.22 times, and $PM_{2.5}$ exceeds the standards by 1.16 times. In the case of summer, it was observed that PM_{10} concentration exceeded the NAAQS during the first three days of monitoring. The highest value was attained on the first monitoring day with ~ 119 $\mu g/m^3$, illustrating the episodic incidence of particulate pollution at the site.

4.2 Chemical Composition

In winter season, the average chemical distribution of PM_{10} was observed to be 18% (21.8 $\mu g/m^3$) of secondary ions ($NO_3^- + SO_4^{2-} + NH_4^+$) and 5% (6.5 $\mu g/m^3$) of

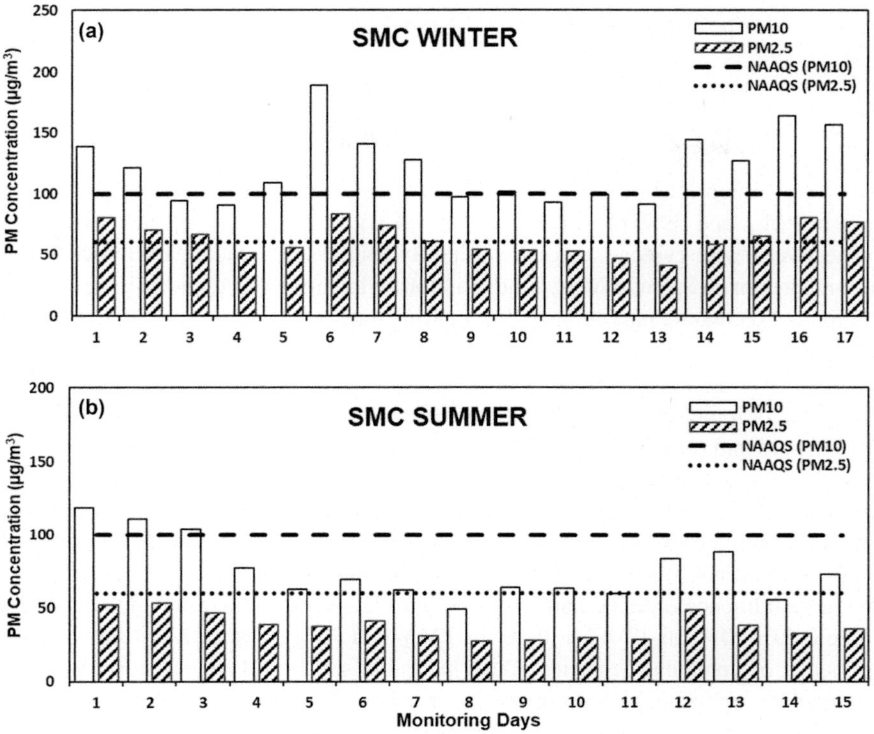

Fig. 2 Temporal variation of **a** PM_{10} and **b** $PM_{2.5}$ concentration over Vijayawada city

other WSISs; 12% (14.7 μg/m^3) of elements (crustal (Ca + Al + Si + Fe) ~ 6% (~ 8 μg/m^3), other ~ 5.5% (6.7 μg/m^3); total carbon species ~ 17% (~21 μg/m^3) (OC ~ 8% (10.3 μg/m^3), EC ~ 9% (10.4 μg/m^3). For $PM_{2.5}$, the average chemical distribution was 25% (17.5 μg/m^3) of secondary ions and 11% (7.8 μg/m^3) of other WSISs; 10% of elements (crustal ~ 4% (2.6 μg/m^3), other ~ 6% (4.1 μg/m^3); total carbon species ~ 25% (17.5 μg/m^3) (OC ~ 13% (9.3 μg/m^3), EC ~ 12% (8.1 μg/m^3). Chemical variation in summer was evident in the result as the average chemical distribution of PM_{10} was 8% (6.1 μg/m^3) of secondary ions and ~ 8% (~ 6 μg/m^3) of other WSISs; 22% (16.9 μg/m^3) of elements (crustal ~ 13% (~ 10 μg/m^3) and other ~ 9% (~7 μg/m^3); total carbon species ~ 17% (12.8 μg/m^3) (OC ~ 9% (~ 7 μg/m^3), EC ~ 8% (~ 6 μg/m^3). For $PM_{2.5}$, the average chemical distribution was 13% (4.7 μg/m^3) of secondary ions and 11% (4 μg/m^3) of other WSISs; 26% (10.2 μg/m^3) of elements (crustal ~ 11% (4.3 μg/m^3)) and other ~ 15% (~ 6 μg/m^3)); total carbon species ~ 27% (10.2 μg/m^3) (OC ~ 15% (5.7 μg/m^3), EC ~ 12%(4.5 μg/m^3)).

4.3 *Water-Soluble Ionic Species (WSISs)*

Secondary ionic species (NO_3^- + SO_4^{2-} + NH_4^+) were dominant among WSISs in coarse and fine particulate matters at the monitoring site. Among overall average of observed WSISs (PM_{10} ~ 20.1 μg/m^3, $PM_{2.5}$ ~ 17.1 μg/m^3), the SO_4^{2-}, NO_3^- and NH_4^+ comprised of 41% (8.2 μg/m^3), ~ 13% (2.6 μg/m^3) and 15% (3.1 μg/m^3) in PM_{10} and 39% (6.6 μg/m^3), 8% (1.4 μg/m^3), 18% (3.1 μg/m^3) in $PM_{2.5}$ respectively, with evident seasonal variability (Fig. 3). The secondary ionic species ascendancy in PM_{10} and $PM_{2.5}$ at the monitoring location may depict the presence of significant secondary particulate pollution from the precursor gases resulting from the atmospheric photochemical reactions. The present study assessed the NO_3^-/SO_4^{2-} ratios in PM_{10} and $PM_{2.5}$ concentration to analyse the potential sources contribution on the major ionic species. In the winter season, the NO_3^-/SO_4^{2-} ratio for PM_{10} was 0.24 ± 0.15 and for $PM_{2.5}$ was 0.26 ± 0.24, respectively. The low NO_3^-/SO_4^{2-} ratio indicates the stationary sources ascendancy during the winter season, which can be attributed to the high usage of coal, wood, and other biomass fuels in the periphery industries due to excess power demand along with low temperature and stable mixing height conditions. Wang et al. [36] reported the NO_3^-/SO_4^{2-} ratio of 0.64, showing the large influence of vehicular activities. During summer, the ratio value estimated to be 0.67 ± 0.41 and 0.47 ± 0.40 for PM_{10} and $PM_{2.5,}$ respectively, showing the influence of vehicular activities. Other probable sources of higher sulphate levels during the monitoring period may be the crematorium activities near the monitoring location.

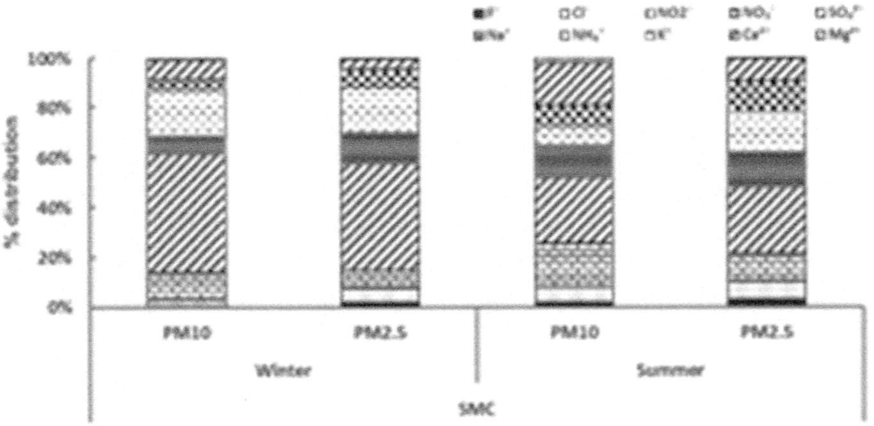

Fig. 3 Ionic contribution in PM_{10} and $PM_{2.5}$ at the monitoring location during winter and summer

4.4 Carbonaceous Species

Organic carbon (OC) and Elemental carbon (EC), collectively known as Total carbon (TC), are typically emitted into the ambient air as the result of vehicular emissions, biomass combustion, and industrial emissions. EC, having light absorption tendencies, shows a positive radiative effect on the earth's atmosphere, has the primary origin, and is emitted from the incomplete combustion of carbon materials [21]. At the same time, OC originated from primary and secondary sources found in various oxidation states can participate in gas to particle reactions in the atmosphere. [26]. Figure 4 presents the temporal variation of OC and EC monitored at the SMC location. The average TC of PM_{10} and $PM_{2.5}$ measured was ~ 21 $\mu g/m^3$ (~ 17%) and 17 $\mu g/m^3$ (~ 28%) in winter and ~ 13 $\mu g/m^3$ (~ 17%) and ~ 10 $\mu g/m^3$ (~ 27%) in the summer season, respectively. The overall average OC and EC concentrations were 8.5 and 8.1 $\mu g/m^3$ for PM_{10}. The average concentration of OC and EC in $PM_{2.5}$ was 7.5 $\mu g/m^3$ and 6.3 $\mu g/m^3$, respectively. In addition, OC/EC ratio was also assessed to determine the potential emission sources at the monitoring site. The average OC/EC ratio for PM_{10} and $PM_{2.5}$ was 1.04 and 1.23, respectively, indicating the influence of vehicular diesel emissions on the ambient PM concentrations. These results are attributed to continuous traffic movement, especially heavy diesel vehicles (HDV) on the NH-16 adjacent to the monitoring location. Cesari et al. [5] reported that lower OC/EC ratios between 0.3 and 1 and 1.4 and 4 indicate the diesel and gasoline vehicular emissions, respectively, whereas OC/EC ratio values between 5 and 12 depict the emission from biomass burning [25].

4.5 Source Identification and Apportionment

4.5.1 Enrichment Factor

Figure 5 illustrates the EF of the crucial elements for the PM_{10} and $PM_{2.5}$ at the SMC site during the monitoring period. The data shows Ag, Cd, Pb, Bi, Zn, As, and Cu exceeded the EF > 10 in PM_{10} and $PM_{2.5}$ in both seasons, indicating the origin of these elements from anthropogenic activities. Maximum EF was observed for Ag, Cd, Cu, Pb, and Zn in $PM_{2.5}$ during the winter, attributed to the high particulate concentration, low temperature, and stable mixing height and anthropogenic activities like construction activities and soil and road dust. Sources of elements like Ga, Cr, and Tl tend to vary seasonally (Fig. 5). For example, Ga is attributed with sources of mixed origin in $PM_{2.5}$ as EF varies between 5 and 10. Similarly, potential sources of Cr and Tl vary between natural and anthropogenic sources following the size fraction and monitoring season. Elements like Mn, Ni, Fe, Mg, and Na have an EF of less than five, which signifies their natural or crustal origin [6]. However, EF > 5 for Mn and Ni in $PM_{2.5}$ during winter also represents anthropogenic origin intrusion.

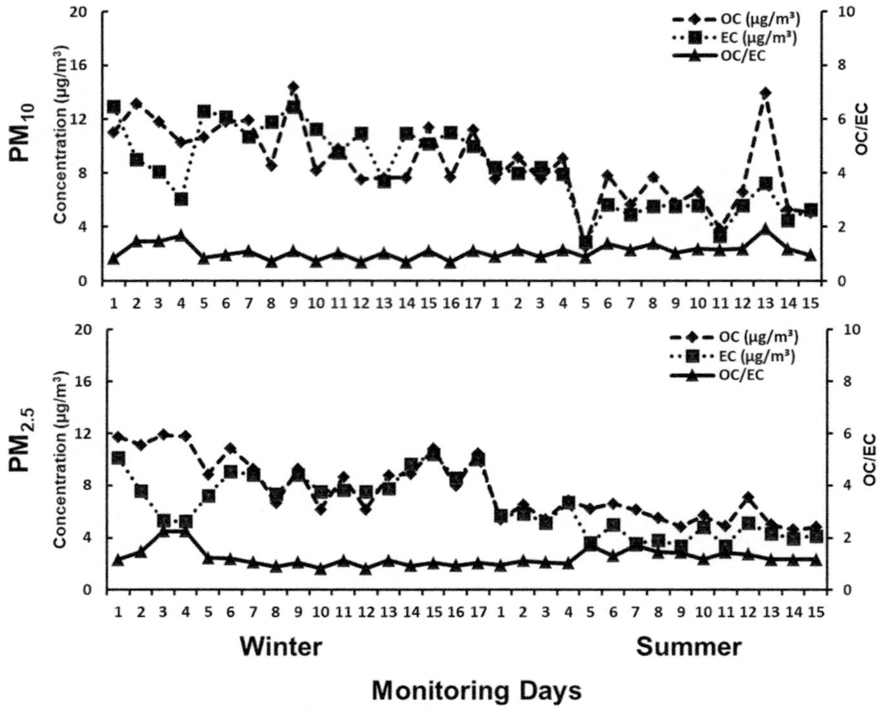

Fig. 4 Temporal and seasonal variation of OC, EC, and OC/EC ratio in PM_{10} and $PM_{2.5}$ at the monitoring location

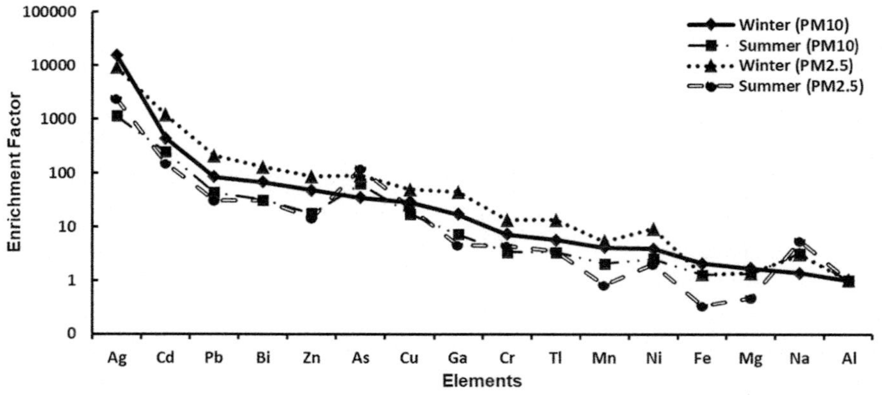

Fig. 5 EF seasonal variation of different elements in PM_{10} and $PM_{2.5}$ at the monitoring location

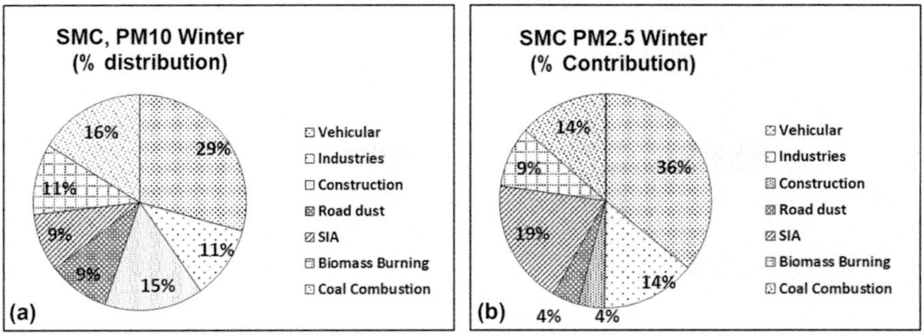

Fig. 6 Source contribution for PM_{10} and $PM_{2.5}$ at the monitoring location during the winter season

Chemical Mass Balance

PM_{10} and $PM_{2.5}$ (Winter Season)

SA results in Fig. 6 describe that the CMB model determined seven sources for both the size fractions during the winter and summer seasons. The model showed vehicular activities as the significant source contributor (29%) at the receptor location, followed by coal combustion (16%) in PM_{10} concentration (Fig. 6a). Chennai-Kolkata highway (NH-16) and one of the important traffic junctions resulted in the heavy influx of vehicles throughout the day at the monitoring location, which attributed to the highest vehicular contribution. The other noticeable source identified by the model was construction activities, contributing to 15% of PM_{10} emissions. Biomass burning (combined contribution of wood-burning, open waste and refuse burning) (~ 11%) emerged as another dominant source during the monitoring period at SMC, followed by secondary inorganic aerosols (SIA) (ammonium sulphate + ammonium nitrate) (~ 9%). High coal combustion and biomass burning are dominant sources at this monitoring location may be attributed to winds from north-east (NE) to southeast (SE) directions, which carry the emission load generated in the upwind direction by the burning of coke, coal, wood, husk in the industries. The close vicinity of industries near the monitoring location also became the influential source (~ 11%), followed by road dust (~ 9%). The source contribution follows a similar trend for $PM_{2.5}$ concentration at the receptor location (Fig. 6b). CMB model resolved the ambient $PM_{2.5}$ concentration into vehicular contribution (36%), followed by coal combustion (~ 19%). During the monitoring period, other dominant sources at the SMC were industrial activities (~ 14%), followed by biomass burning, which contributed 12%. SIA contributed was estimated to be 11%, with an equal contribution from construction (~ 4%) and road dust (~ 4%). The high contribution of biomass burning may be ascribed to the crematoria (~ 1000 m areal distance) near this location and the dominant winds from the north-east (NE) to the southeast (SE), which carry the industrial emissions generated by the burning of coal and wood in the industries.

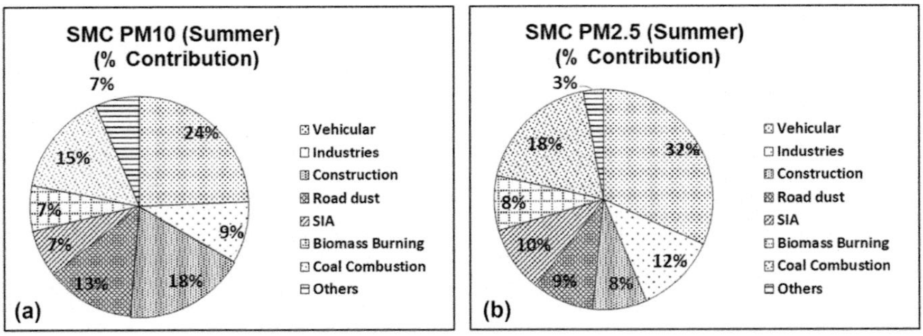

Fig. 7 Source contribution for PM_{10} and $PM_{2.5}$ at the monitoring location during the summer season

PM_{10} and $PM_{2.5}$ (Summer Season)

The CMB model indicates similar trends for the summer season as in the winter season at the monitoring location (Fig. 7). For PM_{10}, the model suggests that vehicular contribution was highest at the site. During summer, the movement of the excessive wind resulted in the dust particles transportation at a larger distance; hence, the model estimated the high contribution from construction (~ 18%) and road dust (13%) (Fig. 7a). The coal combustion and biomass burning contribution was characterized by 15 and 7% shares. In the summer, dominant winds were from the south (S) and south–southeast (SSE) direction, which may carry the industrial coal and wood burning emissions at the monitoring location. The industrial activities' contribution declined and was estimated to be 9% compared to the winter season. Followed by industries, SIA contributes around 7%. It was estimated that other sources (DG sets, fly ash, etc.) also contributed 7% during the monitoring period. Figure 7b shows that the vehicular contribution was 32% for $PM_{2.5}$, followed by a contribution from coal combustion (~ 18%) and industrial emission (~ 12%). Other dominant sources that contributed to the SMC during the monitoring period were SIA (~ 10%), road dust (~ 9%), constructional activities (~ 8%), and biomass burning (~ 8%). Other sources like DG sets, fly ash, etc., also contributed to 3% at the location during summer.

HYSPLIT Model

The air mass backward trajectory analysis was performed by employing the HYSPLIT model at the receptor location; to identify the transboundary origin and movement of airborne particulates. As Fig. 8 illustrates, during the winter season, most of the trajectories collide at the receptor, originating from Afghanistan, Pakistan, Iran, Oman, and Saudi Arabia and traversing through the region of Jammu and Kashmir, Himachal Pradesh, Punjab and Haryana, Rajasthan, Uttar Pradesh, Indo-Gangetic plain (IGP), and Bay of Bengal (BoB). The long-range trajectories possibly traversed from continents and desert regions contain the crustal/mineral atmospheric

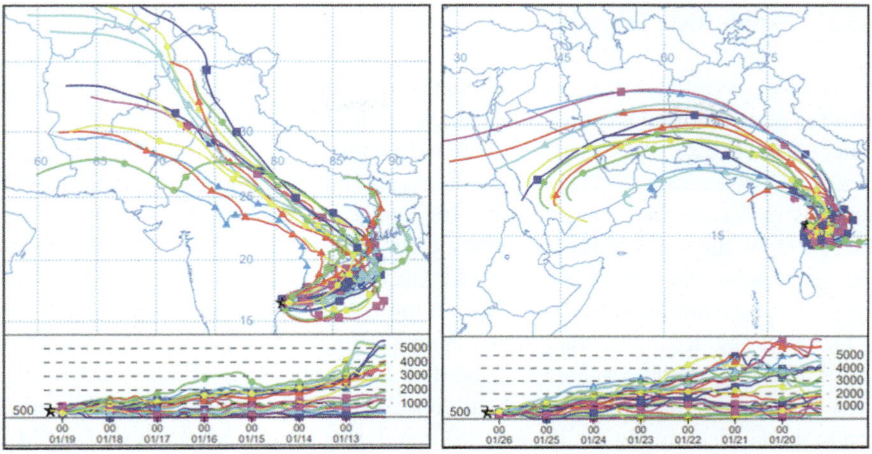

Fig. 8 HYSPLIT backward trajectories plot during the winter season for monitoring location in Vijayawada

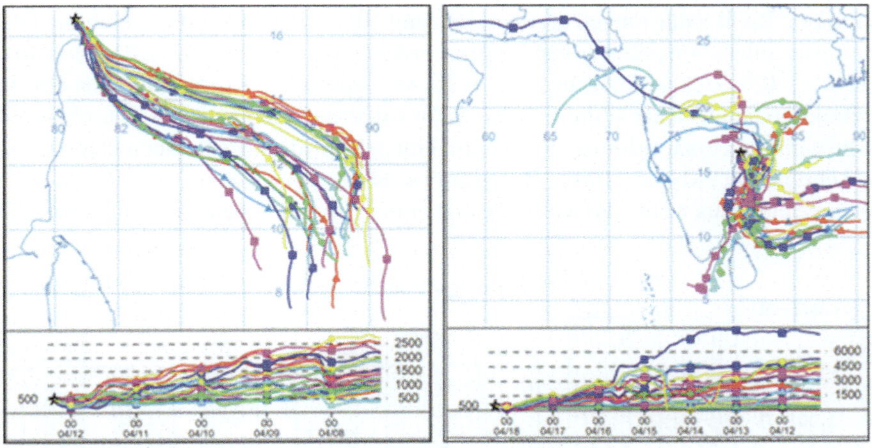

Fig. 9 HYSPLIT backward trajectories plot during the summer season for monitoring location in Vijayawada

particulates. Short-range trajectories originate from IGP, and BoB may characterize the receptor location with local sources and marine aerosols. During summer, almost all the trajectories originated from BOB and traversed to the receptor location, which contains the influx of sea salt aerosols (Fig. 9). However, limited trajectories followed the long-range transport which originated from the desert and continental regions of Iran and Oman.

4.5.2 Health Risk Analysis

This study assessed the non-carcinogenic and carcinogenic risk by estimating selected particulate-bound metals' hazardous quotient (Hq) and incremental lifetime cancer risk (ILCR). PM_{10} and $PM_{2.5}$ bound metals' exposure to the human populace may cause severe health (non-carcinogenic and carcinogenic) toxicity, which depends on factors like concentration, frequency, and duration [1]. Figure 10 and Table 1 show the non-carcinogenic risk of Zn, Pb, Cu, Co, Cd, Ni, Cr, and Mn and the carcinogenic risk of Cd, Ni, and Cr metals through inhalation pathways, respectively. For non-carcinogenic risk (Hq), all the metals hold a value less than 1 (Hq < 1), indicating acceptable risk among both sub-populations, except Mn (Hq > 1). Mn exhibited the highest value,PM_{10} (adults):2.96E+00, PM_{10} (children):1.05E+01, $PM_{2.5}$ (winter, adults:1.63E+00 and children:2.88E+00), predicting significant non-carcinogenic risk among adults and children during winter and summer for PM_{10} and during winter for $PM_{2.5}$. During summer, $PM_{2.5}$ bound Mn was marginally low for adults and marginally high for children (Fig. 10). The non-carcinogenic risk was assessed high among children than adults during both seasons.

The carcinogenic risk (ILCR) of Cd, Ni, and Cr was enumerated in Table 1. Results show that ILCR value ranged between 10^{-5} and 10^{-4} among the sub-population, indicating the inhalation pathway carcinogenic risk of Cr exceeded the USEPA standards, i.e. 1×10^{-6}. The carcinogenic risk of Ni was within the acceptable limits during summer for adults and children. Whereas for Cd, the carcinogenic risk was observed to be within acceptable limits only for children in PM_{10} during summer and for adults and children in PM_{10} and $PM_{2.5}$ during both seasons. During winter, ILCR values for all three elements were observed to be higher than in summer among both adults and

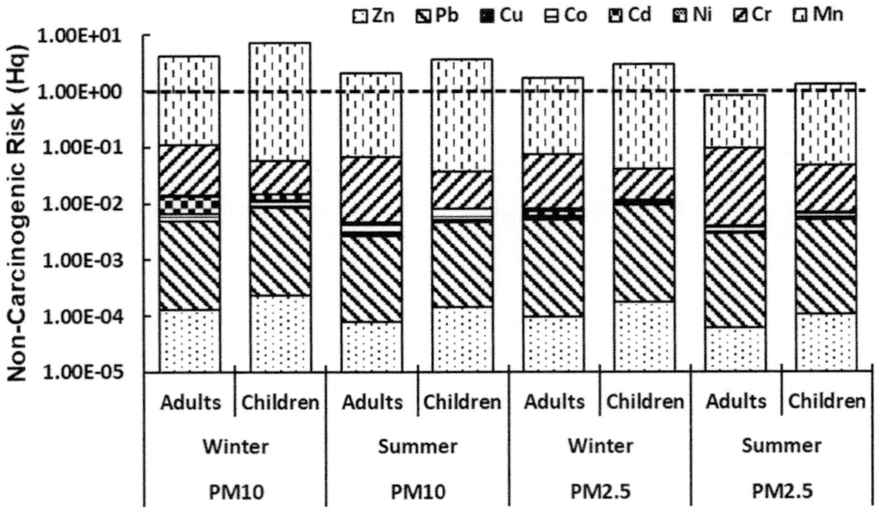

Fig. 10 Non-carcinogenic risk estimated of selected elements for adults and children populace of Vijayawada city during winter and summer

Table 1 Carcinogenic risk of Cd, Ni, and Cr for adults and children population of Vijayawada city

		Cd	Ni	Cr
PM_{10} Winters	Adults	4.04E−05 (4.54E−07 to 3.96E−04)	1.98E−05 (4.34E−07 to 1.63E−04)	1.12E−04 (4.00E−05 to 4.07E−04)
	Children	1.79E−05 (2.01E−07 to 1.76E−04)	8.76E−06 (1.92E−07 to 7.24E−05)	4.98E−05 (1.77E−05 to 1.81E−04)
PM_{10} Summer	Adults	1.22E−06 (1.90E−07 to 1.15E−05)	5.22E−07 (2.22E−07 to 8.04E−07)	7.35E−05 (4.29E−05 to 1.22E−04)
	Children	5.44E−07 (8.41E−08 to 5.11E−−06)	2.31E−07 (9.84E−08 to 3.56E−07)	3.26E−05 (1.90E−05 to 5.41E−05)
$PM_{2.5}$ Winter	Adults	8.80E−06 (5.92E−07 to 6.51E−05)	1.05E−05 (9.15E−−08 to 1.02E−−04)	7.93E−05 (2.09E−05 to 1.93E−04)
	Children	3.90E−06 (2.63E−07 to 2.89E−05)	4.66E−06 (4.06E−08 to 4.51E−05	3.52E−05 (9.27E−06 to 8.58E−05)
$PM_{2.5}$ Summer	Adults	5.95E−07 (3.53E−07 to 1.16E−06)	4.81E−07 (2.05E−07 to 1.65E−06)	1.06E−04 (2.49E−05 to 3.09E−04)
	Children	2.64E−07 (1.56E−07 to 5.16E−07)	2.13E−07 (9.08E−08 to 7.31E−07)	4.69E−05 (1.10E−05 to 1.37E−04)

children, indicating the potential carcinogenic threat in winter. It was also observed that carcinogenic risk was higher among the adult populace than in children, contrary to the non-carcinogenic risk, which was more profound in children than in adults during both seasons. This observation is evident in the study performed by [1], which reported a high non-carcinogenic risk and low carcinogenic risk among children than in adults during the foggy and non-foggy period. Gao et al. [11] also reported the higher non-carcinogenic risk in children than men and women, compared to high carcinogenic risk in adults than children.

5 Conclusion

Understanding the limited studies of airborne particles in the urban atmosphere over the southern region of India, this study was performed to obtain knowledge of PM concentration, chemical components, the contribution of sources, and seasonal and temporal resolution of the Vijayawada city. The 24 h average PM_{10} and $PM_{2.5}$ concentration exceeded the prescribed standards during the winter season, whereas in summer, the PM level was in control. Seasonal variation of chemical analysis showed a similar pattern with the dominance of ionic and carbon content in the city atmosphere. SO_4^{2-}/NO_3^- ratio ranged between 0.78 and 8.8 for both the size fraction, indicating the presence of mixed (mobile and stationary) sources. Although the monitoring location holds a significant influence of vehicles, the presence of industries in its vicinity is attributed to these results. CMB receptor model resolved the vehicular as the primary source with 24–36% contribution during winter and summer seasons for both the size fraction, followed by coal combustion, biomass burning, industrial activities, construction, road dust, and secondary inorganic aerosols. Non-carcinogenic risk (Hq) of PM-bound metals was validated that the health risk is within the prescribed limits except Mn, which poses a severe potential threat to adults and children. Cr imposes the highest carcinogenic risk potential in adults than children among the rest of the elements. Carcinogenic risk poses a more significant threat to the adult population than non-carcinogenic risk, impacting most children.

Acknowledgements This research is funded by Andhra Pradesh Pollution Control Board (APPCB) under the National Clean Air Program (NCAP). Authors also acknowledge the NOAA Air Resources Laboratory (ARL) for the provision of the HYSPLIT transport and dispersion model and/or READY website (https://www.ready.noaa.gov) used in this publication. MS greatly acknowledge the University Grant Commission Junior Research Fellowship (UGC JRF) (Ref: 190510664832) for the Ph.D. scholarship and the Indian Institute of Technology Tirupati, for providing the necessary resources for the research.

References

1. Agarwal A, Mangal A, Satsangi A, Lakhani A, Maharaj Kumari K (2017) Characterization, sources and health risk analysis of PM2.5 bound metals during foggy and non-foggy days in sub-urban atmosphere of Agra. Atmos Res 197:121–131
2. Aggarwal P, Jain S (2015) Impact of air pollutants from surface transport sources on human health: a modeling and epidemiological approach. Environ Int 83:146–157
3. Bai W, Zhao X, Yin B, Guo L, Zhang W, Wang X, Yang W (2022). Characteristics of PM2.5 in an industrial city of Northern China: mass concentrations, chemical composition, Source Apportionment, and Health Risk Assessment. Int. J. Environ. Res. Public Health 19(9)
4. Belis CA, Karagulian F, Larsen BR, Hopke PK (2013) Critical review and meta-analysis of ambient particulate matter source apportionment using receptor models in Europe. Atmos Environ 69:94–108
5. Cesari D, De Benedetto GE, Bonasoni P, Busetto M, Dinoi A, Merico E, Chirizzi D, Cristofanelli P, Donateo A, Grasso FM, Marinoni A, Pennetta A, Contini D (2018) Seasonal variability of PM2.5 and PM10 composition and sources in an urban background site in Southern Italy. Sci Total Environ 612:202–213
6. Cheng X, Huang Y, Zhang S-P, Ni S-J, Long Z-J (2018) Characteristics, sources, and health risk assessment of trace elements in PM 10 at an urban site in Chengdu, Southwest China. Aerosol Air Qual Res 18:357–370
7. Coulter CT (2004) EPA-CMB8.2. User's manual. Research Triangle Park, NC: Air Quality Modeling Group Emissions, Monitoring and Analysis Division Office of Air Quality Planning & Standards

8. Edwards L, Wilkinson P, Rutter G, Milojevic A (2022) Health effects in people relocating between environments of differing ambient air pollution concentrations: a literature review. Environ Pollut 292
9. Fuller R, Landrigan PJ, Balakrishnan K, Bathan G, Bose-O'Reilly S, Brauer M, Caravanos J, Chiles T, Cohen A, Corra L, Cropper M, Ferraro G, Hanna J, Hanrahan D, Hu H, Hunter D, Janata G, Kupka R, Lanphear B et al (2022) Pollution and health: progress update. Lancet Planet Health 6(6):e535–e547.
10. Gajghate DG, Talwar B, Pipalatkar P., & Pustode, T. (2012). Chemical Characterization of PM10 for Metals in Ambient Air of Chennai, India. J Hazard Toxic Radioactive Waste 16(2)
11. Gao Y, Guo X, Li C, Ding H, Tang L, Ji H (2015) Characteristics of PM2.5 in Miyun, the northeastern suburb of Beijing: chemical composition and evaluation of health risk
12. Guttikunda SK, Goel R, Mohan D, Tiwari G, Gadepalli R (2015) Particulate and gaseous emissions in two coastal cities—Chennai and Vishakhapatnam, India. Air Qual Atmos Health 8(6):559–572
13. Han Q, Wang M, Cao J, Gui C, Liu Y, He X, He Y, Liu Y (2020). Health risk assessment and bioaccessibilities of heavy metals for children in soil and dust from urban parks and schools of Jiaozuo, China. Ecotoxicol Environ Safety 191
14. Jain S, Barthwal V (2020). Health impact assessment of auto rickshaw and cab drivers due to exposure to vehicular pollution in Delhi: an integrated approach
15. Jain S, Sharma SK, Vijayan N, Mandal TK (2020) Seasonal characteristics of aerosols (PM2.5 and PM10) and their source apportionment using PMF: a four year study over Delhi, India. Environ Pollut 262
16. Jiang N, Liu X, Wang S, Yu X, Yin S, Duan S, Wang S, Zhang R, Li S (2019) Pollution characterization, source identification, and health risks of atmospheric-particle-bound heavy metals in PM10 and PM2.5 at multiple sites in an emerging megacity in the central region of China. Aerosol Air Qual Res 19(2):247–271
17. Juda-Rezler K, Reizer M, Maciejewska K, Błaszczak B, Klejnowski K (2020) Characterization of atmospheric PM2.5 sources at a Central European urban background site. Sci Total Environ 713
18. Kalaiarasan G, Balakrishnan RM, Sethunath NA, Manoharan S (2018) Source apportionment studies on particulate matter (PM10 and PM2.5) in ambient air of urban Mangalore, India. J Environ Manage 217:815–824
19. Kaur S, Senthilkumar K, Verma VK, Kumar B, Kumar S, Katnoria JK, Sharma CS (2013) Preliminary analysis of polycyclic aromatic hydrocarbons in air particles (PM10) in Amritsar, India: Sources, apportionment, and possible risk implications to humans. Arch Environ Contam Toxicol 65(3):382–395
20. Kaushal D, Bamotra S, Yadav S, Chatterjee S, Tandon A (2021) Particulate bound polycyclic aromatic hydrocarbons over Dhauladhar region of the north-western Himalayas. Chemosphere 263:128298
21. Kaushal D, Kumar A, Yadav S, Tandon A, Attri AK (2018) Wintertime carbonaceous aerosols over Dhauladhar region of North-Western Himalayas. Environ Sci Pollut Res 25(8):8044–8056
22. Li HH, Chen LJ, Yu L, Guo ZB, Shan CQ, Lin JQ, Gu YG, Yang ZB, Yang YX, Shao JR, Zhu XM, Cheng Z (2017) Pollution characteristics and risk assessment of human exposure to oral bioaccessibility of heavy metals via urban street dusts from different functional areas in Chengdu, China. Sci Total Environ 586:1076–1084
23. Murari V, Singh N, Ranjan R, Singh RS, Banerjee T (2020) Source apportionment and health risk assessment of airborne particulates over central Indo-Gangetic Plain
24. Pipalatkar P, Khaparde VV, Gajghate DG, Bawase MA (2014) Source apportionment of PM 2.5 using a CMB model for a centrally located Indian city. Aerosol Air Qual Res 14:1089–1099
25. Ram K, Sarin MM (2011) Day-night variability of EC, OC, WSOC and inorganic ions in urban environment of Indo-Gangetic Plain: Implications to secondary aerosol formation. Atmos Environ 45(2):460–468
26. Rastogi N, Sarin MM (2009) Quantitative chemical composition and characteristics of aerosols over western India: one-year record of temporal variability. Atmos Environ 43(22–23):3481–3488

27. Rathee A, Yadav S (2022) Human and ecological risk assessment : an international health risk assessment using chemical signatures of fine and coarse particles collected at breathing level height during firework display in New Delhi. Human Ecol Risk Assess Int J 1–24
28. Sharma D, Jain S (2020) Carcinogenic risk from exposure to PM2.5 bound polycyclic aromatic hydrocarbons in rural settings. Ecotoxicol Environ Safety 190:110135
29. Sharma SK, Banoo R, Mandal TK (2021) Seasonal characteristics and sources of carbonaceous components and elements of PM10 (2010–2019) in Delhi, India. J Atmos Chem 78(4):251–270
30. Shivani GR, Sharma SK, Mandal TK (2019) Seasonal variation, source apportionment and source attributed health risk of fine carbonaceous aerosols over National Capital Region, India. Chemosphere 237
31. Singh N, Murari V, Kumar M, Barman SC, Banerjee T (2017) Fine particulates over South Asia: review and meta-analysis of PM2.5 source apportionment through receptor model. Environ Pollut 223:121–136
32. Srimuruganandam B, Shiva Nagendra SM (2012) Source characterization of PM10 and PM2.5 mass using a chemical mass balance model at urban roadside. Sci Total Environ 433:8–19
33. Tu X, Fang X, Fang H, Ye C, Liu Z, Jia X, He D, Wang J, Huang H, Zou C, Yu C (2022) OPEN ACCESS Chemical composition characteristics and source apportionment of PM 2.5 in ceramic industrial base during Winter
34. U.S. EPA U.S. Environmental Protection Agency (2009) Risk assessment guidance for superfund assessment. Office of Superfund Remediation and Technology Innovation, Washington, D.C.
35. Vani M, Prasad PRC (2020) Assessment of spatio-temporal changes in land use and land cover, urban sprawl, and land surface temperature in and around Vijayawada city, India. Environ Dev Sustain 22(4):3079–3095
36. Wang Y, Zhuang G, Zhang X, Huang K, Xu C, Tang A, Chen J, An Z (2006) The ion chemistry, seasonal cycle, and sources of PM2.5 and TSP aerosol in Shanghai. Atmos Environ 40(16):2935–2952
37. Watson JG, Cooper JA, Huntzicker JJ (1984) The effective variance weighting for least squares calculations applied to the mass balance receptor model. Atmos Environ 18(7):1347–1355
38. Yadav S, Tripathi SN, Rupakheti M (2022) Current status of source apportionment of ambient aerosols in India. Atmos Environ 274
39. Yuan CS, Wong KW, Tseng YL, Ceng JH, Lee CE, Lin C (2022) Chemical significance and source apportionment of fine particles (PM2.5) in an industrial port area in East Asia. Atmos Pollut Res 13(4)
40. Zhang J, Zhou X, Wang Z, Yang L, Wang J, Wang W (2018) Trace elements in PM2.5 in Shandong Province: source identification and health risk assessment. Sci Total Environ 621:558–577

A Review of Problems and Directions in Creating a Region-Specific Emission Inventory for the Agricultural Sector

Sri Harsha Kota and Thirunagari Baby Keerthi

Abstract The environment and human health are threatened globally by air pollution. Emission inventories are a useful tool for comprehending the sources of emissions and their overall effects on the environment. Most cities and nations do not include entire agricultural operations (fertilizer application, crop burning, equipment use, pesticides, livestock management, and tillage) when developing an emission inventory for the agricultural sector. Additionally, there hasn't been enough work done to improve our understanding of agricultural emissions' effects on the ecosystem. Due to these factors, this study analyses information from current emission inventories for diverse agricultural operations and evidence from other research on agricultural emissions. After a careful literature review, the lack of high-resolution geographical and temporal data, the absence of information regarding activity data, and the unavailability of region-specific emission factors have been identified as the primary challenges in producing a comprehensive emission inventory for the agricultural sector.

Keywords Emission inventory · Agricultural emissions · Tempo-spatial distribution · Particulate matter

1 Introduction

Air pollution has a detrimental impact on both human health and the environment. It is well established that populations who are exposed to elevated amounts of air pollution experience more illnesses, greater reliance on healthcare services, and premature mortality [22]. A recent assessment of the global burden of disease (GBD) has ranked air pollution among the top 10 global killers, and the 6th leading cause of mortality in South Asia. In addition to negative effects on human health, air pollution comes at a high economic cost. According to the World Health Organisation (WHO), "Air

S. Harsha Kota (✉) · T. Baby Keerthi
Department of Civil Engineering, Indian Institute of Technology, Delhi, New Delhi, India
e-mail: harshakota@iitd.ac.in

S. M. Shiva Nagendra et al. (eds.), *Select Proceedings of the 8th Indian International Conference on Air Quality Management (IICAQM 2023)*, Lecture Notes in Civil Engineering 582, https://doi.org/10.1007/978-981-96-2359-4_27

pollution is a concern for wellbeing in all nations, but it disproportionately affects individuals in low- and middle-income nations." The Organization for Economic Cooperation and Development (OECD) predicts that by 2050, particulate matter will be the significant environmental factor responsible for premature deaths worldwide. Exposure to PM with a diameter < 2.5 μm (i.e., $PM_{2.5}$) can result in fatal conditions such as stroke, respiratory issues, lung cancer, heart disease, etc. Many high-income countries including New Zealand, Australia, United States have $PM_{2.5}$ concentrations approx. Within the World Health Organisation's (WHO) allowable range (i.e.,10 μg/m [annual mean]), whereas $PM_{2.5}$ levels in India, central- and northern Africa, etc. are greater than advised [38]. The main sources of primary $PM_{2.5}$ include field burning of agricultural wastes, burning of fossil fuels (oil, coal, etc.), and biomass (wood, charcoal, etc.), etc. Secondary $PM_{2.5}$ is formed when Airborne NH3 reacts with combustion pollutants (such as oxides of nitrogen and sulphates) from industrial processes, power stations, and automobiles resulting in the formation of microscopic solid particles ($PM_{2.5}$), or aerosols. Secondary $PM_{2.5}$ is largely produced by several sectors of the economy, including agriculture, industry, power plants, residential, and transportation. About 40% of world emissions come from livestock, 16% from mineral fertilizers, 18% from burning biomass, and from agricultural wastes. According to a recent study, agricultural emissions dominate all other human sources of fine PM across a large portion of the United States, Europe, Russia, and China.

A database of the quantity of pollutants discharged into the atmosphere is called an emission inventory. The primary method for reporting emissions is through national emission inventories. They must be transparent, comprehensive, precise, comparative, and consistent. The reporting methodology, emission variables, and familiarity with specific sources all have an impact on the accuracy of emission inventories. Agriculture-related atmospheric pollutants are a significant contributor to the quality of the local air. However, prior research on agricultural sources has only concentrated on livestock, fertilizer, or farmland residue burning; a comprehensive agricultural emission inventory identifying critical agricultural operations like the use of farming machinery, tillage, pesticide application, etc. is still lacking, which would limit our capacity to comprehend, anticipate, and regulate the effects of agricultural emissions on human health and air quality. Global emission inventories that consider the actual statistical data from different sources, such as livestock management, fertilizer application, etc., are currently lacking comprehensive bottom-up emission levels from the agricultural sector, particularly for South Asia. For instance, Fig. 1 depicts India's particulate matter emissions from the agriculture sector in 2015 from various global emission inventories such as EDGAR, SMoG, and ECLIPSE. The agriculture sector's emissions are significantly different among these three global emission inventories, which suggests that agricultural emissions are reported inconsistently in various global inventories. To lessen the variability in the modeled emissions of $PM_{2.5}$ and its precursors across several nations, an updated emission inventory based on source activity is crucial for various nations [31].

Figures from Table 1 clearly show the variation in emissions from various inventories.

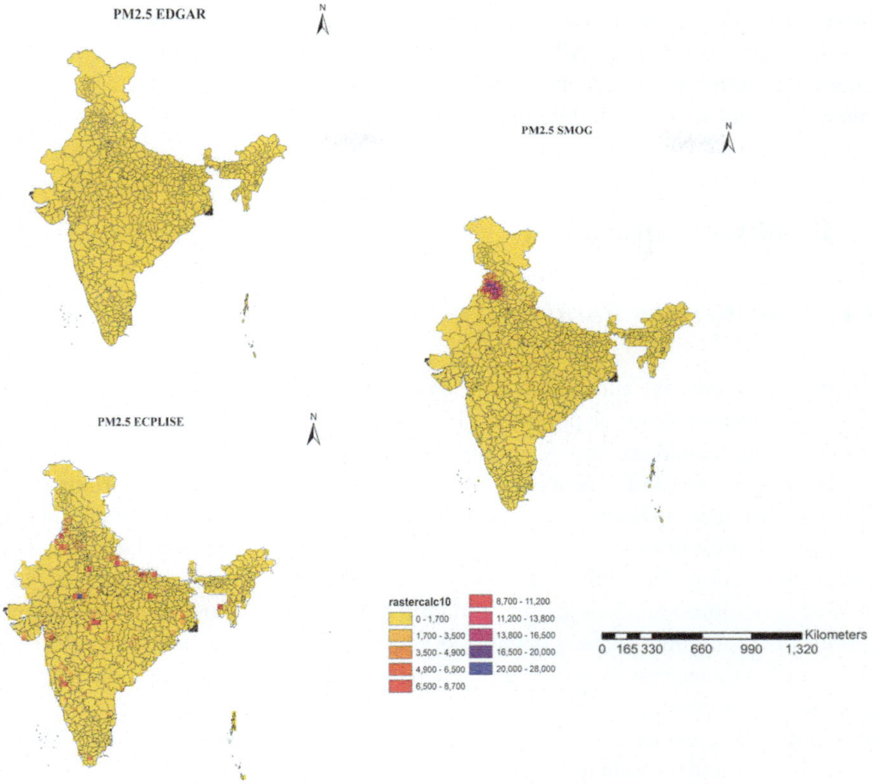

Fig. 1 Emission maps for India showing emissions agricultural sector in various emission inventories

Table 1 $PM_{2.5}$ emissions from Indian agricultural sector for the year 2015 considering three global emission inventories

Inventory	EDGAR	SMoG	ECLIPSE
$PM_{2.5}$ emissions (million t/y)	7.58	1.05	0.56

This article attempts to provide a summary of the major issues surrounding agricultural sector emissions inventories. It draws attention to the necessity of creating an exhaustive emission inventory for the agricultural sector considering various activities such as crop residue burning, agriculture machinery, tillage, fertilizer application, pesticide application, and manure management and points out the places where the accuracy of the emission inventory must be enhanced. This helps in a better understanding of the impact of this source on the environment and aids in the development of suitable mitigation measures to lessen the impact of these emissions from the agricultural sector on the environment and human health. For all these reasons, relevant data from the current emission inventories for various agricultural operations as well as information from many other investigations on agricultural emissions are examined in this research. After a thorough examination of the literature, the key issues with developing a comprehensive region-specific agriculture sector emission inventory that considers all the activities are identified and reviewed.

2 Results and Discussion

2.1 Crop Residue Burning

Apart from industrial and automobile emissions, burning biomass is among the primary contributors to air pollution [17]. According to Akagi et al. [1], biomass burning is the second greatest generator of gaseous pollutants in the troposphere, but it is the major source of primary micro carbon-containing particles. Approximately 730 Tg of biomass is burnt yearly in Asia both from natural and man-made causes, comprising forest fires, grassland fires, and field burning of agricultural residue, with India contributing 18% of this total [34]. According to Venkataraman et al. [39], 18–30% of agricultural residues are burnt in farm fields. The combustion of agricultural residues is prohibited in many developed nations. However, because of inadequate crop residue management, it is still practiced in many developing nations. Burning crop wastes is commonly done for a variety of reasons, including weed eradication, nitrogen rejuvenation in croplands, the preparation of farms for the following season's crop, and the management of agricultural pests and diseases. Major regional variations exist in the quantity of agricultural residues burned in fields, and the crop type is the key determinant. Pollutant emissions and their distribution vary with the seasons, weather, and kind of agro-residues [21]. The cloud microphysics, atmospheric chemistry, and balance in radiation are all significantly influenced by emission levels from burning crop residue in the lower atmosphere. As a result, there is a need to reduce air emissions from burning agricultural residues since they pose a major risk to health [10]. However, many countries' air quality control systems continue to mostly ignore this substantial source of pollution. A suitable air quality standards policy and international collaboration to properly regulate these emissions would be sparked by an accurate estimate of the air pollution caused by burning agricultural crop residue. However, due to issues with uncertainty in emission factors and information on burning activity, there is currently no comprehensive emission inventory for this source. The current EIs for Asia either don't report this open burning emission or offer a rough geographical estimate of it [44]. Several researchers have calculated the emissions of multiple species through burning agricultural residue utilizing IPCC parameters, but they have mainly covered a small number of gaseous pollutants (Nitrous oxide, methane, nitrogen oxides, and Sulphur dioxide), or from a particular location and crop [4]. Due to the absence of local measurements, the preliminary studies mostly on emission inventory from biomass burning in various countries [34] or in specific areas [43] were typically considered emission factors which are developed for foreign nations. However, this method could significantly increase absolute uncertainty in emission estimations due to regional variations in crop kinds and combustion circumstances. High geographical and temporal resolution emission inventories for biomass burning are required.

2.2 Agriculture Machinery Use

In several nations, non-road mobile machinery (NRMM), agricultural-based equipment, has grown to be a major contributor to air pollution. Non-road mobile equipment emits about 25 and 15% of all mobile equipment's particulates [16] and oxides of nitrogen (NO_x) emissions, respectively, in Europe [24]. These emissions account for around three-fourths and one-quarter, respectively, of pollutants from mobile machinery in the US. In 2019 Non-Road vehicular emissions released 0.24 million t of PM, 4.9 million t of NOx, and 0.435 million t of hydrocarbons (HC), based on the annual report of China Mobile Source Environmental Management (2020) [28]. 0.2 million tons of HC, 1.6974 million tons of NOx, and 0.091 million tons of PM were released by agricultural machinery, making up 47.4%, 34.3%, and 37.9% of all emissions, respectively [24]. These pollutants have a negative effect on human health and contribute to global warming. Even though NRMM contributes significantly to overall emissions, a much more comprehensive strategy for collecting and analyzing data is lacking to properly comprehend the health and environmental dangers posed by emission levels NRMM. The statistics demonstrate notable discrepancies among NRMM emission fractions and energy usage. In most nations, NRMM emissions of CO or HC are 2.5–6 times greater than those of motor vehicles. Additionally, emissions of NOx might be up to several times higher in areas where the ratio of NRMM to motorized vehicles is 1.5–3.5 [27]. The worst-case scenario involves PM emissions, which are 5.8 times greater on average in NRMM than in road cars. Germany, which is about 12%, has the highest ratio [27]. It is necessary to create emissions inventories with high-quality data to comprehend and minimize NRMM emissions. However, most nations do not operate in this manner. First off, there isn't a complete list of NRMM. The initial stage in generating an emission inventory is to establish and maintain a reliable repository of NRMM. Unfortunately, a lot of nations lack any kind of NRMM inventory of emissions, and those that have tend to base their inventories primarily on conjecture and scant to no data [42]. Only regulatory bodies can create an emission inventory since a more systematic methodology is required. Second, information on NRMM activity fills in the gap between the quantity of NRMM components and their emission factors. The research indicates that the majority of NO_X and PM NRMM emissions are produced by the constructions, agricultural, and forest machinery, making them the most crucial organizations from which to get this information [14]. To ensure that accurate data are acquired for the inventory, it would be necessary to limit data collection to trained authorities to address the issue of validity [15]. Furthermore, the accuracy of the emission factors for varied working situations, particularly during transient equipment activities, should be given specific attention because they are crucial in estimating the overall NRMM emissions [36]. It is possible to avoid overestimating or underestimating a portion of or all NRMM emissions by using more precise emission data. While the information on fuel usage can be effectively gathered through surveys, more testing and research are needed to determine the emission sources. Finally, the investigations stated above revealed that emission variables evaluated in the course of work under actual conditions differ greatly from those recorded under laboratory conditions [36]. The total comprehension of the impact on both the environment and equipment operators would be improved with more reliable statistics on emission variables recorded while working in actual situations. Therefore, more research is required, particularly on innovative techniques for gathering or evaluating the emission components. The quantity of work would be reduced by using a method that could combine the emission factors for various NRMM models because of the broad range of engine models and fieldwork settings. The lack of information is the underlying cause of all the issues.

2.3 *Tillage*

A prominent source of environmental particulate matter [16] emissions is agricultural activity [18]. According to studies, the main causes of PM are wind erosion, farm tilling, fertilization, the use of pesticides crop harvest, processing of grains, residue burning, and livestock feeding. Secondary aerosols are created in heterogeneous or homogeneous reactivity with gas antecedents, such as ammonia, chemicals, and volatile organic compounds (VOCs), produced from these operations [30]. Additionally, other field procedures like slurry application, pesticide use, unpaved traffic congestion, and fuel combustion by agricultural equipment may also add to PM pollution [2]. The two biggest anthropogenic contributions to farming PM emissions in North America and Europe have been recognized as soil tillage and harvesting of crops [9]. Particularly when mechanical disruptions of farmland, earth fragments with a diameter of less than 10 μm may disintegrate and get suspended in the atmosphere (i.e., PM_{10}). These particles are primarily made up of soil minerals and organic dust produced during fieldwork procedures. Although these activities last only a few days, the number of particulate emissions may be many magnitudes more than what would result from crops being eroded by wind [16]. With comparable cultivation techniques and soil silt percentage distribution, PM emissions connected to soil tillage and agricultural harvest are typically highest in wheat fields. Earlier studies have shown that extreme mechanical disruption (i.e., disking and ploughing), soils with high silt percentage, poor soil moisture, and high wind speed increase the escaping of soil particles to the environment [40]. More particles should escape from wheat fields with smaller root zones and less crop residue. Now, inventories of particulate emissions at national or regional level from farming activities majorly focus on Western nations like USA, Europe, and Canada. In addition to information on the cultivated lands, crop cycles, EFs from field practices, and the proportion of conventional tillage, bottom-up estimations based on emission factors out of these activities are the major approaches for inventory development (CARB 2003). The secret to regional inventory is selecting appropriate EF values. However, these levels are uncommon right now, and they vary a lot amongst various farming regions [11]. Even with the same field activities, variations in soil characteristics, moisture conditions, planting and harvesting procedures raise the uncertainty of EF estimates by several orders of magnitude. According to European Environment Agency (EMEP & Guidebook) standards, using regional EF estimates for inventories is the best strategy for gathering pertinent data. In order to improve the precision of inventory and estimating methods, it is highly beneficial to determine the EF in the major agricultural regions across the world.

2.4 Fertilizer Application

It's critical to anticipate ambient ammonia levels accurately for a variety of reasons. Ammonia is a contributor to secondary aerosols along with gases like NO_x, SO_x, and hydrocarbons. It can interact with the other precursors to generate chemical species like ammonium nitrate and ammonium sulfate. These species make up a large portion of $PM_{2.5}$, which has been associated with negative health impacts. Ammonia leads to the deposition of acids and thereby causes eutrophication [35]. In addition, studies indicate the possibility of ammonia leading to global warming [8]. An accurate emission inventory is necessary for forecasting ammonia concentrations. Animal husbandry is the main source of ammonia in the current inventories [3]. In the most recent inventories, fertilizer application is cited as a significant source of emissions, frequently ranking second. 90% of anthropogenic ammonia emissions in European nations are attributed to animal manure and fertilizer, according to research compiled by Battye et al. [6], and they predict that this is also the case in the US. In the UK, figures of fertilizer's contribution to overall emissions vary from 3 to 21% [35], but estimates for the United States are 7% [33] and 9.5% [6]. Its estimated worldwide contribution is 17% [12]. Undoubtedly, a significant amount of the overall ammonia emissions come from fertilizer, and this category must be carefully considered in order to create an accurate total inventory. Numerous ammonia inventories that take fertilizers into account have been released. These inventories, however, have a variety of shortcomings. Firstly, most of the inventories in use today were created with insufficient spatial resolution, frequently providing a single figure for an entire state. A country-level inventory exaggerates emissions in non-agricultural regions and tends to underestimate emissions from fields since the application of fertilizers is more intensive in agricultural lands. Even inventories that record emissions at the provincial level are insufficient to be used with the most recent atmospheric chemistry models, which employ grid cells with a side length of only 2–4 km. Second, reported inventories sometimes only provide annual average data, which results in insufficient temporal resolution. Seasonal fertilizer use is associated with sowing in the spring and harvesting in the fall. Most ammonia emissions are emitted within the first few days after application [20], hence emissions of ammonia vary seasonally. With a temporal resolution of one year, summertime and wintertime emissions are overestimated while spring and fall emissions are underestimated. Third, inaccurate activity levels are frequently used in published inventories. Numerous activity levels are dependent on fertilizer sales to distributors as opposed to farmers. The dealers might then sell the fertilizer to farms elsewhere, leading to an incorrect distribution pattern of levels of activity. In the absence of real application data, purchases from distributors to farmers provide a more accurate approximation. The differences in emission factors from various types of fertilizer are also not considered by some inventories. For instance, only anhydrous ammonia emissions were included in the National Acid Precipitation Assessment Program (NAPAP) inventory [6]. The main fertilizer kinds and their corresponding emission factors must be taken into consideration to accurately predict ammonia emissions.

2.5 *Livestock Waste*

Livestock manure and fertilizer are the two main sources of ammonia emissions; they account for over 57% of emissions globally more than 80% in Asia [8, 34]. First, they frequently apply uniform, European-based EFs across the board. Studies conducted by Kirk and Nye [23] and the European Environment Agency [13], found that soil pH and ambient temperature have a significant impact on ammonia volatilization, particularly from the application of nitrogen fertilizers and manure management systems. Ammonia volatilization is highly influenced by temperature which fluctuates seasonally. Consequently, a temporal resolution of one year has a tendency to underestimate in the summer and overestimate in the winter. The present inventories are primarily created at coarse spatial resolutions, such as the province level [41], and the 30 min $\times$ 30 min grid [34]. A more thorough inventory is recommended for a modeling analysis of the NH_3 cycles as well as for air quality modeling.

2.6 *Pesticide Application*

Since plant protection products, also known as pesticides, may have harmful effects on human health and the environment, numerous investigations have been done globally to evaluate these effects [19]. The absence of reliable statistical estimates, which is linked to the scarcity of relevant information on pesticide application and, subsequently, emission information, is a significant issue in measuring these impacts [32]. Most European nations do not adequately disclose pesticide use and the statistics that are available typically only show country-level amounts of pesticide Active Substances (AS) used for the major crops without any further spatial breakdown. Many of the active ingredients in commercialized pesticide compositions have a prolonged residence period in various environmental matrices. Relatively longer transport is a critical element for the distribution of these compounds into the atmosphere. They have been labeled as persistent organic pollutants (POPs). Setting up precise inventories of the pertinent air emissions, taking into consideration the compounds' short- and long-distance dispersion, is necessary to comprehend the impacts of pesticides on the environment and human health. Up to 30–50% of the herbicide sprayed on a crop could be lost through evaporation [7]. This could happen because of spray drift, volatilization while being applied, and volatilization through plants and soil after being applied. Organochlorines were the focus of most of the research on pesticide emission inventories that were published [5]. The above inventories accounted for volatilization from soil and plants but does not include drift emissions [25]. Also calculated are the yearly emission factors of a-HCH and b-HCH through sprinkling and tillage, respectively. On a worldwide scale for the years 1980 and 1990, annual emissions to the environment were estimated using these variables with a $1^\circ \times 1^\circ$ latitude and longitude resolution. To quantify both pesticide emissions and residues, a model called the Simplified Gridded Pesticide Emission and Residue Model (SGPERM) was created [26]. A grid with a $0.25^\circ \times 0.33^\circ$ precision was used to calculate emissions to air from 1994 to 2004. Gridded emissions of toxaphene and soil residue inventories for the USA were also made by Li et al. [25]. Although this insecticide was prohibited in 1982, the fact that it was still present in the soil in 2000 emphasizes the significance of these inventories. Bailey [5] determined the global HCB emissions for the middle of the 1990s from a variety of sources, including the usage of pesticides. For the years 1970–1995 in Europe, Pacyna et al. [29] established an emission inventory for HCB. They calculated historical HCB emissions at the national scale using default emission coefficients and national consumption information. Though 80% of the overall HCB emissions come from agriculture, this inventory also includes emissions from many other sources (solvents, production processes, waste treatment, and disposal). A more recent study by van der Gon et al. [37] revealed national emissions of POPs (such as HCB and HCH) and eight potential POPs (such as dicofol and endosulfan, two well-known pesticides used on diverse crops) across UNECE-Europe in 2000. For research assessing the risks to the environment and to human health, it is crucial to create a pesticide emission inventory that contains pesticides from all chemical classes.

3 Conclusion

Following a thorough examination of the literature, it has been determined that airborne particulate matter from agricultural operations exacerbates local, regional, and global air pollution. The release of atmospheric PM through agricultural activities can be understood, predicted, and controlled by using thorough emission evaluation for important crops. It was observed that farmland emission levels and their effects on the environment were not given as much significance as they should have been in previous studies, which tended to concentrate only on straw burning, livestock waste, or fertilizer application at a macro scale. Additionally, it was noted that there is a global research gap in this field when considering different activity levels. First, an analysis of the current situation with these emissions and the corresponding emission inventories is a major requirement at present. The key issues observed in developing an accurate emission inventory considering various activities in the agricultural sector are reported below:

1. There are several issues with livestock waste inventory and fertilizer application. A single number is usually provided for an entire state in most of the inventories still in use because they were produced with insufficient geographical resolution. Since the application of fertilizers is more intensive on agricultural grounds, a country-level inventory tends to overestimate emissions in non-agricultural areas while underestimating emissions from agricultural fields. Insufficient temporal resolution occurs when reported inventories only offer annual average data. Few inventories also do not consider the variations in emission components from different types of fertilizers.
2. Most emission factors for tillage are available for Europe, Canada, and the United States. These emission variables are not appropriate for use in different regions since emissions from tillage activity vary by several orders of magnitude even with identical field operations, soil properties, moisture content, planting techniques, and harvesting methods. Selection of proper EF values is the key to regional inventory
3. In the case of pesticides, many countries do not sufficiently report pesticide use, and the figures that are normally available only show country-level levels of pesticide Active Substances (AS) used for the principal crops without providing any further spatial breakdown. Additionally, the majority of inventory developed for the application of pesticides only considers volatilization from plants and soils, ignoring the drift emissions
4. For the NRMM emission inventory, the absence of information on the complete list of equipment, fuel use, etc. is a major issue in developing emission inventory. Additionally, the emission factors differ significantly between the field and laboratory.

Finally, the lack of high-resolution geographical and temporal data, the absence of information regarding activity data, and the unavailability of region-specific emission components have been identified as the primary challenges in producing a comprehensive emission inventory for the agricultural sector. All these issues must be considered while preparing a comprehensive emission inventory for the agricultural sector for better air quality management and for policymakers to better plan mitigation efforts.

References

1. Akagi S, Yokelson RJ, Wiedinmyer C, Alvarado M, Reid J, Karl T, Crounse D, Wennberg P (2011). Emission factors for open and domestic biomass burning for use in atmospheric models. Atmos Chem Phys 11(9):4039–4072
2. Aneja VP, Schlesinger WH, Erisman JW (2009) Effects of agriculture upon the air quality and climate: research, policy, and regulations. ACS Publications
3. Asman WA (1992) Ammonia emission in Europe: updated emission and emission variations Report 228471008. Bilthoven, The Netherlands7 National Institute of Public Health and Environmental Protection (RIVM)
4. Badarinath K, Chand TK, Prasad VK (2006) Agriculture crop residue burning in the Indo-Gangetic Plains—a study using IRS-P6 AWiFS satellite data. Curr Sci 1085–1089
5. Bailey RE (2001) Global hexachlorobenzene emissions. Chemosphere 43(2):167–182
6. Battye R, Battye W, Overcash C, Fudge S (1994) Development and selection of ammonia emission factors
7. Berg F, Kubiak R, Benjey W, Majewski M, Yates S, Reeves G, Smelt JH, Van der Linden A (1999). Emission of pesticides into the air. In: Fate of pesticides in the atmosphere: implications for environmental risk assessment. Springer, Berlin, pp 195–218
8. Bouwman A, Lee D, Asman WA, Dentener F, Van Der Hoek K, Olivier J (1997) A global high-resolution emission inventory for ammonia. Global Biogeochem Cycles 11(4):561–587
9. Cassel T, Trzepla-Nabaglo K, Flocchini R (2003) PM10 emission factors for harvest and tillage of row crops. In: 12th International emission inventory conference, San Diego, CA
10. Chandra BP, Sinha V (2016) Contribution of post-harvest agricultural paddy residue fires in the NW Indo-Gangetic Plain to ambient carcinogenic benzenoids, toxic isocyanic acid and carbon monoxide. Environ Int 88:187–197
11. Chen W (2015) Research progress in atmospheric particulate matter emissions from agricultural tillage. J Agro-Environ Sci 34(7):1225–1232
12. Ed B, Hans F, Willem A (1987) Anthropogenic NH3 emissions in europe. Atmos Environ 21(5):1009–1022
13. EEA, E (2009) EEA air pollutant emission inventory guidebook. European Environment Agency, Technical 28 Report No. 9/2009
14. EMEP, E., & Guidebook, E. A. P. E. I. (2019) European Union: Brussels. In: Belgium
15. Frey HC, Bammi S (2003) Probabilistic nonroad mobile source emission factors. J Environ Eng 129(2):162–168
16. Goossens D, Riksen M (2004) Wind erosion and dust dynamics: observations, simulations. Wageningen University, Modelling
17. Gurjar BR, Ravindra K, Nagpure AS (2016) Air pollution trends over Indian megacities and their local-to-global implications. Atmos Environ 142:475–495
18. Hinz T (2007) Particulate matter in and from agriculture. Bundesforschungsanstalt für Landwirtschaft (FAL)
19. Humbert S, Margni M, Charles R, Salazar OMT, Quirós AL, Jolliet O (2007) Toxicity assessment of the main pesticides used in Costa Rica. Agr Ecosyst Environ 118(1–4):183–190
20. Jarvis, S., & Pain, B. (1990). Ammonia volatilization from agricultural land. Proceedings-Fertiliser Society,
21. Kaskaoutis D, Kumar S, Sharma D, Singh RP, Kharol S, Sharma M, Singh AK, Singh D (2014) Effects of crop residue burning on aerosol properties, plume characteristics, and long-range transport over northern India. J Geophys Res Atmos 119(9):5424–5444
22. Kaushik C, Ravindra K, Yadav K, Mehta S, Haritash A (2006) Assessment of ambient air quality in urban centres of Haryana (India) in relation to different anthropogenic activities and health risks. Environ Monit Assess 122(1):27–40
23. Kirk G, Nye P (1991) A model of ammonia volatilization from applied urea. V. The effects of steady-state drainage and evaporation. J Soil Sci 42(1):103–113
24. Kubsh J (2017) Managing emissions from non-road vehicles. ICCT Consultant Report

25. Li Y, Bidleman T, Barrie L (2001) Toxaphene in the United States: 2. Emissions and residues. J Geophys Res Atmos 106(D16):17929–17938
26. Li Y, Venkatesh S, Li D (2005) Modeling global emissions and residues of pesticides. Environ Model Assess 9(4):237–243
27. Lončarević Š, Ilinčić P, Šagi G, Lulić Z (2022) Problems and directions in creating a national non-road mobile machinery emission inventory: a critical review. Sustainability 14(6):3471
28. Mee C (2014) Ministry of Ecology and Environment of the People's Republic of China. Announcement on releasing the guidelines for disassembly operation and production management of waste electrical and electronic products. 2015 edition. http://www.mee.gov.cn/gkml/sthjbgw/qt/201412/t20141219_293233.htm
29. Pacyna JM, Breivik K, Münch J, Fudala J (2003) European atmospheric emissions of selected persistent organic pollutants, 1970–1995. Atmos Environ 37:119–131
30. Pattey E, Qiu G, Fiset S, Ho E, MacDonald D, Liang C (2015) Primary particulate matter emissions and trends from Canadian agriculture. J. Wit Trans Ecol Environ 198:143–154
31. Pawar PV, Ghude SD, Jena C, Móring A, Sutton MA, Kulkarni S et al (2021) Analysis of atmospheric ammonia over South and East Asia based on the MOZART-4 model and its comparison with satellite and surface observations. Atmos Chem Phys 21(8):6389–6409
32. Pistocchi A, Sarigiannis D, Vizcaino P (2010) Spatially explicit multimedia fate models for pollutants in Europe: state of the art and perspectives. Sci Total Environ 408(18):3817–3830
33. Roe S, Strait R (1998) Methods for improving national ammonia emission estimates. Emiss Invent Living Global Environ 1:394–405
34. Streets DG, Bond T, Carmichael G, Fernandes S, Fu Q, He D, Klimont Z, Nelson SM, Wang MQ, Woo J-H, Yarber KF (2003) An inventory of gaseous and primary aerosol emissions in Asia in the year 2000. J Geophys Res Atmos 108(D21)
35. Sutton M, Place C, Eager M, Fowler D, Smith R (1995) Assessment of the magnitude of ammonia emissions in the United Kingdom. Atmos Environ 29(12):1393–1411
36. Tan D, Tan J, Peng D, Fu M, Zhang H, Yin H, Ding Y (2021) Study on real-world power-based emission factors from typical construction machinery. Sci Total Environ 799:149436
37. van der Gon HD, van het Bolscher M, Visschedijk A, Zandveld P (2007) Emissions of persistent organic pollutants and eight candidate POPs from UNECE–Europe in 2000, 2010 and 2020 and the emission reduction resulting from the implementation of the UNECE POP protocol. Atmos Environ 41(40):9245–9261
38. Van Donkelaar A, Martin RV, Brauer M, Kahn R, Levy R, Verduzco C, Villeneuve PJ (2010) Global estimates of ambient fine particulate matter concentrations from satellite-based aerosol optical depth: development and application. Environ Health Perspect 118(6):847–855
39. Venkataraman C, Habib G, Kadamba D, Shrivastava M, Leon JF, Crouzille B, Boucher O, Streets D (2006) Emissions from open biomass burning in India: Integrating the inventory approach with high-resolution Moderate Resolution Imaging Spectroradiometer (MODIS) active-fire and land cover data. Global Biogeochem Cycles 20(2)
40. Wang J, Miller DR, Sammis TW, Hiscox AL, Yang W, Holmén BA (2010) Local dust emission factors for agricultural tilling operations. Soil Sci 175(4):194–200
41. Wang W, Lu X, Pang Y, Tang D, Zhang W (1997) Geographical distribution of NH_3 emission intensities in China. Acta Sci Circum 17(1):2–7
42. Zhang G, Sandanayake M, Setunge S, Li C, Fang J (2017) Selection of emission factor standards for estimating emissions from diesel construction equipment in building construction in the Australian context. J Environ Manage 187:527–536
43. Zhang H, Hu D, Chen J, Ye X, Wang SX, Hao JM, Wang L, Zhang R, An Z (2011) Particle size distribution and polycyclic aromatic hydrocarbons emissions from agricultural crop residue burning. Environ Sci Technol 45(13):5477–5482
44. Zhou Y, Xing X, Lang J, Chen D, Cheng S, Wei L, Wei X, Liu C (2017) A comprehensive biomass burning emission inventory with high spatial and temporal resolution in China. Atmos Chem Phys 17(4):2839–2864

Assessment of Crop Yield Loss Due to Surface Ozone Pollution Over Delhi NCR and Indo-Gangetic Plain

Pooja Saroj, Chirashree Ghosh, Siddhartha Singh, and Vijay Kumar Soni

Abstract Economic damage due to surface ozone-induced crop losses affects the Indian economy and food security. Relative Yield Losses (RYL) for wheat and rice crop have been calculated using ozone exposure indices AOT 40 and M7 which is based on observational data of Delhi, Punjab, Haryana, and Uttar Pradesh for the period 2017–2020. RYL based on AOT40 indices have been estimated as 6–9% for rice and 1–12% for wheat while RYL have been estimated as 0.7–3% for rice and 1–6% for wheat using M7 indices. Economic damage due to surface ozone for rice production is more than wheat production in comparison with the results from other studies which may be due to increasing crop prices and more production area. Urgent requirement for surface ozone monitoring in agricultural areas is needed to avoid food security problems in India.

Keywords Surface ozone · IGP · Economy · Crop loss · Relative yield loss · AOT40 · M7

1 Introduction

In India, agriculture is essential for the provision of food, nutrition, and livelihoods. It provides gainful employment for about two-thirds of the workforce and contributes significantly to India's Gross Domestic Product (GDP). Several industries rely on agricultural output for the raw ingredients they need. The country's economy benefits from the agriculture sector's expansion because of its tight ties to other economic sectors. Surface ozone (O_3), which is one of the major air pollutants worldwide, has adverse impacts on natural vegetation and crops [6, 11, 12, 22]. O_3, which is harmful to life on earth's surface, is primarily formed by the photochemical reactions of

P. Saroj · C. Ghosh (✉)
Department of Environmental Studies, University of Delhi, New Delhi, India
e-mail: chirashreeghosh.63@gmail.com

S. Singh · V. K. Soni
India Meteorological Department, New Delhi, India

S. M. Shiva Nagendra et al. (eds.), *Select Proceedings of the 8th Indian International Conference on Air Quality Management (IICAQM 2023)*, Lecture Notes in Civil Engineering 582, https://doi.org/10.1007/978-981-96-2359-4_28

air pollutants released by anthropogenic and natural processes. Numerous Volatile Organic Compounds (VOCs), Carbon Monoxide (CO), and Nitrogen oxide (NOx) are the main air pollutants released into the atmosphere. When exposed to sunlight, these chemicals react to form ozone in the troposphere. Anthropogenic activities like burning biomass and fossil fuels are the main causes of these air pollutants emissions [24, 29]. Urban areas are the major suppliers of ozone precursor gases but rural areas in the downwind could also be affected by ozone, and the balance of VOCs and NOx levels over the rural areas may favor the higher formation of ozone than over the urban areas [8]. It has been suggested that ozone-induced crop production decline could affect food security and cause economic losses [5, 33]. This is a growing concern, particularly in areas like eastern and southern Asia where growing economies have led to higher ozone precursor emissions [1, 4, 34]. With the second-largest population after China, India is a fast-developing nation, and a sizable amount of its economy is dependent on agriculture. Agriculture plays a vital role in India's economy. 54.6% of the total workforce is engaged in agricultural and allied sector activities [7] and accounts for 17.8% of the country's Gross Value Added (GVA) for the years 2019–20 at current prices. Crop yields have a far greater overall impact on the economy since rural demand is intimately correlated with the year's crop yield for a wide range of consumer goods. Numerous studies have used global models to try to estimate crop output losses based on the AOT40 measure over India [5, 10, 13, 33]. Ghude et al. [13] studied the first district-scale high-resolution model study for India to examine and quantify ozone-induced crop losses and discussed important mitigating strategies. They also found that wheat is the most impacted crop with losses of 3.5 ± 0.8 Million tons (Mt), followed by rice at 2.1 ± 0.8 Mt with the losses concentrated in central and north India. Sharma et al. [27] review the ozone-attributable Crop Yield Loss (CYL) using the WRF-Chem model, employing a configuration that has previously been proven to better represent the observed ozone changes over the different chemical environments in India.

Flux-based measures have been created in Europe [15], but they require more weather data and details on the state of the plants. At present, flux-response functions have not been parameterized globally and are available for very few crops [21, 25].

Several metrics have been developed based on findings from the field and open-top chamber experiments to evaluate the effects of O_3 pollution on agricultural yields. M7 or M12 is the average O_3 concentration in the growing season during daylight hours and AOT40 or SUM06 is the accumulated O3 concentration during daylight hours that exceeds certain thresholds during the growing season [19, 20]. Ozone takes into account how plants can detoxify low quantities of oxygen, with crops only experiencing yield losses when the hourly O_3 concentration surpasses 40 (AOT40) or 60 (SUM06) ppb. Ainsworth [3], widely utilized these matrices to assess the global yield losses caused by current O_3 concentrations, which are estimated to be between 2–16 percent for the four main staple crops, i.e., wheat, rice, maize, and soybean. These metrics are simple to calculate, but they often neglect the active reactions of plant eco-physiology to ambient climatic and environmental changes. As a result, they are probably insufficient for analyzing yield losses in a future climate and atmospheric environment [23].

In this study, real-time ground observational data of surface ozone pollutants monitored from Continuous Ambient Air Quality Monitoring Stations (CAAQMS) on the Northwest Indo-Gangetic Plain have been used. The study assesses ozone-related crop yield losses for wheat and rice in the Delhi NCR as well as the North-west Indo-Gangetic Plain region, i.e., the food bowl of India, spread across the states of Punjab, Haryana and Varanasi, Uttar Pradesh for the period 2017–2021. Firstly, the calculation of the cumulative ozone exposure indices AOT40 and the concentration-based Metric indices of ozone M7 have been made.

Also, estimated the Relative Yield Losses and the crop losses for the two major crops, i.e. wheat and rice using the exposure–response functions [20]. Lastly, the calculation of economic losses has been made using crop production losses and the minimum support prices for the crops.

2 Methodology

Yield losses for 3 months of the Rabi wheat (January–March) and Kharif rice (August-October) using ozone exposure indices AOT 40 (accumulated exposure over a threshold of 40 ppb) and M7 (mean daytime surface ozone) using observational data at Delhi, Punjab, Haryana and Varanasi, Uttar Pradesh have been estimated for the period 2017–2020. Data from 9 monitoring stations near the agricultural sites in Delhi have been studied. Rapid urbanization and the growth of trade and industry have reduced the significance of the agriculture sector in Delhi's National Capital Region. Delhi's agricultural sector may not contribute much to India's crop production but being the capital of India, it affects the socio-economic development of this region. Also, one monitoring location has been selected in Punjab for this study namely Punjab Agricultural University, Ludhiana and for Haryana State air quality monitoring station located at Maharshi Dayanand University, Haryana has been chosen due to the non-availability of data from the other monitoring sites. In Uttar Pradesh, there are four monitoring stations in Varanasi and surface ozone data of all these monitoring stations have been analyzed in this study. Short data gaps have been filled in by interpolating the ozone concentrations before and after the data gap. For longer data gaps, the averaged data for consecutive years have been used.

To calculate ozone exposure and its effects on crops, we used two different ozone exposure matrices. Cumulative exposure metrics AOT40 as well as one mean exposure metric are included M7. They are defined as:

$$\mathrm{AOT40} = \sum_{i=1}^{n}([\mathrm{O}_3]_i] - 40]) \tag{1}$$

$$\mathrm{M7} = \frac{1}{n}\sum_{i=1}^{n}[\mathrm{O}_3]_i \tag{2}$$

AOT40 and M7 are in units of ppb. M7 is defined similarly to M12 but uses a 7-h (09:00–15:59) daytime window instead of 12-h for summation.

When comparing these characteristics, M7 gives all data equal weight and accounts for yield losses caused by ozone concentrations below 40 ppb, while AOT40 gives additional weight to high ozone concentrations [32]. Therefore, M7 will be more effective in assessing plant damage and yield losses at low ozone concentrations, whereas the AOT40 will better capture the impact of occurrences with very high O_3 concentrations on plant physiology and yields [14].

2.1 Calculation of Relative Crop Damage and Economic Losses

Semi-empirical equations given in Table 1 have been used for calculating the relative yield for wheat and rice based on AOT40 and M7 data collected from the peer-reviewed literature.

Further, the Relative Yield (RY) is used for calculating the Relative Yield Loss (RYL) and Crop production damage. RYL is defined as the crop yield reduction from the theoretical yield that would have resulted without ozone-induced damages [5]. RYL is given as:

$$\mathrm{RYL}_i = 1.0 - \mathrm{RY}_i \tag{3}$$

$$\mathrm{CPL}_i = \frac{\mathrm{RYL}_i}{(1 - \mathrm{RYL}_i)} * \mathrm{CP}_i \tag{4}$$

The following equations have been considered for calculating the Crop Production Losses (CPL_i) from the Crop Productions (CP_i) and Relative Yield Losses (RYL_i). Here, '*i*' signify the year for which it is calculated. The data of Crop Production (CP) for the study area have been taken from the Directorate of Economics and Statistics, Department of Agriculture and Farmers Welfare, Ministry of Agriculture and Farmers Welfare and the Government of India (https://eands.dacnet.nic.in/).

Economic Cost Loss or Economic Loss (ECL) is defined as the financial loss due to O_3-induced damages in a financial year for any crop. The minimum EL is calculated using the crop production loss (CPL) and the Minimum Support Prices

Table 1 Statistical relationships between relative yield and ozone exposure metrics

O_3 Metric	Exposure-yield relationship	References
AOT40 (Rice)	$RY = -0.0039 \times AOT40 + 0.94$	[20]
M7 (Rice)	$RY = \exp^{-(M7/202)2.47}/\exp^{-(25/202)2.47}$	[2]
AOT40 (Wheat)	$RY = -0.0000161 \times AOT40 + 0.99$	[20]
M7 (Wheat)	$RY = \exp^{-(M7/137)2.34}/\exp^{-(25/137)2.34}$	[19]

(MSPs) for the same fiscal year through the following equation. Here 'i' stands for the year of calculation.

$$\mathrm{ECL}_i = \mathrm{CPL}_i \times (\mathrm{MSP}_i) \quad (5)$$

MSP is defined as the fixed price at which farmers sell their crops to the government. MSP of wheat and rice for respective years were taken from the database of the Directorate of Economics and Statistics, Department of Agriculture and Farmers Welfare, Ministry of Agriculture and Farmers Welfare and Government of India (https://eands.dacnet.nic.in/).

2.2 Study Area

India's agricultural output is highly dependent on the weather and any changes in weather patterns have a significant impact on agricultural production. Northwest Indo-Gangetic Plain (IGP), the food bowl of India, spread across the states of Punjab, Haryana, Delhi and Varanasi, Uttar Pradesh was selected for the study of agricultural vulnerability to surface ozone pollution (Fig. 1). With major crops including wheat and rice, the Indo-Gangetic Plains are the most intensively farmed region in the world. In this region, additional crops like maize, sugarcane, and cotton are also farmed. The IGP is one of the world's most densely populated areas because of its agriculturally productive soil. In terms of vegetation, the IGP is an ecologically homogeneous area. IGP is surrounded by subtropical to moderate temperate climates, which are distinguished by chilly, dry winters and warm, wet summers [31].

In the IGP, rice is usually grown during the wet summer season called kharif (May to November) and wheat during the dry winter season called rabi (November to April).

3 Review of Other Studies

The details of crop loss estimate due to AOT 40 and M7 with other studies over the Indian region in the last decades are shown in Table 2.

4 Result and Discussion

Figure 2 shows the monthly 24-h average surface ozone concentrations for the period 2017–2021 over Delhi, Punjab, Haryana, and Varanasi using observational data.

At Delhi, maximum monthly averaged surface ozone values have been found as 62 ppb in April 2017, 32 ppb in April 2018, 35 ppb in May 2019, 31 ppb in May

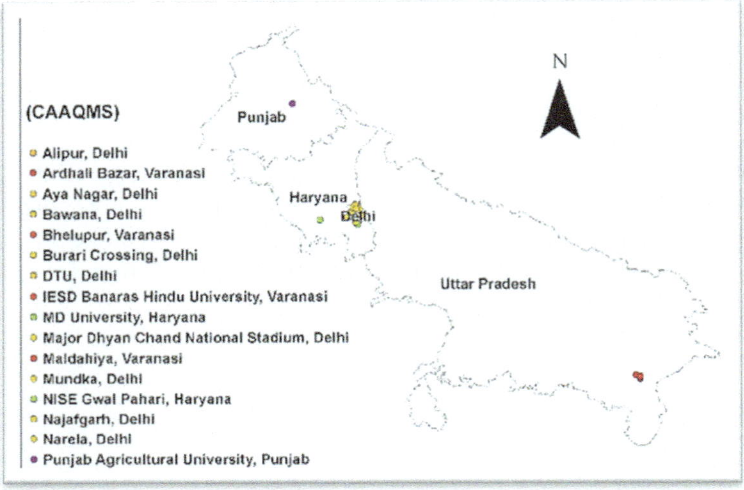

Fig. 1 Sites of surface ozone monitoring stations

Table 2 Summary of estimated economic loss with other studies for India

References	Ozone exposure matrices used	Findings
[27]	AOT40	Estimate Economic damage due to CYL using WRF-Chem model and found over Indian region and found ~5 billion USD for wheat and 1.5 billion USD for rice
[18]	AOT40 & M7	Estimates of the economic loss using surface ozone observational from 17 sites covering different parts of India found that an annual loss of 4.0–14.2 million tons (4.2–15.0%) for wheat and 0.3–6.7 million ton (0.3–6.3%) for rice
[28]	AOT40 & M7	Estimate CPL for wheat, rice, cotton and maize for Punjab and Haryana for the years 2011–2013 using high-quality observational data at Mohali, Punjab. They calculated CPL for wheat amounted to 20.8 ± 10.4, 10.3 ± 4.7 million t and rice totaled 5.4 ± 1.2 and 3.2 ± 0.8 million t in the fiscal year of 2012–2013 and 2013–2014 and, respectively, for Punjab and Haryana took together
[13]	AOT40	Estimate economic damage due to CYL using the WRF-Chem model at a 0.5° resolution for wheat, soybeans, cotton, and rice and found that Wheat is the most impacted crop with losses of 3.5 ± 0.8 million tons (Mt), followed by rice at 2.1 ± 0.8 Mt, with the losses concentrated in central and north India
[10]	AOT40 and M7	Calculate economic damage using surface ozone observational data for the period 2002–2007 covering different parts of India and found that economic loss due to winter wheat accounts for 1,222 to 4,091 million US$ and rice worth 86–276 million

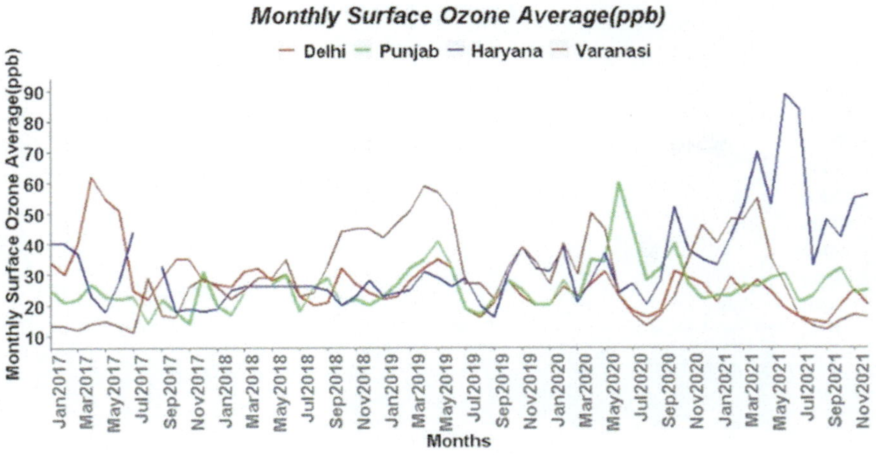

Fig. 2 Monthly averaged surface ozone concentration

and October 2020 and 29 ppb in February 2020. Over Delhi, surface ozone usually peaks in April and May due to strong insolation which promotes the photochemical formation of O_3 and the same pattern of surface ozone can be found in Kunchala et al. [17].

At Punjab, monthly maximum surface ozone values have been found as 27 ppb in April 2017, 30 ppb in June 2018, 41 ppb in October 2019, 60 ppb in May and October 2020 and 32 ppb in October 2021.

At Haryana, monthly maximum surface ozone values have been found as 44 ppb in July 2017, 26 ppb in May and August 2018, 39 ppb in November 2019, 31 ppb in February 2020 and 89 ppb in June, 2021. The summer season in Punjab and Haryana has experienced the highest ozone levels due to the high temperature, low humidity, and intense solar radiation. Also, the post-monsoon months have the highest values due to the large-scale post-harvesting burning emissions of ozone precursors and a lower boundary layer [28].

At Uttar Pradesh, maximum surface ozone values have been found as 29 ppb in August and December 2017, 45 ppb in November and December 2018, 59 ppb in April 2019, 50 ppb in April 2020 and 55 ppb in April 2021. Surface ozone variability over Uttar Pradesh has been similar to Delhi, Punjab, and Haryana showing maximum values in summer and post-monsoon season.

Figure 3 shows the monthly AOT40 values for the period 2017–2021 over Delhi, Punjab, Haryana, and Uttar Pradesh using observational data.

At Delhi, monthly maximum AOT40 values have been found as 11,086 ppb in April 2017, 5082 ppb in June, 2018, 4616 ppb in May 2019, 4601 ppb in May 2020 and 1594 ppb in May 2021.

At Punjab, monthly maximum AOT40 values have been found as 3832 ppb in December 2017, 2432 ppb in April 2018, 6432 ppb in May 2019, 9510 ppb in June 2020 and 2873 ppb in October 2021.

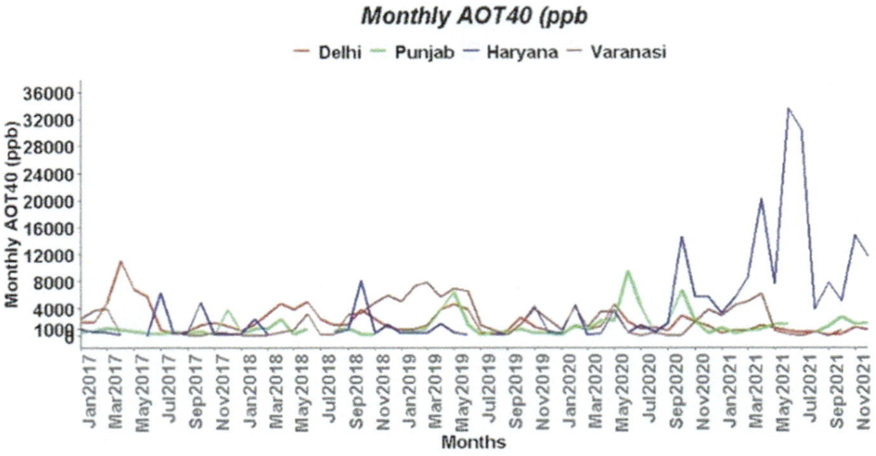

Fig. 3 Monthly accumulated ozone over a threshold of 40 ppb

At Haryana, monthly maximum AOT40 values have been found as 6242 ppb in July 2017, 8063 ppb in October 2018, 4317 ppb in November 2019, 14,486 ppb in October 2020 and 33,590 ppb in June 2021.

At Uttar Pradesh, monthly maximum AOT40 values have been found as 3990 ppb in March 2017, 5891 ppb in December 2018, 7826 ppb in March 2019, 4273 ppb in February 2020 and 6191 ppb in April 2021.AOT40 values are in a range from 10 ppb to 35,000 ppb over Delhi, Punjab, Haryana, and Varanasi for the period 2017–2021.

Figure 4 shows the monthly M7 ozone exposure indices values for the period 2017–2021 over Delhi, Punjab, Haryana, and Uttar Pradesh using observational data. At Delhi, monthly maximum M7 values have been found as 72 ppb in May 2017, 47 ppb in April 2018, 51 ppb in May 2019 and 43 ppb in May 2020 and 38 ppb in April 2021.

At Punjab, monthly maximum M7 values have been found as 50 ppb in December 2017, 41 ppb in April year 2018, 59 ppb in May 2019, 66 ppb in October 2020 and 48 ppb in October 2021.

At Haryana, monthly maximum M7 values have been found as 53 ppb in December 2017, 75 ppb in June 2018, 73 ppb in April 2019, 77 ppb in December 2020 and 76 ppb in January 2021.

At Uttar Pradesh, monthly maximum M7 values have been found as 41 ppb in October 2017, 78 ppb in November 2018, 81 ppb in May 2019, 56 ppb in February 2020 and 51 ppb in April 2021. Except for M7 values, which exhibit a slightly stronger change, the monthly average M7 values resemble the monthly average surface ozone values. M7 indices show maxima during summer (May and June) and post-monsoon (October and November) and minima during the monsoon (July to September) and winter (December to February), however, the difference between the cumulative metric (AOT40), which gives higher weight to high values and low or no weight to low values, and the average-based metric (M7) comes out very clearly.

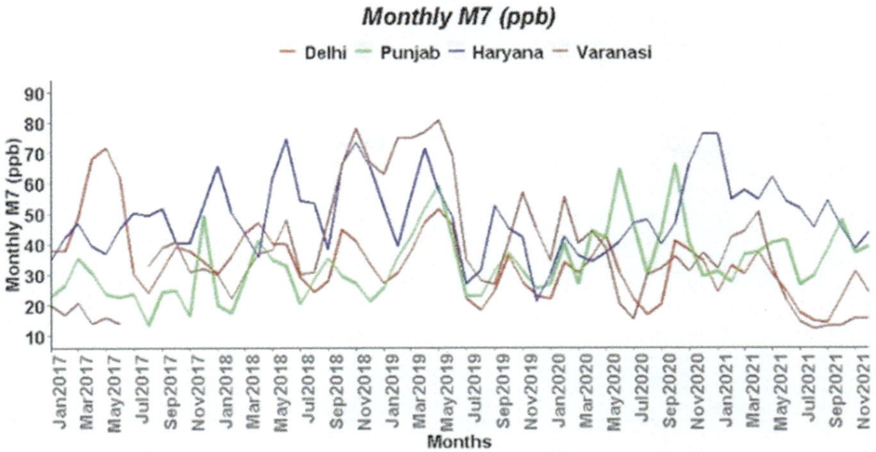

Fig. 4 Line graph of monthly M7 Ozone exposure indices using observational data

4.1 *Scatterplot of Monthly 24 h Average Surface Ozone Versus Monthly AOT40 and M7*

Figure 5 shows the scatterplot of monthly 24 h average surface ozone versus monthly AOT40 using monitored data at Delhi, Punjab, Haryana, and Uttar Pradesh from the year 2017 to 2020. Scatterplot show polynomial fit well for Delhi = 0.78, Punjab = 0.82, Haryana = 0.92 and Uttar Pradesh = 0.74 between monthly 24 h average O_3 versus monthly AOT40. Using observational data, Sinha et al. [28] and Lal et al. [18] also demonstrated that AOT40 fits well with the polynomial curve for northern India.

Figure 6 shows the scatterplot of the monthly 24 h average O_3 versus monthly M7 using observational data at Delhi, Punjab, Haryana, and Uttar Pradesh from the year 2017–2020. Scatterplot show polynomial fit well for Delhi = 0.91, Punjab = 0.8, Haryana = 0.83, and Uttar Pradesh = 0.72 between monthly 24 h average O_3 versus monthly M7. To demonstrate the association between monthly averaged O_3 and M7 data in several other regions of India using observational data, Debaje [10], Lal et al. [18] and Sinha et al. [28] used linear equations.

4.2 *Assessment of the Effect of O_3 on Crop Yield During 2017–2020*

For the wheat and rice crops, for the corresponding crop seasons of January to March (Rabi season) and August to October (Kharif season), AOT40 and M7 values have been calculated. Also, using Eq. (3), RYL has been determined based on this data. Using the Crop Production (CP) for wheat and rice crops, the Crop Production Loss

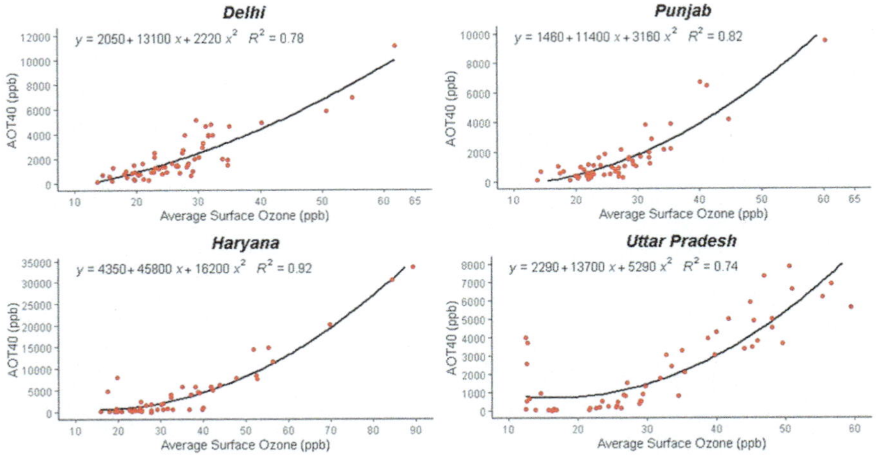

Fig. 5 Scatterplot using polynomial between monthly 24 h average surface ozone versus monthly AOT40 using observational data

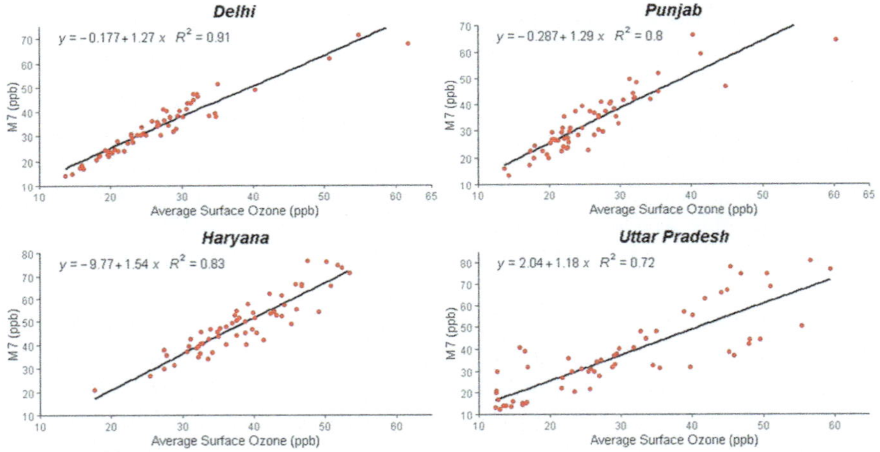

Fig. 6 Scatterplot using linear equation between monthly 24 h average surface ozone versus monthly M7 using observational data

(CPL) has been calculated in Eq. (4). In the end, by using CPL data and MSP, the economic damage has been calculated.

4.2.1 RYL and CPL of Rice Due to AOT40 and M7

Table 3 shows the ozone exposure indices values, i.e., AOT 40 and M7 for the period 2017–2020. Table 4 shows the distribution of rice crop production during the study period.

For AOT40, the average RYL was 6.5% in Delhi, 6.5% in Punjab, 7.07% in Haryana, and 6.35% in Uttar Pradesh, respectively. For M7, the average RYL was 0.27% in Delhi, 0.7% in Punjab, 2.3% in Haryana, and 1% in Uttar Pradesh, respectively. AOT40 shows the highest crop loss density in Delhi, Punjab, Haryana, and Uttar Pradesh in comparison with M7. Tables 5 and 6 show the RYL (%) and ECL calculated using AOT40 and M7.

Figure 7 shows the Economic losses (Million INR) and Relative Yield Losses (RYL %) for rice derived by AOT40 ozone exposure indices for the period 2017–2020. Figure 8 shows the Economic losses (Million INR) and Relative Yield Losses (RYL %) for Rice derived by M7 ozone exposure indices for the period 2017–2020. Ozone levels are significantly lower during the rice-growing season in India following the summer monsoon than they are throughout the winter. This prevents the rice crop during the Kharif season from suffering serious ozone damage. As was previously indicated, around 85–90% of India's annual rice production is produced during the Kharif season. Using observational data from a single site, Sinha et al. [28] calculated a loss of rice of 19–21%, respectively, for Punjab and Haryana in northern India. Rice losses were estimated by Debaje [10] to be between 3 and 16% in three sites in Maharashtra and also six additional locations in other states. AOT40

Table 3 AOT40 and M7 during kharif season

State	Year	AOT40 (ppb)	M7 (ppb)
Delhi	2017	784	32
	2018	2311	32
	2019	1190	26
	2020	1652	26
Punjab	2017	600	27
	2018	764	31
	2019	708	30
	2020	2988	31
Haryana	2017	1865	47
	2018	3126	53
	2019	606	49
	2020	5646	45
Uttar Pradesh	2017	208	38
	2018	2206	48
	2019	869	32
	2020	212	33

Table 4 Distribution of rice crop production during kharif season

State	Year	Crop production (kg)	Area (ha)
Delhi	2017	2880	5854
	2018	2880	5848
	2019	3010	5840
	2020	290	6171
Punjab	2017	4740	258,000
	2018	4450	259,000
	2019	4680	259,000
	2020	4623	258,666
Haryana	2017	2630	48,900
	2018	2730	48,000
	2019	2450	67,000
	2020	2603	54,633
Uttar Pradesh	2017	2550	51,719
	2018	2530	4803
	2019	2660	47,540
	2020	2580	49,090

Table 5 RYL and ECL during kharif season calculated using AOT40

State	Year	AOT40 (RYL %)	MSP	ECL in million Indian Rupee (INR) (AOT40)
Delhi	2017	6.3	1550	17.58
	2018	6.9	1750	21.84
	2019	6.5	1815	22.04
	2020	6.6	1868	24.24
Punjab	2017	6.2	1550	1260
	2018	6.3	1750	1355
	2019	6.3	1815	1473
	2020	7.2	1868	1724
Haryana	2017	6.7	1550	144
	2018	7.2	1750	178
	2019	6.2	1815	198
	2020	8.2	1868	237
Uttar Pradesh	2017	6.1	1550	132
	2018	6.9	1750	157
	2019	6.3	1815	155
	2020	6.1	1868	153

Table 6 RYL and ECL during kharif season calculated using M7

State	Year	M7 (RYL %)	MSP	ECL in million Indian Rupee (INR) (M7)
Delhi	2017	0.4	1550	1.17
	2018	0.5	1750	1.5
	2019	0.1	1815	0.27
	2020	0.1	1868	0.22
Punjab	2017	0.1	1550	249
	2018	0.4	1750	920
	2019	0.4	1815	802
	2020	0.4	1868	918
Haryana	2017	2.2	1550	44
	2018	3.0	1750	70
	2019	2.4	1815	74
	2020	1.9	1868	50
Uttar Pradesh	2017	1.0	1550	35
	2018	2.3	1750	83
	2019	0.5	1815	17
	2020	0.5	1868	22

and M7 estimates for the study area are still much lower than both estimates but calculations are done with a wide network of monitoring sites.

For AOT40, cumulative economic losses for four states are 1553.58 in 2017, 1711.84 INR in 2018, 1848.04 in 2019, and 2138.24 in 2020 million INR. Also,

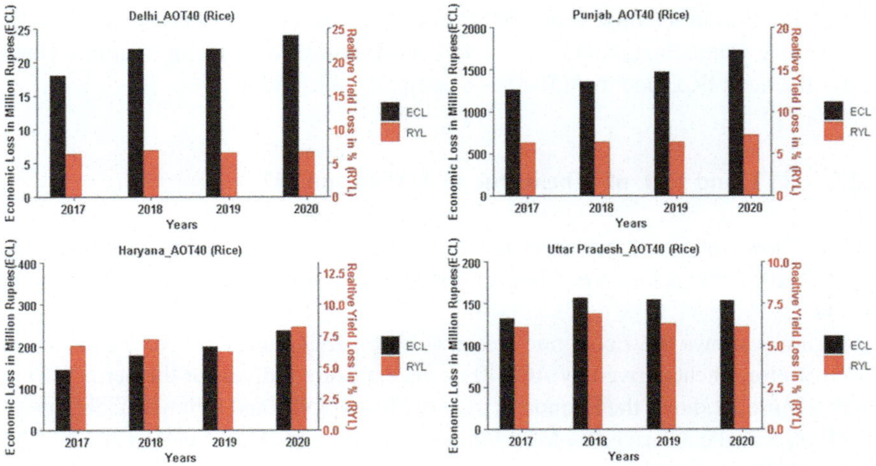

Fig. 7 Economic losses Million Indian Rupee (INR) and RYL (%) for rice derived by AOT40 ozone exposure indices

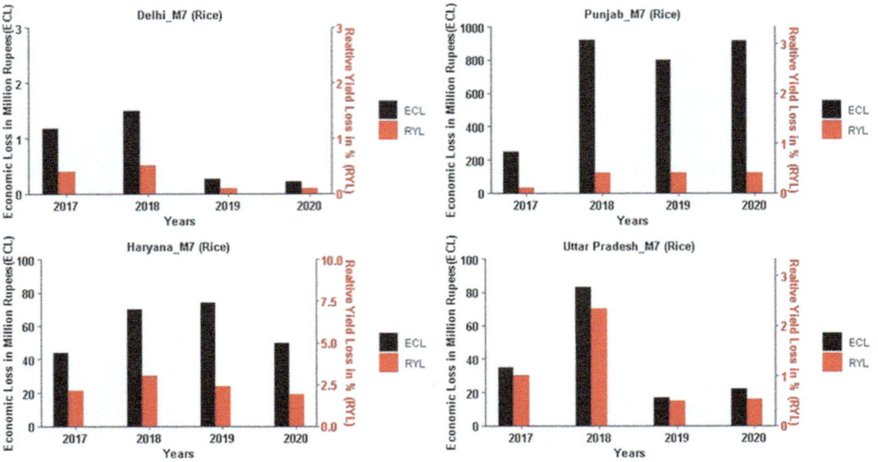

Fig. 8 Economic losses (Million INR) and RYL (%) for rice derived by M7 ozone exposure indices

for M7 cumulative economic losses for four states are 329.17 in 2017, 1074.5 in 2018, 893.27 in 2019, and 990.22 in 2020 in million INR. State-wise, cumulative economic losses due to AOT40 in the four consecutive years were 85.7 in Delhi, 5812 in Punjab, 757 in Haryana, and 597 in Uttar Pradesh million.

Also, State-wise, cumulative economic losses due to M7 in the four consecutive years were 3.16 in Delhi, 2889 in Punjab, 238 in Haryana, and 157 in Uttar Pradesh in million INR. Delhi, Punjab, Haryana, and Uttar Pradesh bear the most crop losses (CPL) due to AOT40. Also, Northwest Indo-Gangetic Plain states account for more than half of all crop production losses (CPLs) for rice caused by ozone exposure nationwide. This is because the region has higher crop production rates as well as elevated pollution loading. Using MSP, economic damage has been calculated for consecutive years. Punjab, Haryana, and Uttar Pradesh being the agricultural states suffer the most ECL due to AOT40 as a comparison to M7.

4.2.2 RYL and CPL of Wheat Due to AOT40 and M7

Table 7 shows the ozone exposure indices values, i.e., AOT 40 and M7 for the period 2017–2020. Table 8 shows the distribution of rice crop production during the study period.

Figure 9 shows the Economic losses (Million INR) and Relative Yield Losses (RYL %) for wheat derived by AOT40 ozone exposure indices for the period 2017–2020. Figure 10 shows the Economic losses (Million INR) and Relative Yield Losses (RYL %) for wheat derived by M7 ozone exposure indices for the period 2017–2020.

For AOT40, cumulative economic losses for four states are 1089 in 2017, 930 INR in 2018, 1356 in 2019, and 1251 in 2020 million INR. Also, for M7 cumulative economic losses for four states are 181 in 2017, 202 in 2018, 744 in 2019, and 686 in

Table 7 AOT40 and M7 during Rabi season

State	Year	AOT40 (ppb)	M7 (ppb)
Delhi	2017	2975	42
	2018	1892	37
	2019	1147	32
	2020	851	29
Punjab	2017	835	28
	2018	855	27
	2019	816	35
	2020	894	31
Haryana	2017	647	41
	2018	1018	53
	2019	352	49
	2020	1684	36
Uttar Pradesh	2017	3416	19
	2018	115	27
	2019	6716	35
	2020	2141	44

Table 8 Distribution of wheat crop production during Rabi season

State	Year	Crop production (kg)	Area (ha)
Delhi	2017	4440	19,350
	2018	4240	19,220
	2019	4310	19,220
	2020	4310	19,180
Punjab	2017	4740	258,000
	2018	4450	259,000
	2019	4680	259,000
	2020	4623	258,666
Haryana	2017	2630	48,900
	2018	2730	48,000
	2019	2450	67,000
	2020	2603	54,633
Uttar Pradesh	2017	3380	69,363
	2018	3500	68,973
	2019	3330	68,369
	2020	3403	68,901

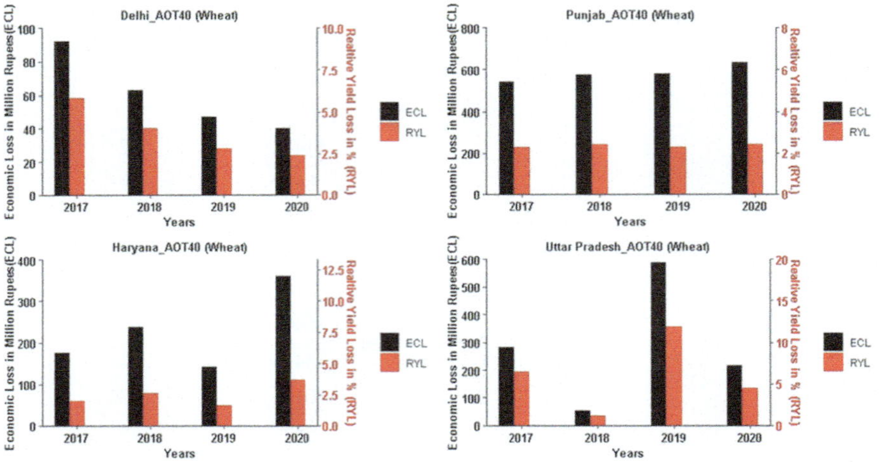

Fig. 9 Economic losses (Million INR) and RYL (%) for Wheat derived by AOT40 ozone exposure indices

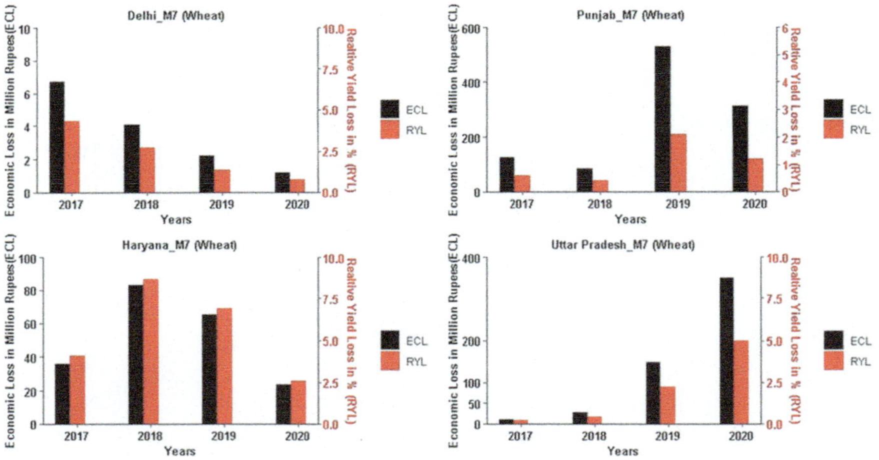

Fig. 10 Economic losses (Million INR) and RYL (%) for Wheat (Rabi) derived by M7 ozone exposure indices for the period 2017–2020

2020 in million INR. Tables 9 and 10 show the RYL (%) and ECL calculated using AOT40 and M7.

State-wise, cumulative economic losses were 241 in Delhi, 2333 in Punjab, 914 in Haryana, and 1138 in Uttar Pradesh million INR. Also, State-wise, cumulative economic losses due to M7 in the four consecutive years were 14 in Delhi, 1056 in Punjab, 208 in Haryana, and 535 in Uttar Pradesh in million INR. The computations in this paper contain some ambiguity, which could be due to errors in observational data such as instrumentation error and data unavailability.

Table 9 RYL and ECL during rabi season calculated using AOT40

State	Year	AOT40 (RYL %)	MSP	ECL in million INR (AOT40)
Delhi	2017	5.8	1735	91.6
	2018	4.0	1840	63.2
	2019	2.8	1925	46.7
	2020	2.4	1975	39.6
Punjab	2017	2.3	1735	540
	2018	2.4	1840	577
	2019	2.3	1925	581
	2020	2.4	1975	635
Haryana	2017	2.0	1735	175
	2018	2.6	1840	237
	2019	1.6	1925	141
	2020	3.7	1975	361
Uttar Pradesh	2017	6.4	1735	282.7
	2018	1.18	1840	53
	2019	11.8	1925	587
	2020	4.4	1975	215

Table 10 RYL and ECL during rabi season calculated using M7

State	Year	M7 (RYL%)	MSP	ECL in million INR (M7)
Delhi	2017	4.3	1735	6.7
	2018	2.7	1840	4.1
	2019	1.4	1925	2.2
	2020	0.8	1975	1.2
Punjab	2017	0.6	1735	128
	2018	0.4	1840	87
	2019	2.1	1925	529
	2020	1.2	1975	312
Haryana	2017	4.1	1735	36
	2018	8.7	1840	83
	2019	6.9	1925	65
	2020	2.6	1975	24
Uttar Pradesh	2017	0.2	1735	10
	2018	0.4	1840	28
	2019	2.2	1925	148
	2020	5	1975	349

However, we employed all the monitoring sites in Delhi near agricultural land, Punjab, Varanasi, and high data quality over MD Rhotak, Haryana which has demonstrated better ozone build-up over India. Furthermore, utilizing exposure–response functions generated from European cultivars may have produced conservative conclusions because Indian wheat and rice cultivars are equally or more sensitive to ozone than European cultivars. Some Asian crop cultivars are just as sensitive to ozone as their European and American counterparts, according to Sarkar and Agrawal [26] research.

5 Conclusion

In this study, an estimation of India's Economic damage to two major crops (Rabi for Wheat and Kharif for Rice due to surface ozone pollution has been made using observational data of surface ozone. For the estimation of crop yield loss, two ozone exposure indices (AOT40 and M7) have been used for the period 2017–2020. For AOT40, the average wheat and rice RYL were 3.75% and 6.5% in Delhi, 2.35% and 6.5% in Punjab, 2.4% and 7.07% in Haryana and ~ 5.94% and 6.35% in Uttar Pradesh, respectively.

For M7, the average wheat and rice RYL were 2.3% and 0.27% in Delhi, 1.07% and 0.7% in Punjab, 5.5% and 2.3%in Haryana and 1.95% and 1% in Uttar Pradesh and Delhi, 0.7% in Punjab, 2.3% in Haryana and 1% in Uttar Pradesh, respectively. For AOT40, the average wheat and rice ECL were 241 and 85.7 in Delhi, 2333 and 5812 in Punjab, 914 and 757 in Haryana and 1138 and 597 in Uttar Pradesh in million INR. For M7, the average wheat and rice ECL were 14 and 3.16 in Delhi, 1056 and 2889 in Punjab, 208 and 238 in Haryana and 535 and 157 in Uttar Pradesh in million INR. Economic damage due to surface ozone for rice production is more than wheat production in comparison with the results from other studies which may be due to increasing crop prices and more production area. During the rice-growing season, surface ozone is relatively higher and which may be because the study area receives more solar insolation and higher concentrations of ozone precursors associated with a stable atmosphere which occurs during the rice-growing season over Northwest India. The main staple crops for the Indian population are wheat and rice and rising surface ozone pollution poses a high risk to their production in India as suggested by M7 and AOT40 indices. To lessen the negative effects of surface ozone on human health and the economy, India should focus on enacting stricter air quality rules and reducing O_3 precursor emissions. According to Kumar et al. [16], the IGP belt, a significant agricultural region, would see the greatest increase in surface ozone levels in the coming years (2045–2044).To minimize the effects of O_3, adaptation to the current high pollution circumstances with tolerant plant cultivars is also required.

References

1. Adams RM, Crocker TD, Thanavibulchai N (1982) An economic assessment of air pollution damages to selected annual crops in Southern California. J Environ Econ Manage 9(1):42–58. https://doi.org/10.1016/0095-0696(82)90005-5
2. Adams RM, Glyer JD, Johnson SL, McCarl BA (1989). A Reassessment of the economic effects of ozone on U.S. Agriculture. JAPCA, 39(7):960–968. https://doi.org/10.1080/08940630.1989.10466583
3. Ainsworth EA (2017) Understanding and improving global crop response to ozone pollution. Plant J 90(5):886–897. https://doi.org/10.1111/TPJ.13298
4. Aunan K, Berntsen TK, Seip HM. Surface ozone in China and its possible impact on agricultural crop yields. AMBIO J Human Environ 29(6):294–301. Available from: https://doi.org/10.1579/0044-7447-29.6.294
5. Avnery S, Mauzerall DL, Liu J, Horowitz LW (2011) Global crop yield reductions due to surface ozone exposure: 2. Year 2030 potential crop production losses and economic damage under two scenarios of O_3 pollution. Atmos Environ 45(13):2297–2309. https://doi.org/10.1016/j.atmosenv.2011.01.002
6. Betzelberger AM, Yendrek CR, Sun J, Leisner CP, Nelson RL, Ort DR, Ainsworth EA (2012) Ozone exposure response for U.S. soybean cultivars: linear reductions in photosynthetic potential, biomass, and yield. Plant Physiol 160(4):1827–1839. https://doi.org/10.1104/PP.112.205591
7. CENSUS (2011) https://censusindia.gov.in/census.website/data/census-tables
8. Chand D, Lal S, Naja M (2003) Variations of ozone in the marine boundary layer over the Arabian Sea and the Indian Ocean during the 1998 and 1999 INDOEX campaigns. J Geophys Res Atmos 108(6). https://doi.org/10.1029/2001JD001589
9. Crutzen PJ (1995) Ozone in the troposphere. In: Singh HB (ed) Composition, chemistry, and climate of the atmosphere. Van Norstrand Reinhold, New York, pp 349–393. https://www.scirp.org/%28S%28351jmbntvnsjt1aadkposzje%29%29/reference/referencespapers.aspx?referenceid=1644670. Accessed 6 Sep 2022
10. Debaje SB (2014) Estimated crop yield losses due to surface ozone exposure and economic damage in India. Environ Sci Pollut Res 21(12):7329–7338. https://doi.org/10.1007/s11356-014-2657-6
11. Feng Z, De Marco A, Anav A, Gualtieri M, Sicard P, Tian H, Fornasier F, Tao F, Guo A, Paoletti E (2019) Economic losses due to ozone impacts on human health, forest productivity and crop yield across China. Environ Int 131. https://doi.org/10.1016/J.ENVINT.2019.104966
12. Feng Z, Hu E, Wang X, Jiang L, Liu X (2015) Ground-level O3 pollution and its impacts on food crops in China: a review. Environ Pollut 199:42–48. https://doi.org/10.1016/J.ENVPOL.2015.01.016
13. Ghude SD, Jena C, Chate DM, Beig G, Pfister GG, Kumar R, Ramanathan V (2014) Reductions in India's crop yield due to ozone. Geophys Res Lett 41(15):5685–5691. https://doi.org/10.1002/2014GL060930
14. Hollaway MJ, Arnold SR, Challinor AJ, Emberson LD (2012) Intercontinental trans-boundary contributions to ozone-induced crop yield losses in the Northern Hemisphere. Biogeosciences 9(1):271–292. https://doi.org/10.5194/BG-9-271-2012
15. Klingberg J, Engardt M, Karlsson PE, Langner J, Pleijel H (2014) Declining ozone exposure of European vegetation under climate change and reduced precursor emissions. Biogeosciences 11(19):5269–5283. https://doi.org/10.5194/BG-11-5269-2014
16. Kumar R, Barth MC, Pfister GG, Delle Monache L, Lamarque JF, Archer-Nicholls S, Tilmes S, Ghude SD, Wiedinmyer C, Naja M, Walters S (2018) How will air quality change in South Asia by 2050? J Geophys Res Atmos 123(3):1840–1864. https://doi.org/10.1002/2017JD027357
17. Kunchala RK, Singh BB, Karumuri RK, Attada R, Seelanki V, Kumar KN (2022) Understanding the spatiotemporal variability and trends of surface ozone over India. Environ Sci Pollut Res 29(4):6219–6236. https://doi.org/10.1007/s11356-021-16011-w

18. Lal S, Venkataramani S, Naja M, Kuniyal JC, Mandal TK, Bhuyan PK, Kumari KM, Tripathi SN, Sarkar U, Das T, Swamy YV, Gopal KR, Gadhavi H, Kumar MKS (2017) Loss of crop yields in India due to surface ozone: an estimation based on a network of observations. Environ Sci Pollut Res 24(26):20972–20981. https://doi.org/10.1007/s11356-017-9729-3
19. Lesser VM, Rawlings JO, Spruill SE, Somerville MC (1990) Ozone effects on agricultural crops: statistical methodologies and estimated dose-response relationships. Crop Sci 30(1):148–155. https://doi.org/10.2135/CROPSCI1990.0011183X003000010033X
20. Mills G, Buse A, Gimeno B, Bermejo V, Holland M, Emberson L, Pleijel H (2007) A synthesis of AOT40-based response functions and critical levels of ozone for agricultural and horticultural crops. Atmos Environ 41(12):2630–2643. https://doi.org/10.1016/j.atmosenv.2006.11.016
21. Mills G, Hayes F, Simpson D, Emberson L, Norris D, Harmens H, Büker P (2011) Evidence of widespread effects of ozone on crops and (semi-)natural vegetation in Europe (1990–2006) in relation to AOT40- and flux-based risk maps. Glob Chang Biol 17(1):592–613. https://doi.org/10.1111/J.1365-2486.2010.02217.X
22. Mills G, Sharps K, Simpson D, Pleijel H, Frei M, Burkey K, Emberson L, Uddling J, Broberg M, Feng Z, Kobayashi K, Agrawal M (2018) Closing the global ozone yield gap: quantification and cobenefits for multistress tolerance. Glob Chang Biol 24(10):4869–4893. https://doi.org/10.1111/GCB.14381
23. Musselman RC, Lefohn AS, Massman WJ, Heath RL (2006) A critical review and analysis of the use of exposure- and flux-based ozone indices for predicting vegetation effects. Atmos Environ 40(10):1869–1888. https://doi.org/10.1016/J.ATMOSENV.2005.10.064
24. Ohara T, Akimoto H, Kurokawa J, Horii N, Yamaji K, Yan X, Hayasaka T (2007) An Asian emission inventory of anthropogenic emission sources for the period 1980–2020. Atmos Chem Phys 7(16):4419–4444. https://doi.org/10.5194/ACP-7-4419-2007
25. Pleijel H, Danielsson H, Emberson L, Ashmore MR, Mills G (2007) Ozone risk assessment for agricultural crops in Europe: further development of stomatal flux and flux-response relationships for European wheat and potato. AtmEn 41(14):3022–3040. https://doi.org/10.1016/J.ATMOSENV.2006.12.002
26. Sarkar A, Agrawal SB (2010) Elevated ozone and two modern wheat cultivars: an assessment of dose dependent sensitivity with respect to growth, reproductive and yield parameters. Environ Exp Bot 69(3):328–337. https://doi.org/10.1016/J.ENVEXPBOT.2010.04.016
27. Sharma A, Ojha N, Pozzer A, Beig G, Gunthe SS (2019)Revisiting the crop yield loss in India attributable to ozone. Atmos Environ 1:100008. https://doi.org/10.1016/j.aeaoa.2019.100008
28. Sinha B, Singh Sangwan K, Maurya Y, Kumar V, Sarkar C, Chandra BP, Sinha V (2015) Assessment of crop yield losses in Punjab and Haryana using 2 years of continuous in situ ozone measurements. Atmos Chem Phys 15(16):9555–9576. https://doi.org/10.5194/acp-15-9555-2015
29. Streets DG, Canty T, Carmichael GR, De Foy B, Dickerson RR, Duncan BN, Edwards DP, Haynes JA, Henze DK, Houyoux MR, Jacob DJ, Krotkov NA, Lamsal LN, Liu Y, Lu Z, Martin RV, Pfister GG, Pinder RW, Salawitch RJ, Wecht KJ (2013) Emissions estimation from satellite retrievals: a review of current capability. Atmos Environ 77:1011–1042. https://doi.org/10.1016/J.ATMOSENV.2013.05.051
30. Tang H, Liu G, Han Y, Zhu J, Kobayashi K (2011) A system for free-air ozone concentration elevation with rice and wheat: control performance and ozone exposure regime. Atmos Environ 45(35):6276–6282. https://doi.org/10.1016/j.atmosenv.2011.08.059
31. Timsina J, Connor DJ (2001) Productivity and management of rice–wheat cropping systems: issues and challenges. F Crop Res 69(2):93–132. https://doi.org/10.1016/S0378-4290(00)00143-X
32. Tuovinen JP (2000) Assessing vegetation exposure to ozone: properties of the AOT40 index and modifications by deposition modelling. Environ Pollut 109(3):361–372. https://doi.org/10.1016/S0269-7491(00)00040-3
33. Van Dingenen R, Dentener FJ, Raes F, Krol MC, Emberson L, Cofala J (2009) The global impact of ozone on agricultural crop yields under current and future air quality legislation. Atmos Environ 43(3):604–618. https://doi.org/10.1016/J.ATMOSENV.2008.10.033

34. Wang X, Mauzerall DL (2004) Characterizing distributions of surface ozone and its impact on grain production in China, Japan and South Korea: 1990 and 2020. Atmos Environ 38(26):4383–4402. https://doi.org/10.1016/j.atmosenv.2004.03.067

Appendix A
Impact of Solid Waste Dump Yards Fire on Air Quality of National Capital Delhi

Monika Sharma[1,*], Mukesh Khare[2], Rajeev Kumar Mishra[1]
[1]Department of Environmental Engineering, Delhi Technological University, New Delhi, India
Email: sharma.monika1646@gmail.com
[2]Department of Civil Engineering, Indian Institute of Technology Delhi, New Delhi, India

See Figs. A.1 and A.2.

Monika Sharma[1], Mukesh Khare[2], Rajeev Kumar Mishra[1]

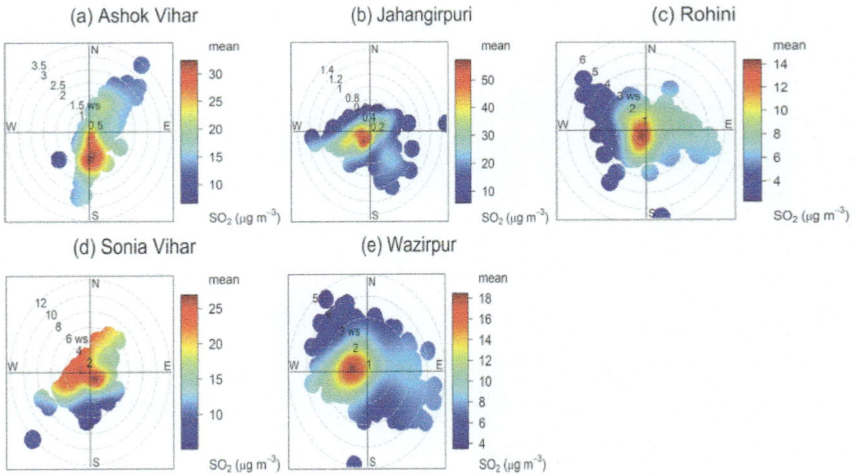

Fig. A.1 **a** Ashok Vihar, **b** Jahangirpuri, **c** Rohini, **d** Sonia Vihar, **e** Wazirpur

S. M. Shiva Nagendra et al. (eds.), *Select Proceedings of the 8th Indian International Conference on Air Quality Management (IICAQM 2023)*, Lecture Notes in Civil Engineering 582, https://doi.org/10.1007/978-981-96-2359-4

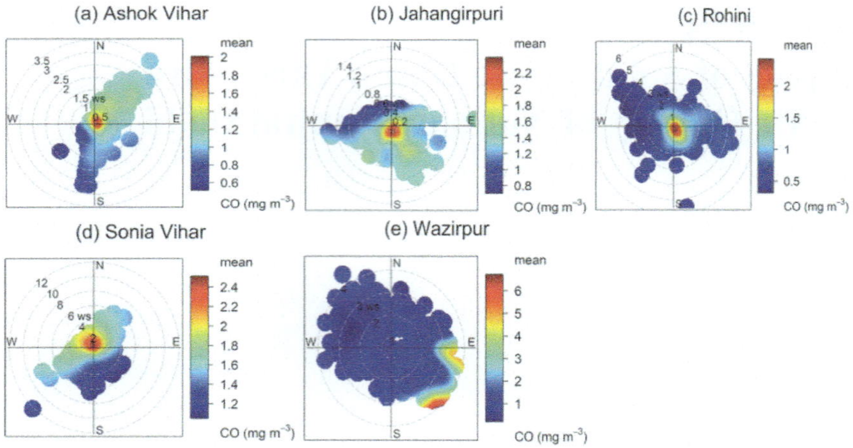

Fig. A.2 **a** Ashok Vihar, **b** Jahangirpuri, **c** Rohini, **d** Sonia Vihar, **e** Wazirpur

Appendix B
Seasonal and Temporal Distribution of Urban Fine Particulates and Associated Health Risk Analysis: A Long-Term Measurement (2018–2020) Analysis in Megacity Delhi

Vignesh Mohan[1], Rajeev Kumar Mishra[1,*], Vijay Kumar Soni[2]

[1]Department of Environmental Engineering, Delhi Technological University, New Delhi, India

Email: vikiviki93@gmail.com

[2]Indian Meteorological Department, Lodhi Road, New Delhi, India

See Tables B.1, B.2, B.3 and Figs. B.1 and B.2.

S. M. Shiva Nagendra et al. (eds.), *Select Proceedings of the 8th Indian International Conference on Air Quality Management (IICAQM 2023)*, Lecture Notes in Civil Engineering 582, https://doi.org/10.1007/978-981-96-2359-4

Table B.1 WHO default values, corresponding to mortality and morbidity with RR of 10 $\mu g\ m^{-3}$ increase of daily averages for $PM_{2.5}$, and PM_{10} and baseline incidence per 10^5

Pollutant	Mortality and morbidity	Relative risk (RR) per 10 $\mu g\ m^{-3}$ (95% CI)	Baseline incident (BI) (per 10^5)
$PM_{2.5}$ [long-term effect (10 $\mu g\ m^{-3}$)]	Natural mortality (30 + years)	1.062 (1.04–1.083)	1013
	ALRI mortality (0–5 years)	GBD 2015/2016 (integrated function 2016)	49
	COPD mortality (30 + years)	GBD 2015/2016 (integrated function 2016)	101
	LC mortality (30 + years)	GBD 2015/2016 (integrated function 2016)	132
	IHD mortality (25 + years)	GBD 2015/2016 (integrated function 2016)	436
	Stroke mortality (25 + years)	GBD 2015/2016 (integrated function 2016)	436
$PM_{2.5}$ [Short-term effect (20 $\mu g\ m^{-3}$)]	HA-RD	1.019 (0.9982–1.0402)	1260
	HA-CVD including stock	1.0091 (1.007–1.0166)	101
	Natural mortality (30 + years)	1.0123 (1.0045–1.0201)	1013

Note *ALRI* Acute lower respiratory infection, *COPD* Chronic obstructive pulmonary disease, *IHD* Ischaemic heart disease, *LC* Lung cancer, *HA-RD* Hospital admission Respiratory Diseases, and, *HA-CVD* Hospitalization for cardiovascular disease

Table B.2 The estimated number of mortality rate (95% confidence interval) due to $PM_{2.5}$ concentration exceeding WHO guideline level from AirQ + Software

Year	Cause of death	AirQ +
2018[a]	All cause	1,44,449[b] (1,13,702[c]: 1,62,891[d])
	ALRI	5,211 (4,047: 6,002)
	COPD	10,346 (7,536: 13,918)
	LC	13,007 (10,605: 14,929)
	IHD	46,055 (33,329: 61,528)
	Strock	42,540 (29,150: 55,730)
2019[a]	All cause	1,26,239 (95,668: 1,46,642)
	ALRI	4,993 (3,864: 5,834)
	COPD	9,678 (6,854: 12,920)
	LC	11,834 (9,277: 13,906)
	IHD	44,584 (31,032: 60,531)
	Strock	40,402 (26,360: 53,552)
2020[a]	All cause	97,398 (70,512: 1,17,520)
	ALRI	4,528 (3,504: 5,383)
	COPD	8,566 (5,784: 11,399)
	LC	9,995 (7,393: 12,231)
	IHD	41,838 (27,456: 57,715)
	Strock	36,731 (21,856: 48,558)
2018[e]	All cause	44,468 (17,664: 67,112)
	HA-RD	79,785 (0: 1,37,697)
	HA-CVD	3,392 (686: 5725)
2019[e]	All cause	34,732 (13,533: 53,380)
	HA-RD	63,301 (0: 1,14,226)
	HA-CVD	2,629 (522: 4,517)
2020[e]	All cause	23,15 (8,834: 36,309)
	HA-RD	42,938 (0: 81,524)
	HA-CVD	1,737 (338: 3,045)

[a] Long-term effect of $PM_{2.5}$. [b, c, d] Centre, Lower, and, Upper values, and [e] Short-term effect of $PM_{2.5}$

Note ALRI Acute lower respiratory infection, *COPD* Chronic obstructive pulmonary disease, *IHD* Ischaemic heart disease, *LC* Lung cancer, *HA-RD* Hospital admission Respiratory Diseases, and, *HA-CVD* Hospitalization for cardiovascular diseases

Table B.3 Summarized standard deviation and mean concentration for different criteria pollutants in each year

Year	Season	PM_{10}	$PM_{2.5}$	PM_1	NO	NO_2	NO_x	CO	O_3
2018	WIN	88.80[a] ± 258.48[b]	58.49 ± 184.26	36.56 ± 66.88	65.56 ± 68.20	13.65 ± 25.19	64.68 ± 97.36	0.94 ± 1.83	5.72 ± 21.71
	SUM	158.32 ± 221.73	90.42 ± 127.75	39.79 ± 42.71	40.71 ± 43.46	19.19 ± 34.91	48.35 ± 73.72	1.23 ± 1.755	10.09 ± 35.70
	MON	135.88 ± 183.16	92.32 ± 114.29	44.12 ± 45.82	21.22 ± 24.11	9.88 ± 22.10	23.34 ± 45.02	0.44 ± 1.11	8.24 ± 24.60
	P-MON	102.42 ± 250.28	55.72 ± 142.58	18.84 ± 46.72	61.20 ± 88.80	12.75 ± 20.71	70.22 ± 108.34	0.80 ± 1.24	7.10 ± 28.32
2019	WIN	115.61 ± 228.98	72.84 ± 141.84	36.75 ± 42.11	37.40 ± 31.24	16.10 ± 18.17	50.37 ± 48.40	1.14 ± 1.08	3.58 ± 22.33
	SUM	74.84 ± 170.97	36.22 ± 83.50	17.81 ± 33.93	30.05 ± 16.16	9.61 ± 7.20	36.54 ± 22.69	0.90 ± 1.01	9.11 ± 37.26
	MON	52.30 ± 84.64	24.76 ± 41.98	12.84 ± 14.28	7.09 ± 10.40	8.03 ± 14.66	13.50 ± 24.52	0.71 ± 1.33	13.53 ± 30.89
	P-MON	120.76 ± 213.83	72.85 ± 126.47	34.94 ± 52.30	31.25 ± 32.17	14.88 ± 25.11	45.55 ± 56.91	0.86 ± 1.82	7.66 ± 28.88
2020	WIN	76.72 ± 155.54	48.74 ± 98.60	25.31 ± 44.23	29.68 ± 28.35	14.02 ± 20.78	41.93 ± 48.86	1.05 ± 1.48	9.06 ± 25.01
	SUM	26.73 ± 74.70	17.28 ± 48.08	6.64 ± 18.58	86.79 ± 70.45	18.10 ± 18.51	94.29 ± 83.74	1.04 ± 0.88	10.55 ± 40.52
	MON	24.30 ± 69.47	14.41 ± 40.63	6.16 ± 16.69	53.48 ± 25.01	13.69 ± 11.61	56.16 ± 33.19	0.19 ± 0.46	12.20 ± 24.32
	P-MON	91.42 ± 192.30	59.12 ± 119.00	27.50 ± 52.83	10.00 ± 19.61	2.13 ± 10.20	10.78 ± 29.78	0.34 ± 0.86	3.56 ± 17.69

[a] Standard Deviation, [b] mean concentration

Note PM_{10}, $PM_{2.5}$, PM_1, NO, NO_2, NO_x, and O_3 unites are $\mu g\ m^{-3}$. The Unit of CO is mg m^{-3}. WIN, SUM, MON, and P-MON stand for Winter, Summer, Monsoon, and Post-Monsoon

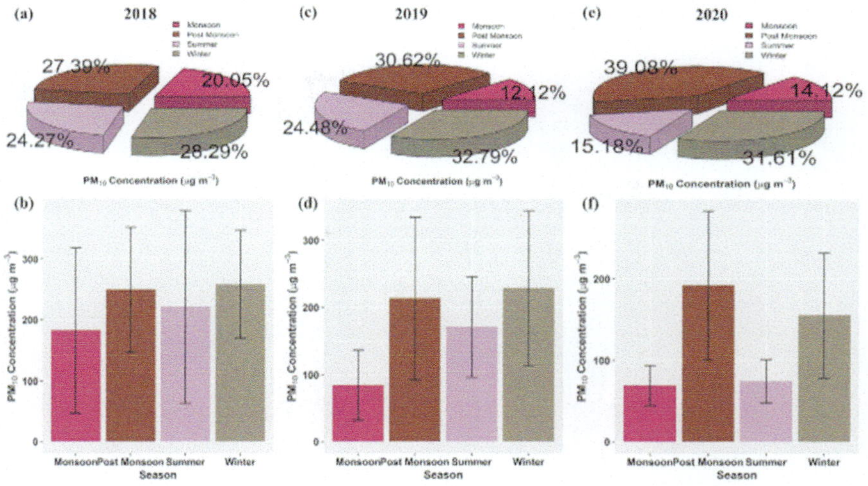

Fig. B.1 Pie chart representing the seasonal percentage distribution of particulate matter PM_{10} in the years 2018, 2019, and 2020. Bar plot representing the seasonal concentration contribution of particulate matter PM_{10} in the years 2018, 2019, and 2020

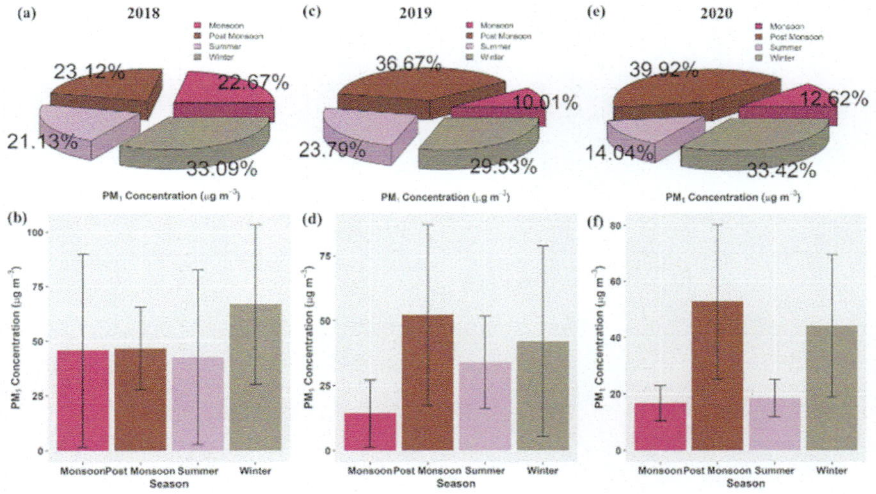

Fig. B.2 Pie chart representing the seasonal percentage distribution of particulate matter PM_1 in the years 2018, 2019, and 2020. Bar plot representing the seasonal concentration contribution of particulate matter PM_1 in the years 2018, 2019, and 2020

Appendix C
Measurements of Size Resolved Nano Particle Concentration and Its Distribution in Polluted Urban Environment During Induced Firework Event

Kanagaraj Rajagopal[1], S. Ramachandran[2], Rajeev Kumar Mishra[1,*]

[1]Department of Environmental Engineering, Delhi Technological University, New Delhi, India

Email: udumalaikanagu@gmail.com

[2]Space and Atmospheric Sciences Division, Physical Research Laboratory, Ahmedabad, Gujarat, India

See Table C.1.

S. M. Shiva Nagendra et al. (eds.), *Select Proceedings of the 8th Indian International Conference on Air Quality Management (IICAQM 2023)*, Lecture Notes in Civil Engineering 582, https://doi.org/10.1007/978-981-96-2359-4

Table C.1 Statistical summary of different size PNC during different periods

S. No.	Parameter	Value	Pre Diwali	Diwali	Post Diwali
1	N_{nuc} (cm^{-3})	Minimum	87.3	27.4	38.8
		Maximum	2062.5	3428.2	2646.1
		Mean ± SD	606.2 ± 462.2	473.1 ± 689	514 ± 495.8
2	N_{satk} (cm^{-3})	Minimum	149.6	136.1	106
		Maximum	2455.9	1429.1	2336.3
		Mean ± SD	878.8 ± 539.3	599.2 ± 396	732.5 ± 551
3	N_{latk} (cm^{-3})	Minimum	284.0	339.2	318.2
		Maximum	2842.7	2233.7	3483.2
		Mean ± SD	1113.1 ± 674.4	1003 ± 572	1074 ± 781
4	N_{acc} (cm^{-3})	Minimum	158.7	506.1	317.8
		Maximum	1505.7	2902.1	2258.1
		Mean ± SD	598.4 ± 278	1111 ± 714	961.4 ± 337
5	N_{total} (cm^{-3})	Minimum	8156	15,120	11,263
		Maximum	66,314	95,073	76,588
		Mean ± SD	32,550 ± 13,957	38,289 ± 23,038	37,203 ± 15,796
6	GMD (nm)	Minimum	35	47	44
		Maximum	117	156	183
		Mean ± SD	73 ± 20	119 ± 27	104 ± 31
7	NO_2 ($\mu g/m^3$)	Minimum	6.80	6.6	6
		Maximum	187.2	40.2	68.8
		Mean ± SD	24.9 ± 31.7	11.7 ± 7	13.1 ± 11.4
8	NO ($\mu g/m^3$)	Minimum	24.6	37.3	30.8
		Maximum	319	145.4	200.4
		Mean ± SD	76.7 ± 45.5	67.6 ± 24.2	77.2 ± 39.1
9	NO_X ($\mu g/m^3$)	Minimum	18.5	25.3	21.5
		Maximum	318.5	108.5	160.7
		Mean ± SD	59.7 ± 47.4	44.4 ± 17.9	50.61 ± 28.3
10	SO_2 ($\mu g/m^3$)	Minimum	5.1	8.1	8
		Maximum	12.5	11	11.8
		Mean ± SD	6.2 ± 1.2	8.6 ± 0.5	8.7 ± 0.5
11	CO (mg/m^3)	Minimum	0.3	1.8	0.9
		Maximum	11.4	4.5	5.6
		Mean ± SD	2.3 ± 1.81	2.8 ± 0.7	2.7 ± 0.9
12	Ozone ($\mu g/m^3$)	Minimum	17	18	18.5
		Maximum	111.3	82	101.4
		Mean ± SD	37.1 ± 25.1	30.6 ± 19.1	32.9 ± 22

GPSR Compliance

The European Union's (EU) General Product Safety Regulation (GPSR) is a set of rules that requires consumer products to be safe and our obligations to ensure this.

If you have any concerns about our products, you can contact us on ProductSafety@springernature.com

In case Publisher is established outside the EU, the EU authorized representative is:

Springer Nature Customer Service Center GmbH
Europaplatz 3
69115 Heidelberg, Germany

Batch number: 08664292

Printed by Printforce, the Netherlands